MORE THAN HOT

More Than Hot

❖ ❖ ❖

A Short History of Fever

Christopher Hamlin

Johns Hopkins University Press
Baltimore

KH

Johns Hopkins University Press
2715 North Charles Street
Baltimore, Maryland 21218-4363
www.press.jhu.edu

Library of Congress Cataloging-in-Publication Data

Hamlin, Christopher, 1951–
More than hot : a short history of fever / Christopher Hamlin.
    pages   cm — (Johns Hopkins biographies of disease)
Includes bibliographical references and index.
ISBN-13: 978-1-4214-1502-4 (pbk. : alk. paper)
ISBN-13: 978-1-4214-1503-1 (electronic)
ISBN-10: 1-4214-1502-X (pbk. : alk. paper)
ISBN-10: 1-4214-1503-8 (electronic)
1. Fever. 2. Epidemics—History. 3. Public health—History. I. Title.
RB129.H36 2014
614.5′112—dc23   2013050172

A catalog record for this book is available from the British Library.

*Special discounts are available for bulk purchases of this book. For more information,
please contact Special Sales at 410-516-6936 or specialsales@press.jhu.edu.*

Johns Hopkins University Press uses environmentally friendly book materials,
including recycled text paper that is composed of at least 30 percent post-consumer waste,
whenever possible.

9/15/17

CONTENTS

Disease we have always had with us. Our ancestors suffered pains in their joints, debilitating coughs and exhausting diarrheas, sore throats and bloody urine, painful and sometimes mortal swellings. Ancient bones tell us that pathological processes are older than written records. And written records tell us that there has never been a time when men and women have not elaborated ways of explaining the incidence and nature of such ills—often in connection with ideas about their prognosis and treatment. Medical and social historians, historical demographers and epidemiologists, all have sought to diagnose past sickness—but rarely in the terms in which past actors understood those ailments. Most older descriptions and terminology do not track easily onto twenty-first-century categories and understandings.

Such terms as *dropsy, continued fever, old age,* and *bloody flux* no longer populate our death certificates. They seem vague, imprecise, even quaint; we take for granted a rather different repertoire of specific disease entities. We have come to think of these disease entities as having a characteristic clinical picture, that picture the consequence of an underlying mechanism of some sort—whether it be a microorganism, a malfunctioning kidney, or a genetic mutation. And we have come to think of these entities apart from their manifestation in the bodies of particular men and women. Of course each of these specific entities has a history and a geographical distribution connected to that history. Thus the rationale underlying the Johns Hopkins University Press's "biographies of disease." The very term *biography* implies a coherent identity and narrative, a discernible movement through time.

But it can in some ways make it difficult to reconstruct the past in the past actors' terms. Men and women in classical Athens or Chaucer's England or Samuel Johnson's London did not share our notions of disease causation or our assumption of specific disease categories. This may seem no more than a statement of the obvious, but retrieving those historical assumptions and linked perceptions is no easy task—and can in fact be made more difficult by our natural tendency to retrieve the past history of disease in twenty-first-century terms. It brings with it the mixed blessing that comes with shining a flashlight into a murky basement. The flashlight's beam serves as a useful kind of sampling device, but one that illuminates only a portion of an otherwise elusive reality.

Christopher Hamlin has embraced a rather different strategy. In writing a biography of "fever," he has chosen a subject that most readers today would call a symptom—and not a disease at all. Yet for most of written history, "fever" was an omnipresent and protean reality, a brutally real reality that threatened incapacity and often death and was an unavoidable part of everyday domestic experience. Fevers were not marked by heightened temperature alone as the twentieth- or twenty-first-century reader might automatically assume. As Hamlin emphasizes, fevers before the mid- to late nineteenth century included heightened sensory perception, often a feeling of burning heat and sometimes subsequent chills, disorientation and even delirium. Some fevers could have a short, intense course, others a long-drawn-out and anxiety-provoking clinical history, before the patient died or recovered; some were associated with skin lesions, others not. Some showed characteristic patterns of what a modern physician might describe as recurrent spiking. And this protean condition was not associated primarily with infectious disease (itself a comparatively modern and thus anachronistic category) and the body's reaction to it, but was primarily an individual, aggregate, multicausal response to a peculiar configuration of circumstances—including, as in certain intermittent fevers, for example, proximity to damp and swampy environments. Fever was very much a *thing* in history, even if we would today call it a symptom, a tactic deployed by the body as it responds to a stressful stimulus or internal disorder.

Hamlin traces a complex history of ideas about fever over

time, from classical antiquity through the nineteenth century and into the twentieth. For much of this time it is necessarily a story of texts talking to texts as concepts of fever were passed on, modified, elaborated, controverted. Such constructions and reconstructions of fever reflected the very omnipresence of the phenomenon they sought to describe and explain. Thinking *about* fever meant thinking *with* fever. What did its incidence reflect about the styles of life and social and physical circumstances in which the feverish body was shaped into being? Was it diet, imprudent work, an inimical physical environment? In the late eighteenth and nineteenth centuries were certain fevers a response to the unnatural and thus pathogenic world of the crowded city and of industrial work?

All this was to change. By the late nineteenth century the capacious and inclusive thing called *fever* was being disaggregated into a series of discrete entities—malaria, typhus and typhoid fevers, influenza and childbed fever. Fevers were by the end of the nineteenth century not phenomena that needed to be understood primarily in terms of the unique characteristics and life events of individual bodies but *things* that could be understood and imagined outside those individual sufferers. Pathology and bacteriology, the tools of the laboratory, had extracted entities defined by mechanism and clinical course from what seemed to later scholars the seemingly formless mass of pre-nineteenth-century fevers. The thermometer and antipyretics had helped narrow and shape the ancient experience of fever into *our* fever—a symptom present in a variety of ailments, defined by the comfortingly objective metric of a column of mercury.

Fever has been in everyday practice reduced to a number, an indicator and not a thing. Generations of mothers have relied on the thermometer and aspirin to tame mild fevers—and to create a functional boundary between those conditions that remain the appropriate focus of household management and those that demand the credentialed physician's attention—and perhaps even hospital admission. But the older meanings of *fever,* Hamlin argues, have not vanished entirely. The cultural residuum of danger, of delirium, of bodily hyperintensity persist in the ways we still employ the terms *fever* and *fevered* to designate states of abandon and loss of control. Christopher Hamlin has written an extraordinarily ambitious biography of a dead cultural entity that lives on in altered and attenuated form.

Charles E. Rosenberg

ACKNOWLEDGMENTS

To imagine and to tell the story of a disease is a peculiar task, and particularly so for fever, at once a familiar experience and a central component of recondite medical theories in many traditions and over long periods. My debts in this book are many.

Charles Rosenberg has been inspirational in many ways but first for welcoming a suggestion for a book on fever, a massive and unruly topic yet one so central to the history of medicine. He has been the finest and kindest of commentators and critics, reminding me of important perspectives, challenging me to think through interpretations and to find balance in the narrative, but most importantly encouraging me to shape the book in my own way.

Many others too have been sounding boards: students at Notre Dame, colleagues at the Institute for Advanced Study, Catherine Cox and her colleagues at the Centre for the History of Medicine in Ireland, and Joel Tarr, Bill Luckin, Margaret Humphreys, Dale Smith, Margaret DeLacy, and Ann Carmichael, supporters of the project and patient listeners to my outlines of it. I thank Christopher Baron for translation help.

I wish also to remember five close friends who died during the years I have been thinking about these matters. Conversations with them have been among the most important intellectual experiences of my life and have shaped my work in ways that are impossible to trace. They are Roy Porter, Elizabeth Hunt, Angela Gugliotta, Harry Marks, and John Pickstone.

I owe a great and joyful debt to Hamlin family members: we have been at times a cottage industry of fever (and sometimes fevered) research. Dr. Elizabeth Hamlin offered insights from modern medicine; Kat, Aang, and Fern Hamlin extracted notes, compiled tables, databases, and bibliographies, and proofread. Their thoroughness, diligence, patience, and good humor not only expedited the project but also vastly enriched the experience of reflecting on the past and present of this curious entity.

I would also like to thank the University of Notre Dame, particularly the Department of History and its chair, Patrick Griffin, and the School of Historical Studies at the Institute for Advanced Study for both financial and moral support. The Hans Kohn membership, established by Immanuel and Vera Kohn, supported my stay in Princeton.

I thank Jackie Wehmueller for her encouragement and flexibility, Joanne Allen for thorough editing, and numerous others on the staff of Johns Hopkins University Press for their long hours on the project and tolerance of my fickleness.

I would also like to thank numerous librarians, digitizers, and archivists, especially staff members at the Barbara Bates Center at the University of Pennsylvania School of Nursing, at the National Library in Ireland, and Harriet Wheelock at the Royal College of Physicians in Ireland.

MORE THAN HOT

CHAPTER ONE

More Than HOT

❖ ❖ ❖

In *The Doctor* (1891), by Sir Luke Fildes, a middle-aged male physician in a rumpled suit sits chin in hand, carefully watching a sleeping, pale-cheeked child. Distressed parents huddle in the background. Although clinical settings have evolved, the painting has continued to reflect the noble ideal of bedside medicine.[1] To its original viewers it would have also represented an experience common to them but less common to us: a long and dangerous fever. Pallor, vigil, worry, exhaustion, a child victim—all suggest a serious fever. The painting is thought to have been inspired by the fatal illness of Fildes's son in 1877.[2]

For most people, fevers, whether dangerous or benign, have been a normal part of life. "No person can live without fever," observed the eighteenth-century commentator Gerard van Swieten (1700–1772), whose writings I shall often cite as representative of a long-lived "classical" tradition in Western medicine. He noted too the ancient adage that fever ended most lives—an exaggeration, but not by so great a stretch.[3]

Innumerable questions arise as we enter this painting.

First, what kind of fever does the child have? To us, diagnosis is critical. It tells us how much to worry. While Fildes's doctor would have used a familiar set of disease names—a century earlier

Sir Luke Fildes, *The Doctor* (1891).
Reproduced with permission, Tate, London / Art Resource, NY.

that would not have been the case—diagnosis would rarely have provided any magical keys to cure or relief. The child's disease may be one of the childhood exanthems (i.e., eruptive diseases), like measles, or it may be a passing generic "flu," the most common form of febrility for those in the medicalized world. Yet at the time *fever* suggested something potentially dangerous. To Victorian viewers, the term often meant typhus or typhoid, familiar in cottages like this one as diseases of deprivation and insanitation. These were long, dramatic fevers lasting weeks. Alternatively (though less likely in 1891 than two centuries earlier) the child may have malaria and may be suffering a brief but complete fever every other day (a *tertian*) or every third day (a *quartan*), perhaps for months.

Usually, no less important than rendering a discrete diagnosis was to interpret each case in terms of a general model of a fever's course. Most important in such models was the "crisis," when, following a troubled period of varying length ("status"), a fever "broke." While many cases failed to fit the model, the pattern structured care and offered hope.

Will this child live? Today, the pathophysiological processes through which infections generate fevers are well enough understood that we are likely to view fever as an orderly, even restorative process. The hypothalamus has been stimulated to elevate the core temperature by molecules (pyrogens) that are either microbial toxins or products of immune response. Fever then can seem to be an accident, if perhaps a beneficial one, an "adaptive response" in modern parlance.[4] In fact the idea of fever as a positive condition is an old one: long before any appreciation of a host-pathogen arms race, healers were seeing fever as a challenging rite of passage between injury and recovery, the body making itself healthy by temporarily making itself ill. They were much less confident, however, that bodies would meet that challenge—it is jarring to find *death* among the modes of fever resolution in old textbooks. Only in the twentieth century would fever, at least in some contexts, come to be represented as a good thing.

What should the doctor do? This one isn't doing anything. Often Fildes's painting has symbolized the futility of prescientific medicine. The doctor watches because there is nothing he can do. There were no antibiotics, and effective antipyretics were still very new. By Fildes's day, timing and restraint had become the watchwords of much medical practice, and particularly so for fevers. The fever would take its course; a body well nourished with digestible aliment (note the mortar and pestle) would survive the ordeal. Thus an expert's monitoring might be the best medicine, preferable to the bleeding or sweating of earlier practitioners.

Although doctors had sometimes aspired to cures, over the centuries they often spoke of the high art of *managing* fevers, which some described as akin to piloting a ship in a storm or riding an unruly horse.[5] Hence the vigil in Fildes's painting. This doctor may have been sitting for a long while, steering the child through the crisis, or he may have just returned for one of multiple visits that day.[6] The fragility of the fevered body at critical stages required frequent reassessments, when the physician had to decide how to nudge the body along the safest channel.

What is he looking for? The modern doctor, with nursing and laboratory support staff, may look less at the sleeping child than at the chart, focusing on the levels of a few key indicators. Fildes's doctor had no useful laboratory tests. By 1891 fever was becoming an objective entity; though none is shown in the painting, thermometers were being used. Like his predecessors, this doctor tracks a rich array of signs and symptoms. Besides hotness, a person's pulse, tongue, respiration, skin, visage, eyes, bowels, urine, and even fingernails gave clues to a fever's character. So too did speech, gesture, and even the victim's posture in bed.

Usually, the focus of Fildes's painting has been its exemplary doctor. *But what of the sleeping patient?* Before it became a thermometric entity, fever was significantly subjective. One felt hot, shivery, weak, achy, or dizzy; light, sound, and touch caused pain; one's mind did not work properly. Of course, the most articulate descriptions are from adults. What sense children made of their troubled bodies is less clear. According to the didactic novelist Harriet Martineau, writing a half-century before Fildes's painting, a child should regard fever as a visit to the anteroom of the afterlife. Did children see it thus?[7]

Nowadays, the thermometer defines fever's beginning and end. But in Fildes's day convalescence was a long, liminal, physician-defined state separating acute illness from full competence. Often it was a period of conflict; thinking themselves well, patients misbehaved. Their error was itself a symptom: mental malfunction lingered when physical symptoms had gone.

What of the parents and others? What does the child's fever mean to family, community, and state? In some times and places, a child such as this one might have been a pariah, a contagious threat to family members, to neighbors, or even, in the case of typhus, to the attending doctor: typhus caught at the bedside would be an idiom of medical martyrdom. The sin or error that this child's fever registers may be individual (laziness, filth) or structural (in-

frastructure, welfare policies). We may wonder about medical institutions. Who pays this doctor? Perhaps he acts as a representative of a contagion-sensitive state. And why is this child at home?
By Fildes's time fever hospitals in many parts of Britain provided
better protection of the public and better treatment for the child.[8]

If in much of the world malaria still rules and new flus and
fevers threaten, the events of Fildes's painting are increasingly foreign to persons in medicalized societies. Vaccination, sanitation,
and rising standards of living have largely eliminated the childhood exanthems (e.g., smallpox and measles) and typhus and
typhoid, the most prominent nineteenth-century "continued"
fevers. The changes are recent. In the second and third decades of
the twentieth century my mother lost two siblings to fevers. I was
in the first cohort of polio vaccinees. My friends and I expected
to get the childhood fevers, but no one died of them. I remember
weeks in bed, vigils of parents (and strangers), and the evening
arrival of a stern, gentle, bushy-browed Fildesian doctor. My children's experience has been quite different. In the time between the
eldest and the youngest even chicken pox, least of the childhood
fevers, vanished from our lives. Now "fever" is usually a brief bout
of "flu-like symptoms." One is "out with a virus." Cheap drugs
lower temperature and ease pain. We self-medicate safely and confidently, knowing the fever will soon run its course.

What fever once was, cancer is now. Long and dangerous, im-
plicating the whole body, potentially deadly, it occasions the reck-
oning with our ephemerality that fever once brought. It is, as fever
once was, central in our clinical, scientific, and political medicine.
Cause and blame, therapy too, are bewildering and controversial.
By contrast, in relatively fever-free parts of the world, *fever*, at
least in the realm of popular culture, has become a euphemism for
sexual desire. Introspection has given way to flirtation. Elsewhere,
of course, serious fevers persist. Far more than in the past, fever
divides—by class, place, and sometimes race.

FEVER AS A HISTORICAL PROBLEM

These issues of medical knowledge and practice, of sufferers' experiences, and of the significances assigned to a form of illness might be raised for any disease. But fever brings four special problems.

First is status. Fever was once a disease but is no longer one. Defined as elevated temperature, it may be seen as a symptom, but traditionally it has not been a single symptom but rather a cluster: where we speak of the chills and aches *of* fever, earlier generations would have treated those and other symptoms *as* fever. It may be best to view fever as a composite condition and the most common form of temporary alienation from normal comforts and capacities. But even that view is misleading. Where malaria is endemic, that alienated state may be the norm rather than an aberration.[9]

Second is how to bound, distinguish, classify, and comprehend the vast and varied domain of febrility. Disease names are the residuum of a long past. Many diseases have been taken to exemplify fever, some with *fever* in their names (e.g., typhus fever, dengue fever), many without (e.g., pneumonia, malaria, influenza, and plague, often seen as the most serious fever). Fever has been seen as a feature of most diseases now labeled infectious, though in tuberculosis, cholera, and dysentery it has usually been a secondary one. Oddly, the exanthems, with their unique rashes, were often marginal to the domain. In addition to these distinct whole-body diseases, there were fevers that followed a local injury, a wound, or a birth. Cancer, scurvy, and heatstroke have been seen to involve fever.[10] Finally, vague and transitory states of disequilibrium might be understood as *feverishness*.

A third issue is how to detect and delimit fever. In modern medicine the thermometer rules. A granddaughter's preschool will tolerate coughing, sniffling, and other contagious toddler horrors, but a minimal elevation of temperature mandates immediate removal. Using that marker, we distinguish fever from a mere cold, though the symptoms overlap and the generic "colds 'n flu" get the same treatment. But for most of history, people have experienced the peculiar physiological state without a disciplining thermometer. Hotness, apprehended by sufferer or healer, was but one of many symptoms. Others, such as a rapid pulse, have often been more important, but rarely was there a single measure. Observers might simply confirm a sufferer's self-diagnosis by agreeing that the person looked or acted fevered.

It is tempting to see fever as independent of language and culture and to assume that persons in the past were identifying the same conditions and features that we do, only in qualitative terms. Assuming universality, we would then translate ancient terms into modern medical truths, for surely vomit is vomit the world round. But caution is in order. For starters, meanings change. For example, in the vocabulary of early nineteenth-century febrility, *low* did not mean low temperature but rather a dangerous "typhoid," or stuporous, state.

It would be naive too to think that some invisible algorithm could, without significant loss, transform a rich register of qualitative assessment into a single variable. It is safer to admit that fever requires analogy. And not only do sufferers and observers seek familiar terms to express bewildering sensations and unusual observations but their usages ripple back into language. Only insiders may fully appreciate the idioms. Consider modern English: *yellow fever* is a disease, *purple fever* has been one, but *gold fever* is not. Knowing that *Saturday Night Fever* is not a malarial variant requires familiarity with 1970s popular culture. Interpretation of the vocabulary of febrility must be a major concern as one moves from culture to culture over a vast period.

Still, I assume here that there is indeed a core concept, a set of conditions that have counted as fever. Its extent, boundaries, and internal organization have varied; so too have the significances assigned to those states (and not merely the theories generated to explain them). Often, the core of that core, the exemplar of fever, has been malaria. Many ancient cultures recognized the distinct periods of malarial fevers. Still, malaria is not always periodic, nor was periodicity always evident to observers—relapsing fever would only be recognized in the 1840s.

Yet as much as possible I will avoid retrospective diagnosis.

An early eighteenth-century attempt to relate the many forms of fever.

From Francesco Torti, *Therapeutice specialis ad febres quasdam perniciosas* (Modena, Italy: Soliani, 1712), Wellcome Library, London, L0025278.

The problems are biological as well as linguistic and cultural. The body's limited repertoire of symptoms, the evolution of infections, and the frequency of coinfection or other comorbidity (e.g., a seventeenth-century "worm" fever) often preclude confident diagnosis. I question too the need for the practice. There are sometimes good reasons for anachronism; more often, we lose much in twisting ancient concepts into our own terms.[11]

Last is the philosophy of fever. Understanding of fever's causes and pathology is recent, but the vast literature on fevers is ancient. In volume upon volume, theorists struggle. Fever, they note, is a unique challenge, a state of body and mind that can rarely be ascribed to any injury to the body's fabric and often leaves no obvious damage. Most explanations fall between two poles.

At one end fever is an extreme version of normality. From time to time, I am hot, worn out, or achy. Perhaps I have eaten unwisely or been in the sun too long (and feel a bit "under the weather"). At some stage I have entered the territory of fever. However mild initially, my fever may become life threatening. At the other end, febrile diseases may be distinct states, exogenous entities, distinct from normal variability, which take over our bodies for a time.

This grand question—whether we are dealing with a single "fever" or multiple "fevers"—structures this book. For millennia, in Europe and much of Asia, the former, "biographic," "phenomenological," or "physiological" view prevailed. Various humoral theories of illness emphasized an individual's departure from healthy equilibrium. Often the descriptive terms *distemper* and *disorder* were used where today we would write *disease*. Usually, the names given to fevers were more descriptive than designatory.

With the late nineteenth-century rise of microbiology, the latter, "ontological" view of specific diseases triumphed. We now see fevers as fundamentally different because we define them by their microbial agents.[12] Still, physiologists and clinicians sometimes protest: knowing which tiny beast has invaded may tell one nothing about the havoc it is causing or how best to treat it. Only rarely has precise diagnosis been essential to getting the temperature down. And circumstances of practice may defy causal knowledge; the impatient patient may demand and be prescribed antibiotics for what is actually a slowly resolving viral infection.

Rarely nowadays is there any fundamental incompatibility between a clinical-pathophysiological orientation and a focus on specific agents. What were often rival approaches now refer to different aspects of medicine. Still, historically and philosophically the divide was often seen as fundamental, the two perspectives being seen to have quite different implications for how individuals should understand their illnesses and how community and state should respond. I may seem here to give undue attention (and sympathy) to the former, physiological perspective. In part that comes with the territory: this is a history of fever, not of fevers. But that focus also helps us balance medical knowledge, illness experience, and societal response, and it helps us avoid interpreting earlier medicine in terms of later—by assuming, for example, that humors and germs represent the same register of explanation.

But I am also concerned with the presumed implications. In a nutshell, the physiological perspective has been seen to highlight individual illnesses, while the disease-specificity perspective has been seen to highlight populations. In a classic 1976 article, "The Disappearance of the Sick-Man," the sociologist Norman Jewson reviewed the successive "cosmologies" of Western medicine from 1770 to 1870, each a union of knowledge, practices, roles, and institutions. Jewson recognized that a premodern holistic, "person oriented," "bedside" medicine had been available only to a small and wealthy clientele, but he lamented its loss. It had been predicated on the uniqueness *and intelligibility* of each sufferer's illness, for in 1770 fevered persons would have understood their illnesses in the same terms as the healers who acted at their behest. Independently, Charles Rosenberg called attention to the distinct therapeutic implications of such medicine: that cures—bleeding to the point of faintness, a warming cordial for chills, or sweat-stimulating camphor—belonged to the same experiential domain as illness itself. One felt physick work.

By contrast, the "hospital" medicine that arose after 1800 was "object oriented," in Jewson's view: alive or dead, the fevered person became a datum from which hospital clinicians inferred general laws of specific diseases. Even before Jewson, the great French sage Michel Foucault had pointed to the new forms of power nineteenth-century states were acquiring in creating medical normality. The organizing of thousands of ill bodies in the reformed hospitals of Paris and the exacting study of the many cadavers produced expanded medical authority while rendering the ill person a case to be counted. Thus it was that fever, an alienating experience in its own right, was alienated from the fevered.

In a later iteration—laboratory medicine, dominant from 1870—data would displace embodiment even further. The temperature chart, hanging at the foot of the bed, out of the patient's sight, confirmed the untrustworthiness of one's own assessment: numbers clarified, personhood confused. According to the dictatorial thermometer, if one is not hot, one is unlikely to be seriously ill. Often such laboratory medicine rested on a militantly physiological view of disease, but no longer did it engage with the subject's feelings or offer an intelligible narrative. Ironically, the paradigm of disease specificity, with its invasions by predatory microbes, might offer a more intelligible narrative than the physiologists' biochemical-feedback loops. The operation of modern drugs likewise often required an act of faith: even if one felt nothing, one trusted in the expert's diagnosis that justified them.[13]

Consider again Fildes's painting as a boiled-down version of this history. We admire this bedside doctor doing his best to manage *this child's unique fever,* precisely what we should want for every child. But few would elect this care. If this case is serious, there should be a clean, well-lighted hospital with a medical laboratory. Commitment to a case should not imply biotic uniqueness, yet here there is no fever chart to link this to the great reservoir of clinical knowledge. However threatening, the Parisian premise of the interchangeability of bodies is the basis of effective, efficient, and *humane* care. Nor, I suspect, would most choose to be stared at by a stranger, however benevolent, when a bottle of ibuprofen is at hand.

In the past the rival views also had distinct implications for explanation. In the older, person-centered physiological medicine, causation had rarely been systematically investigated, since clinical similarity had not implied any common exogenous cause. A fever registered the totality of circumstances that had brought a body to its febrile tipping point. By contrast, the concept of specific fevers led to an "etiological revolution" associated with the germ theory. The discreteness of the agent, not of the clinical pattern, defined the disease. And that agent's presence, demonstrated or presumed, was usually sufficient cause as well.

Linked to causation was accountability. Who was to blame for fever? While the physiological and specificity perspectives provided frameworks for locating responsibility, neither had a single set of implications. Within the older, physiological framework, fever, a product of climate, indulgence, or unjust policy, might warrant fatalism, blaming the victim, or thoroughgoing social reform. The attribution of disease to pathogen exposure may seem apolitical by contrast, but those pathogens lived in bodies, communities, and environments. The allure of controlling fever would be central to the rule that European states and their minion professions would exercise over their own "citizens" and the distant peoples they conquered. Here too the competing frameworks have converged. Fevers may be discrete diseases with unique agents, but qualitative factors affect their incidence and effects. All too often, victims of pathogens have already been made vulnerable by all manner of antecedent factors. Usually effective prevention requires attention to both.

ERAS OF FEVER

My approach is episodic. I examine texts, events, and persons, familiar and obscure, that illuminate broader issues. Others might be included. I bypass some matters well treated elsewhere, but there remains much to explore. Fever has been the invisible elephant in the china shop of the medical past. My hope is that this sketch will encourage more critical reflection on its centrality in human experience and fuller exploration of its history in many times and places.

This book is in four parts. Part I deals with generic concepts of fever that appeared in ancient medico-philosophical systems and persisted over millennia. Chapter 2, "Words," the first chapter of part I, concerns concepts of fever in ancient Chinese, Indian, and Hippocratic medical traditions, the chief "classical" systems of medicine. It also concerns the problem of medical communication—how patient and healer, on opposite sides of hot skin, share knowledge. It closes with Galen's efforts to order febrile phenomena. Chapter 3, "Books," explores the proliferation of models of fever in early modern medical learning. In the aftermath of plague it was important that someone know what was going on. Rarely, however, did the vast ingenuity applied to explaining the disruption of health carry over to effective response.

Part II is about the emergence of fever as a public issue in the eighteenth century. Chapter 4, "Communities," opens with fever as a matter for courtly discussion. One interrogated one's own fevers and commiserated with fevered friends. Fevers also were coming to be recognized as marking class, gender, place, occupation, or race. One might ask what it was about the rich, the poor, women, soldiers, sailors, European travelers, slaves, or, later, eastern European Jews that made them prone to dangerous fever, though that proneness did not invariably represent a public problem. Such recognition did, however, reflect growing awareness of social integration, an awareness amplified by increasing awareness of fever's communicability. Chapter 5, "Selves," treats the feminization of fever and its status as a vehicle of romantic introspection. Fascination with fever in one's own life was coupled with growing empathy for the fevers of all others. Gradually, fever-focused charity would evolve into a mandate for public-health activism.

Part III deals with the remarkable reorientation of the nineteenth century. Chapter 6, "Facts," begins with the efforts of the medical revolutionaries of Paris around 1800 to free medicine from error and obscurity by starting anew with facts only. In the hands of the pioneering clinical statistician P. C. A. Louis, that approach led to recognition of recurring clusters of symptoms that might be linked to a unique internal lesion. An unexpected product of anatomico-clinical inquiry was an interest in disease ontology. What made fevers different became more interesting than what they had in common. Clinical distinctiveness made it possible to sharpen an epidemiological profile, which in turn led ultimately to discoveries of responsible microbes, whose presence would demarcate the species of a fever. Typhoid would be the exemplar, with its microbe confirmed by the mid-1880s. With differential diagnosis came comforting categories of safety or danger, to oneself and to others. These categories might bring access to resources but they also brought rules, duties, and expectations. Chapter 7, "Naming the Wild," concerns the deadly tropical fevers, both epidemic and endemic, which resisted confident Parisian ordering and became sites and occasions of imperial conquest. To represent the "fever" of a place as a set of discrete diseases was to deny its arbitrary unruliness and to represent it as analyzable through appropriate expertise. As fever's symptoms were becoming merely concomitants of invasion by some particular microbe species, fever itself was ceding its complexity, becoming a mere mark on a glass tube. Chapter 8, "Numbers and Nurses," takes up the resurgence of the ancient problem of the dynamics of febrility in late nineteenth-century laboratory medicine. As fever became simply an elevated temperature, its other aspects—debility, sweats, shakes, aches, and wandering mind—became secondary. They were mere nursing matters, and yet in cases of dangerous fevers like typhoid the quality of nursing was increasingly being seen as the key to recovery. But also issuing from these laboratories were the powerful pills that allow us to navigate many fevers at home, free from hospital routine, laboratory objectification, or doctors regarding us from a bedside chair.

The focus of the final part, comprising a single chapter, is the contemporary fragmentation of "fever," the wide gap between the trivial fevers of medicalized societies and the serious epidemic outbreaks that still occur in exotic places, attracting international teams of experts and becoming apocalyptic media sensations. "Machines, Mothers, Sex, and Zombies" explores the status of "fever" in American mass culture. By the mid-twentieth century actual fever had become an illness mainly of young children. Yet by 1960 popular culture was transforming the term into a euphemism for sexual attraction, while in the 1990s epidemic fever would become a popular genre to indulge apocalyptic fantasies—"hotties" and "hot zones," respectively.

In closing with popular culture I may appear to trivialize the serious illnesses many still suffer, the problems with which researchers, clinicians, hospital administrators, and public-health officials still struggle. Often I find these extensions of *fever* to be profoundly disturbing, as well as in poor taste. But even deadly disease does not escape culture. Perhaps fever is particularly fungible: woozy, temperature-destabilized, and ill-functioning bodies invite interpretation. Just as fever-causing microbes will continue to adapt, we should expect meanings of *fever* to evolve too.

In the early twenty-first century, the most frightening fever may not be one occurring in single human bodies. "The planet has a fever," former vice-president Al Gore told a congressional committee in 2007.[14] Under the heading "Curing a fevered planet," another writer calls for "geotherapy": "The Earth is sick. If the planet were a person we should say: she has a rising fever."[15] The well-known Keeling curve, showing the rise of heat-trapping CO_2 in the earth's atmosphere, has become a fever chart.

Mere metaphor, this "fevered" earth? Not necessarily. Seeking to understand fever in terms of overall physiological regulation, early twentieth-century researchers explored how the mechanisms of climatic stability operated within bodies. L. J. Henderson's classic 1913 title, *The Fitness of the Environment,* reflects the orientation. That we now ponder macrocosmic instability in terms of the microcosmic should not startle. Will some global immune response kick in, returning us to normal? Will earth's body stabilize at a higher temperature? Or, as in hyperpyrexia, which so worried late nineteenth-century physiologists, will its temperature continue to rise, causing its control mechanisms to fry?

PART I

The Fevers of Classical Medicines

❖ ❖ ❖

The two chapters in this first part explore concepts and philosophies of fever. By concepts of fever I mean the sorts of circumstances that occasion use of the vocabulary of febrility within a culture. Such usage may be contested, but acknowledgment of febrility requires some degree of shared perceptions by victims, societies, and practitioners, not just of whether fever exists but of what having fever means. Philosophies of fever—learned explanations of what is happening within the body and what to do about it—may be more univocal: they are interpretations imposed on the sick. Considering these elements together offers a general way to think about fever in societies remote in time and culture from modern medicine.

I focus on common elements and orientations in concepts and philosophies of fever in the classical medical traditions of China and South Asia, but mostly in the Mediterranean tradition associated with Hippocrates and Galen. In Europe there would be a radical rejection of the classical approach beginning around 1800, yet many people today still manage their health, at least in part, through classical Chinese or Ayurvedic approaches, finding these more satisfactory than the biomedical alternative. And while it is easy to represent modern medicine as the antithesis of these tradi tions, many aspects of ancient orientations have resurfaced there too. The idiom may have changed, but the practical conclusions are much the same.

In all these medical traditions, while the domain of *fever* may be familiar, conceptual coherence and underlying rationale are likely to be much less so. The distinction we now make between a feeling of feverishness and having a specific disease (one of the fevers) was less sharp before 1800. What is now objective, a matter of response to infectious agents, was more subjective, a transitory state involving temperature maladjustment and other changes in sensation, comfort, and function.

Although fevers generally were distinct states, often they were functions of place, time, and activity. Usually fever registered the power of the environment, broadly construed, to disrupt one's internal equilibrium. Thus, the author of the Hippocratic text *On Regimen* warns of fever that comes from overdoing a morning workout. "Some have rigors as a result of their exercises, that is to say, from the time they put off their clothes to the time they finish, and the rigors are renewed on cooling down." The result is "high fever . . . with heavy delirium."[1] Such fevers need not come from a single episode of overtaxing; they might represent accumulated assaults, natural, social, and sensory, for words and whips might contribute to fever just as heats and airs did.

This may seem an odd understanding of fever. Access to good shelter, central heating, and clothing made of miracle fabrics that sequester moisture from our skin probably desensitizes us to the profundity of compensatory physiological responses to combined coldness and wetness. Or to the effects of heat: in some places, a significant number of "fever" cases were probably heatstroke, known briefly in the nineteenth century as *thermic fever.* And whereas many of us enjoy the illusion of our skin as a secure border, separating the autonomous body from external nature, many classical theorists emphasized permeability. To them fever was primarily a disorder of perspirability.

If some fevers might become serious, even deadly, here what is called *fever* is a condition continuous with ordinary life; one may call it a normal abnormal. That all circumstances of the victim's life might contribute to generating fever made it artificial to isolate any "cause" beyond a general account of the patient's course of life leading from health to illness. Such views often did bring suggestions of how to prevent (and cure) by better management—*diet* in the broadest sense. For example, Hippocrates's exerciser is to cut back the workout and replenish lost moisture with moist foods (which for the writer includes puppies' flesh).[2]

Such concepts of fevers persisted for a very long time. Five hundred years after Hippocrates, Galen used the example of a sweaty runner who gets chilled to introduce the simplest form of fever.[3] Hippocrates' overtaxed athlete introduced John Huxham's work on fever twenty-two hundred years later. His contemporary, the Dutch physician Herman Boerhaave, would use the example of an upwind skater.[4]

Exercise heats us. Evaporating sweat still chills. Perhaps these sensations overlap with feverish symptoms, but to us these temporary discomforts have nothing to do with the diseases that cause real fever; there has been no infection by a discrete agent. Classical writers knew that some feverish conditions fit less well than others into such frameworks. The occasional visitations of an unusually hostile season or a mobile "plague" or "pestilence" might be ascribed to processes loosely akin to what today we call *infection*. These might be seen to involve both obscure states of atmosphere and the spread of the seeds of the disease from one person to others. But these causal factors were more placeholders than explanations. Such fevers would come to be called *idiopathic,* distinct diseases without evident cause. To Gerard van Swieten in the mid-eighteenth century such fevers, arising from "an epidemic or poisonous stimulus," challenged the coherence of medicine. One might learn to treat them, but being incomprehensible, they had no place in rational medicine.[5] For us, of course, infectious epidemic fevers are the true ones; center and periphery have reversed.

Believing that medicine does and should change continually and progressively, we may take the mere longevity of the fever-from-exercise idea as confirmation of a view of the medical past as uncritical stagnation. In fact, much changed—concepts, mechanisms, techniques, vocabularies. New diseases appeared. There were revolutionary discoveries. The circulation of the blood, systematic anatomy, and the microscope all left their marks on concepts and philosophies of fever. In part the stagnation verdict only reflects comparison with the extraordinary advance in the last two centuries. But then as now, most discoveries and anomalies were accommodated, folded into the dough.[6] In the mid-eighteenth century, van Swieten was still reconciling the views of the moderns, particular those of his mentor Herman Boerhaave, with those of the ancients; learned physicians in China were doing much the same. In part, the long absence of any great mandate for change was due to the extraordinary achievements of ancient medicine. A well-traveled practitioner of broad experience would never see all the illnesses, in all the places, in all kinds of persons, or reproduce the scientific foundation of Galenism. Art *was* long, and life short.

But not only did Hippocrates and Huxham share a concept of fever, they had similar understandings of how medical explanation fit into society and, more broadly, cosmology.[7] The term *classical* is apt not only because each of these systems of medicine was ancient and long-lasting but also because each was well integrated into its culture. One key criterion of their explanations of fever was intelligibility, another was moral practicability. Much fever represented error, reversible or avoidable at least in principle. Balance and temper needed to be considered. Managing a body was not so different from governing a nation or making beautiful music.[8]

A consequence was that patient and practitioner shared common expectations of explanation, prevention, and cure; the practitioner was simply more experienced in managing the many determinants of health. Since technical terms were lay terms too, learned knowledge was never wholly detached from the experience of the fevered body. Today, by contrast, the depth of medical expertise is often measured by its unintelligibility. Our symptoms are immediately translated into a foreign numerical language. We expect the physician to know us in a different way than we know ourselves. Reading ancient fever texts from that modern perspective is understandably frustrating. In using terms that engage with the layperson's familiar world, authors rely on analogy. But they are saying what something *is like,* when we want them to say precisely what *it is and does.*[9]

Exploring these ancient and varied traditions gives us a sense of the enormous malleability of *fever.* It also helps make clear how much changed with the coming of modern biomedicine.

CHAPTER TWO

Words

❖ ❖ ❖

The Indian medical sage Susruta called fever the "lord of ailments."
He regarded it as "king of all bodily distempers inasmuch as it can
affect the whole organism at one time." It bounded human lives
and was "perhaps an indispensable condition under which a creature can come into being or can depart from this life." Recovery
marked one's incarnation status—only gods, and some humans,
could beat this heat.[1]

The most familiar form of alienation from the normal self, the
experience of fever can help to bridge the several classical medical philosophies. In ancient China, India, and Greece doctors
recognized the distinct malarias, fevers that repeat in one-, two-,
or three-day cycles, what would become in the Latin West *quo-
tidians, tertians,* and *quartans,* respectively. In Indian medicine,
"fever" is a primary organizing concept. Chinese texts discuss a
serious generic fever, but that discussion is dwarfed by interest
in pathophysiology. The authors of the Hippocratic Collection,
seventy or so texts composed in the century after 430 BC, were
concerned with the peculiar fevers of certain seasons, with fevers
arising from injuries, and with fevers of ordinary living (like those
due to overexercise), which usually pass quickly.

In the first part of this chapter I consider aspects and issues important later in the book: the problems of what fever terms mean and of how to distinguish types of fevers, the range of pathological theory, efforts to punctuate the courses of fevers in terms of "critical" days, and the relation of fever to daily life. In the second, I review central elements of the Galenic fever synthesis, a comprehensive explanatory framework that would reach across millennia and continue to set the agenda of fever inquiries into the nineteenth century.

THE VOCABULARY PROBLEM

"How do you feel?" the healer asks. "Woozy," I answer. But this term that captures my feverishness so well is an Americanism that appeared in 1897, with no clear antecedents, and initially was used more often for drunkenness than for febrility.[2] What does fever talk *mean*? Even if a term translatable as "fever" indicates a universally recognized domain, it may nevertheless highlight only certain aspects of that condition and imply certain things about it.

Fever strains language. Yet the fevered need words both to narrate their illnesses for themselves and to secure care. Vocabulary is the central problem in comparing fever traditions and, equally, in exploring the relation between sufferers' experience and theoreticians' knowledge.[3] Translators and interpreters must juggle their own technical vocabulary, which provides units into which earlier diseases can be rendered; the condition being experienced by the sufferer and witnessed by the healer; and the language that the sufferer and healer use to express that experience. All these are entangled. It is tempting to separate experiential from theoretical terms, the subjective from the explanatory—*hot* and *woozy* from the four humors of Western classical medicine (blood, phlegm, and yellow and black bile), or yin and yang, or the three *doshas* of Vedic medicine, or the salt, sulfur, and mercury of early modern alchemical physicians. Alas, it won't work. *Bile* may be taste, smell, substance, or theoretical conceit; *perspiration* may refer to real sweat or some theoretical effluvium; even *pulse* is laden with theory.

Just as we find *woozy* expressive, ancient authors relied on readers' familiarity with terms rather than formal definitions. There are no pre-Hippocratic medical dictionaries, and even if there were, we would have difficulty defining the referents. Medical vocabulary is full of fossil terms that made sense in long-abandoned theories—e.g., *typhus*—as well as neologisms suggested by clever analogies. While the ideogrammatic character of the Chinese language made terms essentially analogical,[4] Greek and Roman writers too were avid analogy makers. Burke complains that their terms were "imaginative" and "imprecise," but it is hard to see how it could have been otherwise.[5] Fever healers and sufferers were trying to express unfamiliar qualitative bodily states unrelated to visible injury. By contrast, their technical, ostensibly non-analogical terms are often untranslatable, like *qi* in Chinese medicine or the names of the three *doshas* in Indian medicine—*kapha, pitta,* and *vaya.* And, notwithstanding the guidance of grammar, it may not even be clear what parts of speech are being deployed. In English, *fever* is both a noun and a verb, reflecting the recognition of fever as an active process or state of being, and *fevered* is an adjective.[6] Nor can translators assume consistency of usage.[7]

The problem may be even greater than we admit, biased as we are by faith that common biological experiences should be expressed similarly and by the expectation that ancient medical knowledge will simply be a more primitive version of our own. Notably, most translations of ancient works into modern languages were done *after* fever's definition stabilized in the late nineteenth century. Authors who had written long before 98.6°F were thus understood to have been referring to that critical temperature; named fevers were presumed to be identical to the modern diseases whose agents were then being discovered.[8]

Yet the meaning of *fever* itself has varied; and we cannot assume that references to fever imply a raised temperature at all. Unusual hotness has not always been a defining or even an essential feature of fever; references to fever may not imply raised temperature at all.[9] Susruta's definition had four elements: interrupted perspiration, heat, pain, and numbness in the limbs.[10] A general distinction in Chinese medicine was between cold-damage disorders, *shang-han* and *wenbing*, and *warmth factor disorders*, a term that would gradually be reduced from describing a wide range of conditions to describing the dangerous infectious fevers of modern medicine. Other fever terms circulated at various times and in various contexts: *ribing*, referring to warm or tepid diseases, *yibing* for epidemics, and *hanre* for the exhaustion that might accompany fever, particularly malarial. *Yao* covers much the same domain, while *chiech* refers to less serious fevers.[11] Even familiar Greek and Latin roots are problematic. The derivation of the Greek *pyretos*, the most generic fever term in the Hippocratic Collection, from *pyr* (fire) might seem clear, but the analogy need not be solely to heat. Both fire and feverish disease consume, spread, and dry. Nor did the term clearly refer to fever in general; in popular usage it may have referred solely to malaria.[12] A reference to "sharp burning heat *and* powerful fevers" suggests at the very least that "fever" was not reducible to heat.[13] The Latin *febres* was also problematic. Robert Christison in 1842 noted that *fervere* referred both to heat and to glowing—the flushed skin that was a characteristic of many fevers. Van Swieten suggested a derivation from *februare*, which referred to "a *purification* or *cleansing*" since "by a fever the body is frequently depurated [purified]."[14]

Even if we stick to hotness, anomalies arise in translating a subjective sensation—perceived by the patient, the examining healer, or both—or a theoretical construct into a thermometric entity. For, as the phrase *latent heat* reminds us, heat need not translate directly into perceived or measured hotness. And hotness itself is not unambiguous. In Chinese medicine, among the most important correspondences of yang and yin are "hot" and "cold." Though central as theoretical opposites, they do not imply sensations. Yang is hot, but more than heat.[15]

The great value of the texts of the Hippocratic corpus, composed in the decades before and after 400 BC, is that they allow us to explore the usage of fever terms in multiple contexts. While some texts may be the work of a historical Hippocrates, the "Father of Medicine" admired by Plato and glorified by Galen, many are viewed as works of associates in the medical sect of Cos, which he presumably headed, and a few are likely the works of a rival sect centered on the nearby Anatolian mainland at Cnidus. Unlike the great Chinese or Indian compendia or the comprehensive Galenic synthesis, the Hippocratic corpus is a motley. It includes the famous professional manifestoes but also theoretical texts on pathological processes, how-to texts for diagnosing and prognosticating, treatises on distinct issues of surgery or practice, lists of diseases, manuals of hygiene and regimen, and finally, seven books of *Epidemics*, which combine observations about seasonal diseases with seemingly random case histories.

These texts suggest how much terms and concepts may evolve. When now we say "I'm burning up," we are saying that our fever is a bad one, using a common idiom to proclaim severity. The Hippocratic texts contain numerous references to a malady, *kaûsos,* whose key attribute is that it is a "burning" fever. Today, burning can be an attribute of any fever, but ancient authors seem clear that *kaûsos* was a particular disease.[16] Perhaps because it had a Hippocratic pedigree, authors were still trying to diagnose cases as *kaûsos* in the eighteenth century, though there was little agreement on which illnesses to apply the diagnosis to.[17] Usually *kaûsos* was left untranslated or was translated literally into Latin as "febris ardens," then into English as "ardent fever." That translation does sharpen the question whether *ardent* or *burning* is meant as an adjective (as we use it) or is part of a compound noun. That distinction is predicated in turn on the distinction between symptom and disease and, behind that, on whether the user of the term believes in a generic fever or in specific fevers.[18] In modern English, of course, *ardent* is no longer about burning but about commitment or depth of desire. But there are still residues: think of a burning love.[19]

Other references to burning appear to refer to drying. In *Epidemics IV,* Thersander's wife suffers an acute fever associated with breastfeeding: "Her tongue, as [the] other parts were burnt up, was likewise burnt at the same time, and became rough like thick hail." (She also brings up worms.)[20] Finally, the adjective *caustic,* which shares the root of *kaûsos,* reminds us that the burning may be more chemical than thermal. Acridity would be associated with heat. Thus another sufferer has a heavy head and "burning heat" in the hypochondrium, which sounds like what today we call *heartburn*.[21] Here, the confusion is one commonly experienced at American dinner tables: spicy hot or hot hot?

Even commonplace adjectives may be difficult to convert into a modern lexicon of bodily experience. Consider the puzzling *Epidemics VI* 1.14, here in Wesley Smith's translation: "Fevers: some are pungent to the touch, some gentle. Some are not pungent but increasing. Some are sharp but decreasing to the touch, some are straightway burning hot, and some are faint throughout. Some dry, some salty, some with blisters dreadful to see. Some damp to the touch. Some are red, some livid, some yellow. And so on."[22] Whatever this once meant, it now seems a cacophony of peculiar comparisons. Particularly problematic is *pungent.* For us it applies to taste or smell, not touch. It is the opposite of neither *gentle* nor *increasing,* just as *hot* is not the opposite of *faint.* The colors presumably refer to skin—or do they?[23] Émile Littré's early nineteenth-century standard edition is no help: Littré has *mordicante* (acrid) and *douce* (sweet) for Smith's *pungent* and *gentle.* He notes too that the passage baffled Galen, who worked in medical Greek.[24] In the early modern period, "sharpness" would often be ascribed to invisible sharp particles. Sporadic references to "pungent heat" or "biting heat" (or, later, to *calor mordax*) would persist into the early twentieth century. It would come to refer to an examiner's perception that in some fevers the skin grew hotter the longer one touched it. (Ultimately one had to pull one's hand away, perhaps to plunge it into cool water.) The great nineteenth-century Leipzig clinician Carl Wunderlich hypothesized that a chemical secretion was burning the examiner's skin.[25] *Pungent* heat, along with other smells, tastes, sights, and tactile experiences once part of the practitioner's lexicon, would ultimately succumb to the triumph of quantification. Long before that, pungent heat was probably more a Hippocratic holdover than a central element of evaluation.

The mysterious *kaûsos* was one of the four great acute diseases with which Hippocratic physicians were concerned, the others being pleurisy, peripneumonia, and phrenitis. These were exceptionally dangerous, said to kill more than "all the others together." This domain—"acute" diseases—corresponds roughly to what today we would recognize as the domain of serious fevers. (The English *ague*, merely a translation of the French *aiguë*, "acute," initially comprehended all fevers, only gradually narrowing to refer to endemic malaria, which, curiously, is not included in the Hippocratic list.)[26] While what the Hippocratic physicians called *peripneumonia* is usually what we call *pneumonia*, so too, Mirko Grmek believes, was pleurisy, a far more important diagnosis then than it is today.[27] What we may see as a symptom, the delirium of the fevered brain, was a distinct disease, a "phrenitis" (or possibly a "lethargy," often an even more serious condition of mind and body).[28] These latter conditions were seen as organic diseases, not mere mental states or complications. The *phrenes*, the organ implicated in phrenitis, was located not in the brain but near the diaphragm.[29] It has gone; *frenetic* and *frenzied* are no longer technical terms; *lethargy* too has been demedicalized.

A final example of linguistic anarchy is *typhus*. We know it as a dangerous louse- or tick-borne fever caused by spirochete. In *Epidemics IV* we read of a slave cured of "a bilious ardent fever [a *kaûsos*] (of that kind which is called τυφος [typhus])." When John Redman Coxe thus rendered the passage in the early nineteenth century, *typhus*, reintroduced into European medicine a century earlier, was becoming a common English term, though it was still used very loosely.[30] Evidently deciding, as had his Latin source, that we did not know what the author had meant, he left τυφος in Greek. A generation later, the authoritative Hippocratic translator Émile Littré would render the term as *stupeur* (stupor). Littré was following Sauvages, who had revived the term in the eighteenth century. More recently, Wesley Smith has it as "delirium"—not quite the same.[31] Complicating things further, this stuporous typhus resembled none of the five diseases called *typhus* in another Hippocratic text, *Internal Affections*.[32] Nor are we apt to see stupor as the main characteristic of modern typhus or of typhoid, the very different disease that has stolen its adjective. *Typhoid* would come to designate a serious gastrointestinal disease that sometimes involved a restless "muttering delirium."

THE FEVERED

Beyond sensitizing us to the tangle of terms whose residue infects our own thought and speech, the Hippocratic texts inform us of the practice of fever medicine. The seven books of the *Epidemics* present the illnesses of about 450 people, most of them residents of Thrasos and Perinthus, in northern Greece, that occurred over a few seasons around 400 BC.[33] The descriptions of the cases, probably the work of several authors, reflect no single principle of selection or organization. They may represent student notes or contributions to an enterprise of epidemiological generalization. Many entries seem more concerned with the natural histories of diseases than with treatment. Sometimes the message is what not to do, such as treating fever with running, wrestling, or baths—"trouble on trouble"—which is how "Herodicus killed fever patients."[34]

During the period that these cases reflect, the Aegean populations may have experienced deteriorating health and prosperity. Homer had not sketched a fever-filled world; nor do Hesiod's (seventh-century BC) Boetian peasants seem to suffer. The seasonal disease described in the *Epidemics* may have been a new disease, malaria, ignited by a critical mass of parasites, people, and Anopheles mosquitoes that had accumulated by the fifth century BC. In 1907–9, following recognition by Ronald Ross of the mosquito's role in malaria transmission, he and the classicist W. H. S. Jones would explore the role of this apparently new malaria in ending a golden age and undermining the Greek nation for the next two and a half millennia: "Gradually the Greeks lost their brilliance, which had been as the bright freshness of healthy youth. . . . Their initiative vanished; they ceased to create and began to comment," wrote Jones. "Patriotism, with rare exceptions, became an empty name. . . . Vacillation, indecision, fitful outbursts of unhealthy activity followed by cowardly depression, selfish cruelty and criminal weakness . . . [set in]." Malaria was also implicated in "abortion, infanticide, universal libertinism, drunkenness, want of religion, gross superstitions," and murder, especially assassination by the cowardly means of poisoning.[35]

Nothing so dramatic seems to be occurring in *Epidemics IV,* from which I take my examples. The least polished of the seven books, it may come closer to constituting a survey of disease.[36] There one meets about one hundred people—men and women, children and aged persons—of varied classes and occupations, including slaves, artisans, even a teacher. Patients are identified by name, home, or familial relation. They suffer a variety of ailments, but fevers are far the most common. Some of these are associated with trauma or wounds;[37] others appear to represent five seasonal epidemics that occurred over roughly a year and a half.[38] Sufferers experience chills, sweats, and delirium; practitioners expect fevers to pass through many stages. The fevers may be localized at some stages, systemic at others. They will resolve in a "crisis" whose arrival may be predicted by studying the body on certain "critical days." The terse notes on a "young man" capture the practitioners' approach: "feces mucous, bilious, ripe, slimy, frequent. Continuous fever. Dry tongue. Crisis on the sixth day. It seized him again on the seventh, but departed the same day with trembling. Thick, sticky flow from the left ear on the sixth day."[39] As we do today, they use *fever* both for disease and for symptom.

There is no typical fever in *Epidemics IV,* but the following examples capture the range:

- A wound on a man's lower leg develops "a blackness." After this spreading ulcer becomes "clear," pain arises in the chest and in the ribs on the same side. The patient "[grows] feverish" and dies (§1).
- Timenes's sister (or niece) has had difficulty breathing. Her flanks become distended; possibly she is pregnant. She has "slimy, bilious" vomiting. On the eleventh day the practitioner notes an inflamed digit, along with diarrhea and vomiting. After brief improvement, on the sixteenth day

fever returns and she dies. The commentator notes: "She was fevered before the apostasis [abscess]." (§26).

- Thestor's servant has a caustic fever with purging and tightness in the hypochondrium (diaphragm). She sweats and has a crisis; then after a brief relapse the fever moves off (§9).

- Cyniscus's acquaintance has coughed up some concocted matter, had a nosebleed, and suffers from a heavy head and weakness. His feet remain warm (§53).

- Aristees, delirious, with a blanched complexion and pale green feces, at the onset of his disease has trembling fingers and trembling lips, though they were usually "rather quick and ready speakers" (§45).

- A woman from Boulagoreus, wasting away with swollen spleen, sheds tears and begins to sweat and shiver on the fifteenth day. She begins to have pain in her spleen-side (left) ear and in her thorax (§35).

- An amenorrheic servant experiences abdominal hardness and distension, then dysentery and tenesmus, followed by a week-long fever, but it is either a "slow," (Coxe), "mediocre" (Littré), or "light" (Smith) fever. She recovers full health (§38).[40]

Other texts reveal the development of a self-conscious observational agenda:

First we consider the nature of man in general and of each individual and the characteristics of each disease. Then . . . the patient, what food is given to him and who gives it—for this may make it easier for him to take or more difficult—the conditions of climate and locality both in general and in particular, the patient's customs, mode of life, pursuits and age. Then we must consider his speech, his mannerisms, his silences, his thoughts, his habits of sleep or wakefulness and his dreams, their nature and time. Next we must note whether he plucks his hair, scratches or weeps. We must observe his paroxysms, his stools, urine, sputum and vomit.

Yet another passage suggests that the patient may know his or her fever best.[41]

As the practitioner's interest in "speech, . . . mannerisms, . . . silences, [and] . . . thoughts" indicates, the Hippocratic fever was a condition of mind as much as of body; the distinction itself is anachronistic and problematic.[42] Observers were interested both in unaccustomed behavior and in the ideation leading to it. Recovering from intoxication, a youth becomes delirious. He presents with a rough tongue and a burning fever, along with a tumid belly and sleeplessness, then begins behaving "very wantonly, getting up, fighting, and talking very obscenely, contrary to his usual manner." A woman suffering chills and heaviness "threatened her child irrationally," then becomes quiet, then "sat up and rebuked those who were there."[43] Delirium, sometimes understood as phrenitis, receives much attention. The patient feels "loathing," avoids others, seeks darkness, and may see the dead.[44] One writer describes the hallucinations generated by rising bile as follows: "There seem to appear before his eyes reptiles and every other sort of beasts, and fighting soldiers, and he imagines himself to be fighting among them; he speaks out as if he is seeing such things, and he attacks and threatens, if someone will not allow him to go outside."[45] Sometimes, such details had prognostic significance in an aphorism-based medicine. Still recognized are *crocydismos* and *carphologia:* picking at walls or one's bedclothes augers death.

The Hippocratic writers appear to have been more interested in delirious speech and behavior than their Chinese and South Asian colleagues were, or than Galen and his followers would be. Not until the eighteenth century would there be similar fascination with delirium. Interest in the patient's feelings sometimes translated into a remarkable recognition of the importance of sensitive nursing. A dark room and the "softest bed-clothes" might be a major factor in recovery, since in some fevers the patient must remain still for a long period.[46] Discussion of involuntary seminal emissions during fever includes recognition of the patient's embarrassment.[47]

DIFFERENTIATING FEVERS

In Hippocratic and classical Chinese medicine, naming and classifying diseases are less important than understanding the internal events of illnesses. The *Epidemics* writers record the sequences of symptoms in individual cases. Others imagine the dynamics of the febrile process, but there is no diagnostic mandate, no belief that cure rests on getting the name right. The few Hippocratic nosological works, which are detached from the rest, are among the oldest texts in the collection. Some likely represent Cnidus rather than Cos and an Egyptian or Mesopotamian orientation toward list making over system building.[48] For the most part these works describe localized afflictions, usually presented in order from head (the host to very many ills) downward. Besides noting symptoms and courses, writers address pathology, prognosis, and treatment. There is much overlap, and rarely are descriptions adequate for diagnostic use. In one of the earliest texts, sections 12–75 of *Diseases II*, many diseases are not even named, distinguished only as "another disease." But others are named, and there are instructions for treating tertians and quartans, the most distinctive forms of malaria (§§42, 43). *Kaûsos* (§63) already exists, as do pleurisy (3 types) (§§44–46), pneumonia (§47), lethargy (§65), and phrenitis (§72). There are also a "malignant" disease (§67) of "fever and chills," of overhanging eyebrows with pains in the neck, head, and groin, and nonsensical speech (deadly on the seventh day); a "livid" disease, which is a dry fever in which the patient cannot look up (§68); and an unnamed fever of the head, detailed with remarkable sympathy: "If anyone moves him, he vomits copiously and easily; his teeth are set on edge, and he is numb. The vessels in his head are raised and throb, and he cannot bear to be still, but is beside himself and frenzied from the pain." Pus from ears or nostrils will relieve it (§16; see also §19).

Another early text, *Internal Affections*, lists the five typhuses (none featuring stupor *or* corresponding to the modern typhus). One, with edema, follows a tertian or quartan; in another there is joint pain; a third is characterized by emaciation and hunger. The final two typhuses are a late summer typhus with diarrhea and distension and a generic summer fever resembling *kaûsos*, characterized by "sharp burning heat and powerful fevers; weakness and loss of control of limbs from heaviness. . . . violent colic in the guts, foul feces" (§§39–44).[49] *Internal Affections* also lists three ileuses, which would be febrile diseases: a winter disease characterized by chills, leaden countenance, headache, trembling, and panting; a summer disease of swampy areas with similar characteristics; and an autumnal "sanguineous" ileus (§§44–46). Diseases called "jaundices" also often included fevers.[50] The texts say nothing about how these designations were arrived at, what causes the different conditions, nor how common they are, where they are rife, or who gets them. Rarely can these diseases be equated with modern species, but the lists do seem to be the legacy of laborious generalization at some remote time.

Susruta's nosology, by contrast, is formulaic. Theory dictates all possible fevers. There are eight—one for each of the three single *doshas (kapha, pitta,* and *vayu),* one for each possible pair, one for all three, and one for fevers from other causes. These may be subdivided further by parts of the body affected. Thus any fever may be designated by class and species. Concerned with symptomatic treatment, classical Chinese medicine emphasized the changing internal location of the evolving fever.[51] The great Chinese texts were concerned more with active properties and processes, more with explanation than with denomination.

PATHOLOGY

What is occurring within the fevered body? Analogies abound. Their intended status is rarely clear—fact? theory? something looser? For Hippocratic authors, fever is variously a matter of heating and cooling, of large and small structures, and of the actions of biles.[52] In traditional Chinese medicine yin-yang and five-phase explanations coexist. A Hippocratic author begins the text *Ancient Medicine* by rejecting a view of fever as a mere reflection of the body's ambient environment. Instead, feverishness involves overcompensation: a hot flush follows a cold bath; one shivers on leaving the fireside; chilled feet burn as they thaw; the greatest heat alternates with the strongest chills. Exposure to heat or cold can have the opposite effect, the author explains, because other factors are operating: a thing may be "bitter and hot, sharp and hot, salt and hot, and countless other combinations both with heat and cold." Sufferers from a common cold will notice an unusually acrid nasal discharge, for example.[53]

Regimen II explains how these effects are produced. "Things acid, sharp, harsh, astringent" affect the body, "some by drying, others by stinging, others by contracting, make the moisture in the flesh shiver and compress itself into a small bulk."[54] The body's interior architecture was a key factor. To these writers, what are for us "organs" were mainly containers, varying in shape and hence in function. Some saw the abdomen—"the (lower) cavity"— merely as a large hollow space, a passageway between the body's two ends.[55] The combination of shape and surface determined the proper distribution of wetness, the key to sound functioning. Heat and cold, the *Regimen III* author suggests, are simply proxies for moisture regulation. A morning walk may bring on feverish trembling and heaviness of the head, because "the body and the head are emptied of their moisture."[56] Container-shaped organs attract moisture, explains the *Ancient Medicine* author. Women, with their container-shaped wombs, are moister than men. Chinese authors agree.[57]

The author of *Regimen in Acute Diseases (Appendix)* uses moisture-based reasoning to address the contentious issue of when and what to feed the fevered patient. Since inappropriate or mistimed food may nourish the fever rather than the resisting body, he pays great attention to the proper uses of the many gruels, or ptisans. Practitioners are warned not to feed a person when his or her feet are cold. Cold feet are both a consequence and a cause of fever. Eating enough to reheat the feet will cause fever to be "kindled" in the thorax, in turn overheating the head, whose subsequent cooling will then draw heat from the extremities, resulting in cold feet again.[58] Sometimes such thermal vacillation was attributed to the mixing of bile and phlegm with blood. Initially that would chill and congeal the blood, causing shivering as blood vessels contracted. As the blood was rewarmed, the bile and phlegm would be heated too; the resulting mix would be "many times hotter than normal; . . . fever follows of necessity because of the overheating of the blood after the chill." That initial surfeit of bile and phlegm might be due to diet, exertion, wounds, external heat and cold, or even "seeing and hearing, but least of all by these."[59]

Susruta envisions something similar. Their heat pushes the altered *doshas—kapha, pitta,* and *vaya*—into the "Rasa (lymphchyle)," obstructing circulation and perspiration, "impair[ing] the digestive fire and expel[ling] the inherent heat *('Ushman')* out of its seat" and over the body. Susruta also cites similar precipitating factors: a blow, a prior illness, a suppurating abscess or ulcer, fatigue, diet (and unwise conduct), grief, but also poison, change of season, astrological factors, and "the smelling of any kind of poisonous herb or flower."[60] Whatever the mechanism, the attraction of familiar dynamic processes involving combustion or boiling is plain. If in a loose sense, distillation—moisture-filled cavities and condensing heads—will long supply the master analogy; the fevered body will be seen as a thermodynamic system.

Closely associated were idioms of cooking and digesting, seen as aspects of a single process.[61] For the Hippocratic writers, the entire fevered process is a "coction," akin to digestion. There are three stages: the cooking, or *pepsis,* of some entity must be followed by a separation, *apostasis,* and then by an expulsion. In this process, as in distillation, some components (of blood or phlegm) become rarefied, or gassy, while others are left as residues. The former move, bubble, and cross membranes; the latter stay and stick and may rot and infect what is around them. In a successful resolution, "there melts away from the phlegm and bile . . . the finest part," *Diseases I* explains. Harmful material that has been "burnt up" or "thinned," that is "weak," is boiled off, transpired across skin in a welcome warm sweat, or exhaled in breath. By contrast, a cold sweat indicates that there is too much "peccant" material or that it "has not yet [been] brought to maturity, or thinned or burnt up." It is a thicker and "more ill-smelling" sweat portending a longer and more dangerous illness.[62]

Therapeutic judgment then involves monitoring these three processes and deciding whether the cook, Nature, requires any assistance. Thus *Regimen in Acute Diseases (Appendix)* reminds the doctor to give no gruel until "their urines or sputa have become mature," that is, properly cooked. (For self-diagnosers, mature sputum resembles pus, while urine should have the reddishness of vetch meal.) Death occurred if coction and expulsion failed or overtaxed the patient.[63] Often there is a distinct internal geography to these processes, and we can track a fever's course. A part of the body (e.g., an organ or a side of the body) that is hard becomes soft. Always the healer must consider not only the parts that are affected but also how far the fever has progressed in its likely course and how to push it on toward resolution and exit.

For any of these traditions, we should be wary of textbook suggestions of a formulaic humoral medicine or any grand dichotomy between pathologies of the fluids (humoralism) and of the structures (solidism).[64] Instead, it is the dynamic interaction of fluids with structures that produces the sequence of events that make up disease. Generally, what is imagined to occur within a body is but a version of familiar events outside it. But sometimes authors must look beyond these prosaic explanations. Usually, structures are characterized by passivity, yet they may also have active properties (e.g., moisture retention) or what we view as chemical properties that challenge those frameworks.[65]

Especially important for the Hippocratic writers, and even more so for their successors, are interactions within vessels. Thirsty travelers suffer a summer fever from the drying of the small vessels, whose contraction attracts "sharp bilious sera," causing "great fever," with prostration and belly pain.[66] The vascular emphasis, along with a complex geographic view of the body in which fever is movement through a landscape, is most fully developed in Chinese medicine. Unlike the interior of the Hippocratic "hollow man," in which spaces, heat, moisture, and gravity determine febrile events, the Chinese interior is a collection of organ states linked by multiple blood- and *qi*-carrying channels. In his "plain conversation" with Huangdi, the "Yellow Emperor," the physician Qibo explains the common course of "cold attack" (fever), which moves day by day from one domain to the next. The region of the great yang is attacked first, with head and neck pain. Next affected is the channel of *yangming*, the eyes and nose, with hotness, dryness, and sensitivity to light. The route of *shaoyang*, or lesser yang, links the ears to bones along the flanks; there the cold attack brings hearing loss and pain to the chest and hypochondrium. Then, in *taiyin*, the stomach and the throat are affected. On the fifth day, the fever has moved on to the lesser yin: the kidneys, the lungs, and the tongue are implicated, resulting in thirst. Transfer to the *jueyin*, linking the genitals and the liver, results in a shrunken scrotum but also depressed affect. This model is subject to multiple variants, Qibo explains. Multiple channels may be affected simultaneously; organs may be affected; the qualities of *qi* may change—all dangerous possibilities. Alternatively, if cold damage is limited to a single channel, it will usually disappear within a week or two.[67]

Beyond the direct effects of an imbalance of yin and yang and unexplained qualitative impairments of *qi*, these pathological problems are effectively structural: obstructed blood vessels, stagnant flow, or reversal of the direction of flow. The Chinese preoccupation with channels is of a piece with the centrality of pulse diagnosis and needling or moxibustion therapies. What Qibo's account makes clear, however, is that fever is less a state than a process: the whole has become destabilized, setting off gyrations of cause and effect.

CRITICAL DAYS AND CRISIS

Today, the hallmarks of medical authority are diagnosis, an explanation of what is wrong, and a scientific therapy. Course and prognosis are bound up in diagnosis. But most classical authors writing on fever sought authority in their assertions of the schedule according to which the disease would unfold. The failure of real cases to conform might warrant comment but did not detract from the essential validity of their rules for the resolution of fevers. I have noted already a Chinese view of fever's ideal course. The complexity of the interior landscape in Chinese medicine accommodated considerable flexibility, even in the malarias, exemplars of regularity. Susruta, by contrast, is rigid. The time of day a fever begins and its length are key elements of diagnosis. Fevers of kapha come in the morning, of pitta at noon, and of vayu in the evening. Vayu fever worsens on the seventh day, pitta fever on the tenth, and kapha fever on the twelfth.[68]

Hippocratic authors too were preoccupied with the schedules of fevers. But they concentrated on relating a person's condition on particular "critical" days to his or her condition on later critical days and to overall outcome. Thus, "if a continuous fever patient suffers distress on the fourth and the seventh days, and he has no crisis on the eleventh, he generally dies."[69] The approach may well reflect the prominence of malaria in the ancient Mediterranean world, with its well-defined cycles.[70] Sometimes the rationale was therapeutic: knowing when a change was due allowed one to anticipate and divert the fever onto a safer track. The lore of critical days figures prominently in books of aphorisms, and it is tempting to see the writers of the *Epidemics* as data collectors for the aphorism makers, noting when in a disease particular events occur in a grand effort to discover the natural laws of disease. Yet no records of any such data-processing project have survived, and others, noting the peculiar concentration on fours and sevens (events on the fourth day predict those on the seventh) and the preoccupation with odd or even, emphasize magic—Pythagorean number theory—rather than empiricism. And doubters such as the first-century commentator Celsus would observe that critical-days theorists differed and that most of the days up to about the twentieth had been identified as "critical" days.[71]

Whatever their origins, the preoccupation with course and critical days gave structure to illness, helping to distinguish diseases and possibly guiding practice. The scheme may have served as a sort of clinical decision tree. Notably, as the period of the disease lengthens, the intervals between critical days lengthen: they go from occurring roughly every other day to every fourth or sixth day and then to every ten or twenty days. By the fourth day many fevers will have resolved or will have killed their victims, the author of *Crises* explains. Hence those fevers that remain are a natural set, as are those persisting on the seventh day.[72] After about three weeks, the character of febrility shifts again: the fever may take years to resolve or may remain for life. None of this is clear at the outset.

Accordingly, Langholf supposes that critical-days schemes functioned as a "sort of ideal 'time-table' [with the physician] paying most of his visits on the days he regarded as potentially critical" and revising that schedule as the fever evolved. Celsus would emphasize the downside of the practice: doctors saw patients when their fevers *should be* shifting, not when they *were.* And indeed the Hippocratic author of *Crises* had admitted that "none of [the customary] periods can be numbered exactly in whole days, any more than the year or the months . . . consist exactly of whole days."[73]

The most important event on a critical day was "crisis" itself, that great juncture that led either to death or to recovery. Thus, for example, "patients with ardent fevers have their crisis in fourteen days, which relieve them or carry them off."[74] Since the outcome often depended on the regimen in place when the crisis came, accurate prediction was important. Physiologically, crisis was expulsion: what had been cooked exited "through either the mouth, the cavity, or the bladder," according to an early text, which notes that "sweating is a form of resolution common to them all."[75] (Many added hemorrhage, including nosebleeds and menstruation, which had a profound resolving power.) Such an event having occurred, the healer could declare that the patient was over the worst. The fever and delirium would subside, and a long, peaceful, strengthening sleep would follow. Recovery would proceed steadily so long as the patient followed the right regimen.

While crises would become emblematic of fevers, Hippocratic writers applied the concept generally—again, an indication of fever's status as an exemplar of disease.

As an empirical entity, "crisis" was ambiguous. In a disease of oscillations, which expulsion was the true crisis?[76] In succeeding centuries, physicians would note "imperfect," "incomplete," or deceptive crises. They would struggle with diseases in which relapses (by definition, postcrisis events) were more dangerous than the original fevers and acknowledge that some fevers (e.g., typhoid) often resolved without an obvious crisis.

As with *critical days,* the meanings of *crisis* went beyond the secular. This physical transformation was intertwined with spiritual redemption. In the context of popular religiosity, *miasma* (a term that would become familiar in nineteenth-century epidemiology) signified a state of "individual or collective impurity," and the *kátharsis* that resolved that impurity was, like crisis, a purgation. *Phármakon,* the term for the means by which it was accomplished, refers to a "charm or ceremony" but also to a "drug," notes Langholf.[77] "Fever" too, and particularly malaria, involved divinities; in Rome the cult of Dea Febris was ancient, and Tertiana and Quartana were deities for the associated malarial fevers. Antifever amulets and spells, including the famous "abracadabra," remained important in Europe well into the nineteenth century.[78]

Though the importance of crises and critical days rose and fell, mid-nineteenth-century physicians were still understanding fevers in these terms, a reflection of the ongoing authority of Hippocrates, the persistence of coction-expulsion concepts of fever, and also the understanding of lives in terms of decisive and cleansing events, whether conversion experiences, sacramental transformations, or even good and peaceful deaths. Critical-days lore entered popular culture. In Elizabeth Gaskell's *Ruth* (1853) a rural Welsh innkeeper questions a practitioner's assertion that a fever will peak that night: "It is four days since he was taken ill, and who ever heard of a sick person taking a turn on an even number of days; it's always on the third or the fifth, or seventh, . . . He'll not turn till to-morrow night, take my word for it."[79] Mrs. Jones may not reflect the latest learned opinion on which days were truly the critical ones, but everyone knew that fevers operated by such laws. By that time, however, critical-days concepts had little practical significance and were ceasing to be a measure of medical authority. Increasingly, keeping track of the days on which febrile events occurred was merely the mapping of the disease's course.

Of course, crises still occur in many diseases, but febrifuges and antibiotics have made time sensitivity less important: if you have the ill, take the pill. Yet much of classical fever therapy was directed at facilitating a resolution. A preoccupation with critical days reflects the complexity and protractedness of fevers, their delicacy, the frequency of relapse, and the importance of *timely* intervention. And if modern medicine is no longer fixated on sevens and fours, it is on nines to fives, nights and weekends.

ON TO GALEN

Despite profound changes in medicine in the half-millennium that separates them, Hippocrates is often linked to Galen (128–200).[80] Together they are seen as cofounders of the infamous mantra of the four humors, the spell that would enchant the great Islamic physicians and the learned Latin masters for another millennium and a half.

That foreshortening not only overlooks an extraordinary record of medical exploration in the intervening Hellenistic era, it underemphasizes differences between the two in three critical areas. First, Galen was dedicated to the textbook ideal of a coherent and comprehensive system of how the body worked, what went wrong with it, and how disorders might be put right. Second, there were significant conceptual differences related to fever. Third, considered in the next chapter, there were enormous differences in the contexts of medical inquiry and practice.

In pursuit of a learned medicine, Galen did filter, arrange, and unify the motley Hippocratic writings. Others had appreciated the Hippocratic achievement, but Galen would be the most successful of the appropriators. Via systematic commentary, he made Hippocrates confirm both Galen's own views and Plato's. That coherence relied, however, on the elevation of a single work, the Hippocratic text *On the Nature of Man,* as definitive of Hippocratic views. To others, including later historians, it has seemed a simplistic overview expressing views inconsistent with other Hippocratic texts. But from it Galen welded a theory of four humors to an Empedoclean-Aristotelian cosmology of four elements, giving us the familiar humoral theory.[81]

The artificiality of a Hippocrato-Galenic common front would only begin to be apparent following the Renaissance humanists' retrieval of many more texts by both.[82] In the eighteenth century, Hippocrates's admirers would seek to restore the purity of the Hippocratic heritage. Galen, doctrinaire and insufferable, became the villain: he had profited from the Hippocratic achievement while distorting it. Yet the restorers too were selective. In fact, they and Galen were doing what any recipient of past knowledge does: extracting what was useful from the illustrious predecessor, overlooking embarrassing errors, and reinterpreting obscurities in terms of contemporary knowledge.[83] And however much latter-day Hippocrateans might glorify the ideal of medical empiricism, a full extrication of pure Hippocrates from Galenic corruption was impossible; Galen's conceptual mortar still bound texts together and underwrote medicine's intellectual authority.

As in the Hippocratic writings, fever would be the paradigmatic disease for Galen and for centuries of successors. It dominates the explanatory part of medicine, as well as nosological, diagnostic, prognostic, and therapeutic discourse. It is also at the center of a model of a patient-practitioner relationship based on a healthful regimen. While Galen, like Susruta, strove for a system in which all phenomena had a clear relation to all others, he offers no comprehensive treatment of fever.[84] This reflects both the breadth of febrile phenomena and the circumstances in which Galen wrote. However much he aspired to comprehensiveness, as an omnispecialist Galen often addressed topics incidentally and in polemical contexts. Many of his books read as incisive, sometimes savage review articles. Rhetorical tactics interfere with strategic, systematic exposition.[85]

I focus here less on Galen himself than on the elements of what would become a composite "Galenic" fever theory, to which predecessors and successors contributed as well. It will be evident that these are elements, not doctrines; single authors, including Galen, often handle aspects of them in incompatible ways. They include

- an appeal to the balance and quality of four humors to explain fever (and other diseases) and development of a nosological framework that will distinguish diseases pathologically as well as clinically, thereby allowing a distinction between disease species and individual case;
- the emergence of a "black bile" as the universal bad boy of pathology;
- an interest in internal putridity, even recognition of "putrid" fever;
- a focus on vessels over other structures;
- a focus on the pathological properties of the heart, blood, pneumas (spirits) in the vessels, and chiefly the arteries;
- an emphasis on preternatural heat as the defining feature of fever;
- a fixation on the pulse as the chief indicator of the internal state;
- a greater emphasis on bloodletting as the primary therapeutic modality; and
- a comprehensive approach to causation that addressed the disjunct between individual illness and general disease.

These elements take us in a circle.

Humors, Black and Putrid

The sheer simplicity of standard humoral theory is beguiling. The ancients, we learn, believed in four humors: blood, phlegm, and yellow and black bile. In a healthy body these are balanced in keeping with the person's (sanguine, phlegmatic, choleric, or melancholy) "temperament." The four humors and temperaments are readily linked to four elements (earth, air, fire, water), four qualities (hot, cold, moist, dry), and four fevers (continuous fever [blood] and three we now call malarial: quartan [black bile], tertian [yellow bile], and quotidian [phlegm]).[86] Those for whom pathological anatomy was the real route to medical progress would later complain that concentration on the body's fluids ("the humors") deflected attention from injuries to structures ("lesions"). Yet many diseases do involve altered fluids or flow, especially if "fluid" includes whatever nerves transmit. For fever, a sharp dichotomy between humoralism and solidism is unhelpful and often anachronistic.

No less remarkable than humoralism itself is the selection of the familiar four. Other fluids would be important to fever theorists: chyle, pus or "ichor," milk, urine, semen, perspiration, lymph, a nervous fluid, and a mysterious "serum."[87] Some Hippocratic texts used a pathology based on two humors, phlegm and bile; in *Diseases IV,* which uses a pathology based on four humors, they are blood, phlegm, bile, and water (associated with the spleen).[88] Susruta's three humors include an inner wind, along with bile and phlegm. In Chinese multivascular medicine distinct humors may not be named, though the quality of what flows is central.[89] Within the Galenic heritage there was ambiguity concerning whether blood was a single humor or a mix of all the humors. Nor were their attributes fixed. Phlegm, once a hot humor associated with inflammation, became a cold one.

Standard humoral theory assumed two biles, yellow and black, which ultimately would be seen as separate substances. *Bile* and its adjective, *bilious,* are among the most frequently used and most frustrating terms in Galenic medicine; they are variously subjective, objective, and theoretical, used to refer to stains, smells, and sometimes tastes of vomit (or feces) and the interior doings they presumably indicate. The infrequency of rising bile, its association with illness, and its bitterness, hotness, or sharpness made it more distinctly pathological than blood or phlegm and invited physical models of bile as sharp (acrid) particles that pricked one from within. But why did standard theory assume two biles? Plato notes red, white, and bile-colored (yellowish) bile. He doubts their common essence, yet here too noun-adjective ambiguities arise.[90]

Exploring the utility of such schemes is more fruitful than seeking their rationale. The four-humors model reflects a characteristic Galenic approach of creating a logical map of pathological possibility. Thus, there are three possible locations for a disease: in an organ; in the body's physical frame; or in its *homoiomeres,* matters that pervade it, such as fluids, flesh, or connective tissue. In each, a limited number of things may go wrong. Damage to an organ, for example, will be to its size, its shape, or its structural integrity.[91]

Most fevers are homoiomeric diseases, affecting the whole body without obviously damaging its fabric.[92] Usually, they are manifestations either of *dyscrasia,* the excess of a humor, often from ill-advised eating or drinking, or of what would come to be known as *cacochymia,* a bad, "corrupt," or "putrid" humor. Galen's fever theory had two further dimensions: determination and degree. Like the proverbial loose cannon, the pathogenic humor(s) coursed through the body, moving from site to site and perhaps getting stuck; one spoke of their "determination" to a particular part, resulting in local inflammation that might lead to systemic feverishness.

While the approach may seem to allow infinite variability, it allowed Galen to demarcate any disease by classifying it in all the necessary dimensions. In the next chapter I will outline the dominant Galenic fever classification; here it is enough to note that Galen's approach, which nicely united the agenda of specificity with physiological explanation, proved an excellent way to organize knowledge. In contrast to the motley lists of the Hippocratic nosological texts, here diseases are systematically laid out.[93] Aberrant cases can then be represented as instances of coexisting diseases or as consequences of unusual contingent or constitutional factors.

Galen's framework would reify the distinctiveness of black and yellow biles, each with its own organ and diseases. The black bile would be the most problematic conceptually as well as the most dangerous. It would prove difficult to isolate such a substance and to theorize any physiological function for it. Retrospectively, we

may understand much blackness as the residue of internal hemorrhage.

Multiple maladies beyond the obvious melancholia would be
attributed to this black bile. Often it would be particularly associated with corruption (rather than mere imbalance) of humors and
with "putrid" fever. Here too, noun-adjective ambiguity arises. As
a theoretical term, *putrid fever* would refer variously to any of a
large class of fevers, including all intermittents and many forms of
continued fever, which were due to corrupted (putrid) humors. It
would refer also to a disease species (putrid fever). As an observational term, *putrid* usually referred to necrotic processes in parts of
the body. Sometimes the index of putridity was tissue destruction:
plague buboes would be the exemplar. Foul smells—of breath,
body, or excreta—might also indicate putridity, as would mul-
ticolored discharges or great oozing reddish or blackish blotches
on the skin. Sometimes cases of diseases not inherently putrid
might become such, as in a "putrid" smallpox.[94] Usually, putridity signaled danger. There is a broad evolution of usage. The first
type, putridity in a theoretical sense, while characteristic of early
modern Galenism, is rare after 1650. Though it has been tempting to link putrid disease to exposure to rotting filth, and thus to
see an anticipation of the sanitarianism of the nineteenth century,
usually putridity is a pathological rather than an environmental
concept, albeit one resting on the familiarity with destruction of
organic matters by decay.[95]

Galen associated putridity with a smoky residue from the combustion of the "vital spirit" of the arteries. Among successors there
was no single theory of the underlying process beyond a general
association of putridity with heat and moisture.[96] *Putridity* would
go out of fashion as a common descriptor of fevers during the first
half of the nineteenth century, though late-century writers would
still sometimes apply the term in reference to tropical diseases.

Vessels, Heats, Pneumas, Pulses

How do bad humors make fevers?

Galen defined fever as preternatural heat. He and his successors had difficulty defining preternaturality in terms other than the disruptive heat of fevers, however. Moreover, inasmuch as heat might be a substance rather than the condition of hotness, the differences between good heat and feverish heat might be qualitative, differences of kind rather than merely of degree.[97]

In the centuries after Hippocrates, anatomists had disclosed a far more complicated human interior. With the recognition of each new structure came the question of what it did. What of the multiple vascular systems, for example: veins, arteries, and nerves? Galen held that the veins, radiating from the liver, carried nutriment. The nerves communicated an animating (animal) spirit. The pulsing arteries, linked to the heart and lungs, distributed a vital spirit, or pneuma, extracted from inhaled air and burnt to generate the special heat of life. Though Galen's soot-based explanation could combine a surfeit of food (contributing to incomplete combustion and putridity) with vascular constriction, perhaps from cold, the pneuma theories that flourished in post-Hippocratic medicine strained the earlier, easily visualized kitchen analogies.[98]

Galen and his successors located the center of systemic (as opposed to localized inflammatory) fever variously in the heart, the vena cava, the arterial blood (full of the soot from pneuma combustion), or, later, the arteries themselves. A controversial issue over the long career of Galenic medicine would be whether terms like *pneuma* or *spirit* were sufficient or required an ulterior physical explanation in term of particles or flow. The sharpness of their particles might explain the properties of humors, for example, as they did for Plato and some Hippocratic authors. Or perhaps particles got stuck, creating back pressure; or rapidly flowing humors eroded vessel walls, allowing contents to leak out and cause mischief.

Artery-based pathology coevolved with pulse-based diagnostics. Whether caused by the properties of the flowing pneuma, the impelling heart (especially important after William Harvey's famous seventeenth-century research on blood circulation), or the vitality of the vessels themselves, a pulse was felt. Galen would make the pulse the chief point of access to the fevered interior. Hippocratic writers had ignored pulse variations, while their Chinese contemporaries developed a system of subtle discrimination of the body's many pulses. Galen would emphasize qualitative characterization of diastole and systole, estimation of the periods following each, and other quasi-spatial dimensions. Some predecessors had urged more complex pulse taxonomies; and Galen's system, a twenty-seven-node matrix, would be too complex for many of his successors. Even experienced fingers could not distinguish all the pulses the theorists could imagine, some complained.[99]

Bleeding

Humoral imbalance was almost always due to surfeit rather than deficit. The usual response was humor removal, and particularly, in fever, phlebotomy. In principle, Galen's position—"bleed in acute diseases except . . ."—was identical to Hippocrates's. But while the gruel-dispensing Hippocratic authors had been reluctant bleeders, seeing many exceptions, Galen saw few. Only those who could stand it should be bled, all agreed, and usually only at an early stage of fever. Youths, young adults, males, and the plethoric could stand it; Galen—who had, after all, had a gladiator-based training—tended to calibrate to that norm, at least on this issue. His defensiveness reflects competition with the milder regimen favored by Methodist rivals, who found in Hippocrates authority for a dietetic regimen. But Hippocrates too, Galen insisted, would have favored bleeding—judiciously, but copiously (and usually to a faint), but always with regard to the patient's constitution.[100]

Etiology and Epidemiology

Constitution, in turn, would become a part of a causal analysis. Galen's approach to causation was as important as Koch's famous postulates would be eighteen hundred years later. Pathological theories might come and go, but Galen's view of what a causal explanation should entail persisted until around 1850. Galen and

Koch addressed different causal questions. Galen sought to explain the differential courses of individual illnesses rather than the necessary agent of a disease. Indeed, the revolution of the germ theory represented as much a transformation of appropriate causal questions as of correct answers.

Galen defended causal inquiry against those who saw it as futile, even dangerous. They worried that in a medicine in which cure was often the reversal of cause, an erroneous assessment of cause might lead to harmful therapies. His chief opponent was a predecessor, the third-century BC Alexandrian anatomist Erasistratus. To Erasistratus, causation implied necessity. A true cause always produced the same effect. A test case was fever arising in a few members of an audience in a crowded, overheated theater. From an Erasistratean standpoint, unless all attendees took fever, the conditions in the theater could not be the cause.

Erasistratus's error, Galen explained, was to assume that causes were singular: he should have known, as Celsus (who had made the same argument) put it, "that nothing is due to one cause alone." For Galen, *cause* meant whatever "by its nature contributes" to an effect.[101] He included as causes differences in individual susceptibility. His stock example was the difference between athletes, who "endure prolonged and very violent movements . . . without being fatigued," and "ordinary people, [who,] even if we toil only slightly more than is customary, are immediately distressed."[102] Accepting the Erasistratean view would also undermine health, since it was predicated on the absurd position that whatever did not directly incite fever was harmless. Were Erasistratus right, a physician should not

> prevent anyone from exercising all day and all night through fear that they might suffer from exhaustion; nor . . . forbid them late nights, nor shield them from the summer sun, nor shall we provide clothing for the servants, or require shoes for ourselves. For since no harm is done by hard work, late nights, heat or cold, we shall view these things as empty superstitions. Nor shall we rebuke those who have overeaten or got drunk.[103]

For Galen, knowledge of cause had therapeutic implications. Those who ignored cause waited for symptoms to appear before belatedly treating them. By contrast, those who understood a fever's trajectory could anticipate and perhaps forestall dangerous symptoms. Thus the wretch with an Erasistratean doctor "must wait until the third or fourth day [for relief]. And though he could have recovered by the third day [under Galen's care] he must contract some more serious disease."[104]

By *cause* Galen meant both the internal correlates of manifest symptoms *(proximate cause)* as well as external events that presumably precipitated a disease *(remote causes)*.[105] Remote causes included factors external to the body, such as environmental heat or cold, miasms or contagia, as well as a body's predisposition (or diathesis) to suffer certain harms, a consequence of both contingent and hereditary factors. None of these need be a necessary cause. Any of them, singly or in combination, might bring on active disease, in the same way that a match ignites tinder, an image well suited to fever.[106] Some of these factors, partially manageable, would gain fame as Galen's six things non-natural: diet, exercise, exposure to weather, excretions and secretions, sleep, and the passions of the mind. Their management would be the nexus of medical advice in much of the world for the next seventeen centuries.

What Galen systematizes was implicit in other classical medical systems. The differences between persons explain both the variable presence of disease and its remarkable variety.[107] The Hippocratic authors had not dwelt on cause, yet they knew that the diseases of a season brought some unity and much variety. If there is a unique aspect to Galen's approach, it is the legalistic context in which he raises etiological questions. He attacks Erasistratus in an imagined courtroom. The Erasistratean advocate will deny that the beating of an elderly woman could have caused her death, since a similar pounding of a healthy youth does not produce a similar effect.[108] But with prosecutor Galen's help, any court will see through the ploy.

With regard to the social determinants of fevers, the analogy opens the door to fundamental issues of public health and of justice. Ultimately it will raise questions about where, whether in the name of obligation or efficiency, a society should insert the pry bar to shift the causal nexus toward healthier humors and on behalf of what model of person—crone or strapping gladiator? Galen himself overlooked such matters. The fevers of his aristo-cratic patients came from bad life choices. They could relocate, moderate their partying, and avoid worry, if they only would.

CHAPTER THREE

Books

❖ ❖ ❖

"Galenism," syncretic and shifting, would become the medical philosophy of Byzantium, Islam, and later the Latinate world and the European empires. Though often remembered only for the dogma of the four humors and an erroneous map of the circulatory system, Galen was more important as an architect of medical reasoning, governor of vocabulary, and populator of the domain of objects, agents, and properties. "Even down to modern times [Galenism] has had its partisans," a mid-nineteenth-century medical author would declare.[1] Over the centuries that ancient framework would flex to embrace new febrile diseases—plague, the "nervous fevers," eventually yellow fever, cholera, and others. Even more impressive than the breadth of Galen's writings over a long life was their depth.

In search of precursors, historians have often pounced on bits of modernity in Vesalius, Paracelsus, Harvey, or Fracastoro. But these authors retained more than they rejected; their novelty was overwhelmed by antiquity.[2] For in many and sometimes subtle ways Galen's philosophy still dominated the agenda, dictating what needed to be explained and in what ways.

Galen succeeded because he won a battle of books. Until the early nineteenth century, the literature of fever was more a matter of tracts and treatises than of hospitalized cases, epidemiological statistics, or dissecting-room or laboratory observations. Rarely are Hippocratic texts conspicuously literary. Only a few declamations to a wider public appear among manuals, lists, notes, and aids to memory.[3] But the context of authority changed as medicine became centralized in the imperial capitals of Alexandria and then Rome. There, would-be healers, uncredentialed by any state or professional body, competed for elite clients, who could pick from among the multitude "clustering round the bedside of the sick to offer their independent advice."[4] Galen flourished in that environment. Starting out with good connections (leading to a public appointment to care for gladiators in Pergamum), he quickly rose to become chief physician to the emperor Marcus Aurelius and then to his heir Commodus. While singular cures, properly presented, might impress, so too did philosophy, and Galen was a first-rate philosopher. One got the best in medical arguments by marshaling the medical past and confidently claiming to know most about the body. There Galen excelled.

Similar factors operated in the other classical medicines. Indian practitioners must learn to debate, counseled one author, adding that medical debates might be friendly explorations of an obscure question or no-holds-barred struggles between professional rivals. Often fever—that mysterious condition of body and mind varying in time, place, and character and in responsiveness to the healer's skills—was the focus of these debates. Whoever could comprehend and conquer it would win such trials.[5]

In such academic medicine the way to authority was to be an author. Only some of Galen's hundreds of works have the sneering tone of the attack on the cause-denying Erasistrateans, but collectively they represent his response to a landscape of medical theory that, like philosophy itself, had become significantly sectarian in the post-Hippocratic centuries. Best known of the sects are the Empiricists, who eschewed theorizing about processes within the body's mysterious interior in favor of generalized experience. While they sympathized with such concerns, the great third-century BC Alexandrians Herophilus and Erisistratus would base medicine on anatomy and even experimentation, including vivisection. Methodists, followers of Asclepiades of Bythnia (first century BC), attributed diseases to vessels that were too taut (hot and dry) or too loose (wet and cool). Emulating the dietetic orientation of the Hippocratic writers, they favored gentle interventions, supplementing diet and exercise with massage and bathing. Another group emphasized the dependence of health on the qualities of a spirit, or pneuma, that ramified throughout the body.[6]

Galen presents himself as looking down on this raucous nonsense from the arbiter's chair, which he rightfully occupies. Yet amid his ridicule are tactical assimilations, as he uses one tradition to expose the weaknesses of another. Enlisting the Hippocratic heritage secured his flanks against the Empiricists, for example. Later medical history reflects his success. So thorough was his housecleaning that it quickly became almost impossible to appreciate Galen's rivals on their own terms, for in his view, he had assessed all and incorporated whatever was salvageable. Some early modern writers would simply equate this Galenic pastiche with the medicine of "the Grecians." Yet there were many other respected authorities in these centuries. In many cases their works have not survived. Caelius Aurelianus, writing in the fifth century AD and focusing on fevers alone, alludes to *On Fevers,* by his fellow Methodist Soranus; to works with the same title by Diocles and Erisistratus; and to *Fevers and Inflammations,* by Antigenes, all now lost.[7]

While Galen was the big winner in the struggle to leave one's mark on later medicine, others did to him what he had done to Hippocrates, for systematization invited condensation and oversimplification. From the fourth century to the seventh, the Alexandrian and Byzantine educator-encyclopedists Oribasius, Alexander of Tralles, and Paul of Aegina repeatedly distilled his texts to extract the practical. Thus Paul's seventh-century medical encyclopedia was plucked from Oribasius's fourth-century catalog.

Few of Galen's hundreds of treatises would be regularly read. To engage fully with his enormous achievement was unrealistic.

Frontispiece of Galen's *Methodus medendi* (Paris: Simon Colin, 1530). This image, which was used in many of Galen's works, traces Galenism from Hippocrates through the important consolidators Oribasius and Paul of Aegina.

Wellcome Library, London, L0007039.

Critics might nip at the margins, reconcile Galen with their own experience via commentary, or render his abstractions more orthodox (some Jewish and Islamic writers found his "spirits" too materialistic).[8] The great *Liber Canonis,* by the eleventh-century Persian philosopher-physician Ibn-Sina (Avicenna), the fountainhead of Latinate medical teaching in the High Middle Ages, incorporated both philosophy and practice, but the shorter handbooks used by students rarely included more than a half-dozen texts: Hippocrates's *Aphorisms* and *Prognostics,* Galen's short book on practice, and the *Isagogue,* an introduction by Galen's first Islamic translator, Hunain ibn Ishaq.[9] This latter work was a condensation both from the Alexandrian curriculum (comprising sixteen of Galen's works) and from the usual Islamic curriculum, which comprised these sixteen, plus four from Hippocrates and Aristotle's logic.[10]

The post-Gutenberg publication of many Galenic texts would reflect the rising tide of humanism more than a felt medical need, authority more than utility. While many texts continued to be neglected, Galen's books on fevers, crises, and pulses did command interest. They had been among those Ibn Ishaq had translated into Syrian and were among the first printed in Latin. A Paris edition of the *Differentiation of Fevers* appeared in 1519.[11] In 1525, the year the iconoclastic physician-alchemist Paracelsus allegedly burned Avicenna's *Canon* in the town square of Basel to signify the end of Galen's reign over medicine, the first five-volume edition of Galen's "complete" works was published in Venice. The next century would see "Arabists," loyal to medieval interpretations, pitted against the new humanist Galenists, some of whom, like Paracelsus, were also interested in Hippocratic texts.[12] Much of that contest would occur within universities, but all the while medicine was being transformed from without—by alchemists like Paracelsus, by experimenters like William Harvey, and by uppity surgeons, but also by exotic therapies, like the Andean cinchona, the source of quinine and seemingly a magic bullet for some intermittent fevers. Even as Galen himself was increasingly a foil, a target for modern authors, Galenism remained fluid, an important rhetorical and conceptual resource. Hippocrato-Galenic terms, theories, and practices mixed easily with other forms of knowledge.

But the medium is no less important than the message. I will suggest here and develop in subsequent chapters the idea that prevailing genres of fever presentation are inseparable from the evolution of concepts of fever and approaches to its management. In this chapter I review three phases of the philosophy of fever: the oversimplified Galenism that dominated until the 1640s, the free-for-all of brilliant obfuscation that would follow the application of chemical and physical models/metaphors to physiology and pathology (roughly 1640–1720) by those who sought to liberate medicine from Galen's chains, and the reconciliation efforts beginning around 1690 as medical teachers sought to compress that wildness into textbook coherence. But first I will consider briefly two intersecting factors that reinforced medicine's trend toward bookishness, travel and plague.

THE FEVERS WE FIND (AND THAT FIND US)

With a few exceptions, notably their famous treatise *Airs, Waters, and Places,* the Hippocratic writers concentrate on their own place of practice. Either they assume that what is true there applies everywhere or they see no point in speculating about the geographic distribution of disease. But on their foundation later writers would build a medicine to counsel travelers on how to avoid danger and to find health. Increasingly, that geographic sensibility would apply to one's home as well, for sometimes survival might require one to become a traveler. The *Epidemics* authors certainly take note of the coming of seasonal epidemic fevers, but they seem curiously resigned to them, no matter how deadly. To wealthy Romans, by contrast, it was obvious that one should leave town during the malarial summer.[13]

Medical counsel might help one make routine summer travel plans, but pestilence or plague, the irregular arrival of deadly disease, was quite another matter. These "visitations," whether of disease, famine, or both, as was often the case, were occasions of terror and panic. Knowing that plague was stirring or that a harvest had failed might allow one to anticipate disaster but rarely allowed one to prevent it. Indeed, such events marked the limits of a medicine based on individualized dietetics. Powerless to cure the stricken or to prevent the epidemic or guarantee the crops, Hippocratic writers ignored events like the famous "plagues" of Athens in the years after 430 BC. They focused instead on what they could do, suggests Vivian Nutton.[14]

The seventh-century physician Paul of Aegina tried to reassure readers that in many cases diseases experienced in common simply reflected the commonality of habits—diet, work, or reliance on a bad water source. The Indian physician Agniveśa questioned the very possibility of epidemic disease, asking how people of such different natures could have the same disease.[15] But it was hard to deny that the normal course of events included these terrifying, random demographic disasters. The pathogenic winds of Chinese medicine or Agniveśa's winds and waters were beyond human control, but they came all the same.[16] Notably, accounts of epidemics and famines are often from civil sources; these are geophysical or providential events, matters for chroniclers, not doctors.[17]

But explanatory expectations were rising during the Middle Ages. How much fever was intrinsic to place and how much was accidental was a key issue of learned controversy. Was fever-ridden Egypt inherently unhealthy? Responding to an accusation by the tenth-century Tunisian physician Ibn al-Jazzār, the eleventh-century Egyptian physician Alī ibn Ridwān admitted that Egypt's climate did predispose all life to weakness and putridity. Drinking the water and eating rank fish from the overheated Nile indeed induced biliousness. But urban insanitation exacerbated that predisposition: the droppings (and bodies) of pets and latrine drainage went into the streets or the Nile. Psychical and political causes, such as "fear of a ruler" (which brought sleeplessness, bad digestion, and an altered heat), contributed to epidemics. So, too, the exhausted bodies of other travelers infected the air with epidemic disease. And during famines, the "rotten vapor of the sick" af-

fected even the well-off. Geography mattered, Ibn Ridwān admitted, but it was these contingent factors, non-naturals in a broad
sense, that made Cairo a fever zone. Ibn Ridwān wrote at a time
of war, political instability, high prices, unusual variations in the
level of the Nile, and pestilence, which together killed a third of
Egypt's population.[18]

Well before the infamous Black Death, a generic "plague" was
understood in terms of fever; indeed, it was seen as the worst of
all fevers. In plague, declared Rufus of Ephesus in the first century
AD, "there is everything which is dreadful, and nothing of this kind
is wanting as in other diseases. For there are delirium, vomitings
of bile, distension of the hypochondrium, pains, much sweatings,
cold of the extremities, bilious diarrhoeas, . . . thin and flatulent;
the urine [is] watery, thin, bilious, black, having bad sediments,
and the substances floating on it most unfavorable; . . . [there are]
blood from the nose, heat in the chest, tongue parched, thirst,
restlessness, insomnolency, strong convulsions, and many other
things . . . unfavorable."[19]

The bubonic plague that repeatedly roamed over Europe and
Asia from the fourteenth century on would likewise be conceived
as a form of fever. The buboes, black swellings in the armpits or
groin, might simply be putrid symptoms of a severe fever. Plague
in fact has multiple presentations, and often in epidemics it was
combined with other fevers. Yet it could matter greatly whether
the true plague had come or a town was merely suffering from
a "pestilential fever," which, as the term suggests, was more an
epidemiological than a clinical or pathological entity. For plague
challenged the predominantly naturalistic explanatory repertoire
of Hippocrates, Galen, and the Arabs. It could not be due to bad
management of the non-naturals, because it affected animals too:
embodying nature, they couldn't err. Accordingly, such visitations
might call for an utterly different kind of explanation. They might
be due to corruption of the air, perhaps from some geophysical
aberration, or to astral, demonic, or divine operations.[20]

Plague also terrified in a way that made other fevers seem mild,
even "desirable," by comparison.[21] By the end of the Renaissance,

plague was a public matter in most European towns. The Italian city-states led the way.[22] Yet it did not acquire that status all at once or in a uniform or comprehensive way. That plague came despite people's best efforts to manage their humors did not mean that recourse to individualized learned medicine should be abandoned, for perhaps some doctor might know the true antidote or the right regimen, for example, drinking one's own urine.[23] The economic and social disruptions were inescapable, but the needs for nursing care, funerals, and burial might or might not be met. Considerations of communicability were a further wrinkle, tied up as they were with determinations of blame. In Renaissance Italy, communicability, considered in broad terms as the power of the diseases and doings of others to undermine one's own health, went far beyond contagion to include suspected poisonings, the toleration of filth or sin, and the presence of strangers or Jews. All might somehow foster plague. But even then, public causes did not necessarily imply a public response: one might simply leave.[24] Galen had managed not to be in Rome when the Antonine Plague hit in 166. Still, by the end of the sixteenth century, towns and states were generally acting against plague, forcefully if sometimes futilely: sequestering, succoring, and purifying.

WRITING FEVER

Plague fed print. There was learned authority to be secured, detailed advice to be delivered, hope to be nurtured, order to be maintained. Authors chronicled and commemorated; they wrote claiming the ability to cure, to prevent, or at least to account for plague in moral or in natural terms. Plague tracts, often written in the vernacular, have been seen as "the first form of popular literature in the West."[25]

With plague tracts came a flood of other medical publication, including fever manuals. Publication remained a route to prestige, but it was much more than that. One could profit by selling one's learning directly; by translating and harvesting the works of some august authority, either an ancient like Galen or a modern like Daniel Sennert (and perhaps increase sales by offering that wisdom in bite-size bits); or even by selling books that told readers of the wonder cure—a pill or a spa—that the author could, perhaps exclusively, provide. The books were not only a commodity themselves but also central to the selling of other commodities in a busy medical marketplace in which practitioners, differentiated by training, specialization, and therapeutic modality as well as cost and class, competed for custom. A golden age of fever writing would develop only in the eighteenth century, with the decline of plague and growing concern about the fevers of the Indies and newly fashionable nervous fevers. Sixteenth-century works on fever were mostly glosses of Galen's two books on the differentiation of fevers, expanded by Avicenna. A good example is the pioneering 1566 fever tract by the English physician John Jones, *A Dial for all Agues* ("agues" here including all fevers).[26] The university-educated Jones was well read. Though he hinted at his success as a practitioner, he was more conspicuous as an author. He wrote also on mineral waters and child rearing.[27] Jones sought to order learned fever lore for his reader, reconciling Hippocrates and Galen (and Aristotle) with one another and with their successors, both medieval (e.g., Avicenna, Ali ibn al-Abbas al-Majusi [Haly Abbas], Isaac Israeli, and Abu Bakr Mohammad ibn Zakariya al-Razi [Rhazes]) and modern (e.g., Benedictus Victorius [Faventinus], Jean Fernel, Girolamo Cardano, Leonhard Fuchs, Georg Bauer [Agricola], and Jacob Dubois [Sylvius]). There were no serious disagreements among them, he maintained.[28]

Following Galen, Jones explained that ague was "nothing else but an unnatural heate." Over-heating from exercise, "as it often chaunceth to dauncers, runers, laborers and leapers," remained the touchstone, though rarely did such activity blossom into full fever (ch. 1). In typical Galenic fashion, Jones mapped possibility. One understood fevers that existed in terms of those that might exist. While fever per se was heart-centered, the several fevers were distinguished by their location in particular fluids and parts. Mild ephemeral fevers were located in one of the three spirits: the liver-based natural spirit, the heart-based vital spirit, or the brain-based animal spirit. Collectively they were known as *synochus non pu-*

trida. More serious fevers were located in the humors or in the solids. Fevers due to "rotten" humors (technically, "putrid" fevers) might be intra- or extravascular. Intravascular fevers were continuous, while extravascular fevers were periodic. Galen had associated three of the four main humors (phlegm, yellow bile, and black bile) with the three distinct intermittent fevers (quotidian, tertian, and quartan). The thorny questions of what humors actually were and what went wrong with them (disproportion, unsound condition, or both) taxed Jones's mediating abilities. Most sources agreed that the four primary humors derived from and were quickly affected by food and that health was balance. But Jones also suggested that blood represented the proper mix of humors, while other humors were pathological aberrations of it. Thus, "Melancholy natural is the dregs of the blud" (ch. 3). And he saw humors as compound. A natural phlegm became unnatural by the inclusion of "other humors": "watrye, slimye, glassy, plastry, salt, sowre, harshe, & stypticke" (ch. 3). If the composite humor putrefied uniformly throughout the vessels, the fever was a *synochus sive contiens putrida,* something like a general inflammation of the blood.

Last were fevers of the solids, "hectic fever" and marasmus. These slowly consumed the body's moisture and its substance, a "consumption" that might be associated with phthisis of the lungs (known today as pulmonary tuberculosis) or with starvation. Jones, like many Renaissance writers, invoked the lamp metaphor, namely, that once the fire has burnt the oil, it consumes the wick itself, the body's fabric.[29] Hectic fever might be treatable; marasmus was not. There were also "accidental agues" stemming from ulcerations or inflammations of particular parts.

As so often with Galen's framing, the practice of bringing *most* fevers within a single scheme helped one make sense of the anomalies. Thus the infamous semitertian was a mix of a continual and an intermittent tertian. There were also a "bastard tertian," an epialic fever (in which one felt hot and cold at the same time), the Hippocratic *kaûsos* (or "burning ague"), and a new disease, the "English sweat," or "stoupgalante," which Jones saw as a stomach-centered fever linked to the Hippocratic "tiphodes" (chs. 14–15). Even then, there was space for "erratic" fevers wholly outside the scheme. Plague, because "it spryngeth and groweth by . . . venomous ayre," was one such. Effectively, it was an extramedical calamity akin to "earthquekes, fluds, windes, beastes wild and venomous, war, famine, and pestilence." Valid responses included prayer, avoidance of fruit, improved sanitation, and flight, though one must take care not to run into plague while fleeing from it, as Jones had: "Unawares I lodged by chaunce with one that had it running upon him" (ch. 8).

Sometimes Jones ties these fevers to specific abuses of the non-naturals. For Jones as for Galen, the exemplary fever was a putrid synochus from plethora, common in youths who partied too much—"them which abound with blud, and of sanguine complection, replenished with humors, fat and corpulent, solemners of Bacchus feastes, gorge upon gorge, quaf upon quaf" (ch. 5). Alternatively, those who lived "an idel life," who partook of "deinty fine meates, and delyte[d] in drynkyng and bathyng, chiefly immediately after meat," risked a phlegm-based quotidian (ch. 11). Readers were to respect even minor fevers: "If thei neglect the wel using of themselves . . . & the Phyisitions counsel" (ch. 1), these might evolve into states of feverish putridity or consumption.

Well into the seventeenth century Jones's colleagues were reassuring readers in exactly the same terms. Edward Edwards (1638) condensed fevers into a single chart, albeit one covering fifty-three pages.[30] Nicholas Culpeper, the most prolific of the English medical writers, presents the same matter in a bulleted list but adds the planetary alignments responsible for plague outbreaks. For agues (increasingly restricted to malarial diseases) he has a sure cure: cinquefoil gathered under the influence of Jupiter and the moon. He corrects Galen, that "good old soul," on the matters of vomiting in tertians and the incurability of marasmus; Culpeper has had marasmus and cured himself.[31] The French pharmaceutical writer Brice Bauderon adds exotic remedies (e.g., potable gold) for each fever.[32] But even sophisticated theorists like Daniel Sennert follow the scheme, even as they embellish it.[33]

Such texts represented an extraordinary oversimplification of ancient knowledge. With regard to the Hippocratic heritage, they marginalized *kaûsos* and ignored the other acute diseases, peripneumonia, pleurisy, and phrenitis. With regard to the Galenic, they passed over critical days, pulse interpretation, and procatarctic causes. Marginalized too were the exanthems, like measles and smallpox.

As with Jones, publishing was one way of making a medical living. Books might sell regardless of whether persons sought one's services. Culpeper was as much a publisher as he was an author. *Sennert* was effectively a brand name. Decades after his death, English publishers (including Culpeper) were selling chop-shop bits of his writings and those of other Continental medical superstars. But what were buyers getting? Some writers, like Bauderon and Culpeper, did offer advice concerning medication and prevention, but these theory-heavy tracts are not mainly self-help guides. Sensitive to the accusation that broadcasting professional knowledge was unprofessional, Jones insists that his fever book is no substitute for expert care by a learned and licensed physician like himself. He must publish, he explains, precisely to combat the power of print. Easy publication, together with the commonsensical character of humoral medicine, had led to the mistaken belief that anyone could master this "Physicke of the kitchen." It was because "so many . . . ignorant wives bableth, . . . pelters prateth, . . . Rogers and makeshifts marreth: to the derogation of an arte moste worthye," that Jones must write. The reality, he says, is that medicine is complicated. There was no "Cooke so cunning that had the perfect knowledge of all the sixe thinges not natural."[34]

Unlike many of his successors, Jones does not discuss cures, regimen, or the applicability of general knowledge to the diseases of England. His emphasis on theory and classification suggests that the key issue was authority itself. The tracts seem intended to reassure. Men of learning *really do know* what is going on in one's fevered body; fevers, unlike plague perhaps, belong to the realm of order.[35]

A Discourse of all sorts of Fevers. 1

First what a Fever is.

A Fever is an unnaturall heat kindled either of the

1 Spirits,
  1 Being inflamed without putrefaction of matter.
  2 Causing either Diaria
    1 Simplex i. dier.
    2 Plurimum lasting
      1 24. houres.
      2 Some dayes.

2 Humours causing a prefixed or rotten Fever, they rot either
  1 In the vessels, and that three wayes, for either
    1 All the humours rot equally, as in Synochus putrida.
    2 Humours rot; and that either,
      1 Equall.
      2 Inequall.
    3 One humor alone causing a Fever continuall according to the nature of the humour putrified.
  2 In the vessels causing an intermitting fever according to the nature of the humour putrified in the missenteries, &c.
  3 In and without the vessels, both causing a compound fever both intermitting, and continuall.

3 Solid parts of the body, causing a Fever Hectick.

The first page of Edward Edwards's fever chart gives the standard breakdown between fevers of the spirits, of the humors, and of the solids used by early modern Galenists.

From Edwards, *The cure of all sorts of fevers both generall, and particular* . . . (London: Thomas Harper, 1638).

During the half-century after 1660, that stability gave way to theoretical chaos. Indeed, later reformers would so despair over the "systems" of seventeenth- and eighteenth-century medicine that they would reject theory altogether in favor of a disciplined commitment to fact. One branch of Galen's challengers, the iatrochemists, imagined bodily processes in terms of the alchemical laboratory. Others, iatromechanists, would treat the body in terms of the laws of mechanics, making fever a consequence of vascular friction. Each oriented practitioners toward certain observations (e.g., qualities of blood versus qualities of pulse), but both were enterprises of analogy more than programs of empirical inquiry or theory testing. And rarely did the rival frameworks have direct therapeutic implications, notwithstanding the insistence of authors that right theory was the guide to correct practice.

Despite their iconoclasm, these reformers were still operating within classical medicine.[36] Like Galen, theorists sought the one plausible analogy that would explain all. And rarely were their analogies wholly new. The chemists' fermentation-putrefaction dyad resembles Hippocratic notions of appropriate coction; the mechanists' concern with obstructed flow recalls the ancient interest in altered pulses and blocked pores. And what seemed to be the novel achievements of the scientific revolution were readily assimilated. In showing the heart to be merely a pump, William Harvey's pioneering research directed attention to two ancient themes: vascular performance and the characteristics of hot blood. In therapeutics too, purging might overtake bleeding, but the strategy of re-equilibration did not radically change. Despite flirtations by some iatrochemists with hypothetical toxins as presumed causes of distinct diseases, most still saw fever as a cascade of internal events in which a delicate balance between inner and outer had been disturbed. Some of those who became prominent after 1700, notably Herman Boerhaave, had little trouble accommodating the rival schemes (treating them as heuristics or as referring to differing levels of explanation) or reconciling ancients with moderns. Such mediations could yield strange theoretical bedfellows, such as Newtonian Hippocrateans, and led sometimes to incoherent terminology as older terms persisted long after theories that had generated them became obsolete.

In recognizing the chaotic results of their theorizing we should not lose sight of the critical spirit of these fever theorists. They asked harder questions, challenged vague concepts, and sought ulterior levels of understanding even at the risk of losing contact with clinical practice. But however much they might be infused with the spirit of the scientific revolution, individually and collectively they could not deliver its benefits. Whether in terms of investigative techniques (anatomy and microscopy), discoveries (blood circulation), or approaches to knowledge making (empiricism, experimentation, communal skepticism), these excited advances in all directions had the effect of undermining professional unity and thus scientific progress.

PHILOSOPHERS OF FEVER

Of the eight philosophers of fever considered below, five—the Hippocratic experimentalist Santorio Santorio, the iatrochemist Thomas Willis, the mathematician Lorenzo Bellini, the Cartesian Friedrich Hoffmann, and the empiricist and skeptic Thomas Sydenham—were prominent theorists and/or exemplified distinct approaches to reconciling theory and practice. Two others, the Helmontian and Puritan James Thompson and the Baconian con man Robert Talbor, were marginal figures but their careers highlight the close connections fever theory might have with religion, politics, or commerce. Last, and as a transition into the quite different fever regimes of the eighteenth century, I consider the great consolidator Herman Boerhaave.

Santorio

The Paduan medical professor and physiologist Santorio Santorio (1561–1636) studied phenomena of pulsation and bodily temperature, but he is best known for experiments concerning bodily inputs and outputs measured in subjects suspended in a chair hung from a balance. Much loss, he found, came as "insensible perspiration" expelled through pores or in exhaled breath.[37] Although "insensible perspiration" was not sweat, but usually inversely proportional to it, it resonated with longstanding concern with the movement of some unknown essence across the body's permeable borders, especially during the expulsive "crisis" of feverish diseases.[38] There was much appreciation of the variability of perspiration in fevers. Writers often expressed more concern about the *dryness* of hot skin than about its hotness. A few, both before Santorio and after, saw fevers as maladies of the pores, which had failed to work.

Published in 1614, *Medicina Statica,* Santorio's presentation of the chair-balance experiments, did not become a bestseller until a century later. The popularity of this experimental, inductive, and

quantitative work anticipated the physics envy that would periodically infect biomedical science in later centuries. Yet Santorio
presents his work as Hippocratic hygienic aphorisms: the balance
chair teaches "rules for health." Six of the eight chapters explore
the relations of insensible perspiration to the non-naturals—
airs and waters, eating and drinking, sleeping and wakefulness,
activity and rest, secretions (sexual), and state of mind. Healthy
humans, he found, voided roughly five pounds each day, mostly
during sleep (one was to sleep clothed and not move about).[39]
Maintaining perspirability was a key component of a healthful
regimen. Drinking cold water decreased perspiration; eating mut-
ton, which vaporized quickly, enhanced it. "Chearful and angry
persons . . . perspire more healthfully" than "the fearful and pen-
sive," he asserted. Joy itself came from pore-opening foods; ob-
struction presented as grief. Failure to perspire brought on fever,
while therapies that inadvertently subverted perspiration exacerbated it. Much of fever's seasonality was due to changes in the
perspiration-facilitating capacities of the air.[40]

Insensible perspiration fit easily into many frameworks. It
brought an additional perspective on what happened to the body
in the bed, reinforcing concern about diaphoresis and proper coction. If it fostered interest in meteorological determinants of fevers, such as heat and humidity, that impact was initially more
conceptual than experimental: not until the mid-nineteenth century would there be significant follow-up research.

Willis

Or perhaps decomposing blood was the key to fever. If we
must classify him, the Oxford physician-physiologist-anatomist
Thomas Willis (1621–1675) was an iatrochemist. But while post-
Paracelsian chemical medicine is often linked to an underground
of alchemists and mystics, as well as to heterodoxy and radicalism,
Willis was an establishment Anglican royalist.[41] In *Two Medico-
Philosophical Diatribes* (1658–59), his first major publication, Willis developed a general concept of fermentation and applied it
to fevers. In his view, fermentation reflected an innate tendency

of materials to disassemble into their components—earth, water, and the alchemists' *tria prima* of salt, sulfur, and mercury (or "spirit"). A ferment was the initiator or an accelerator of such separation. Willis offered homey farmstead analogies: pickling and the making of bread, beer, wine, vinegar, butter, and cheese all involved fermentation. As the ageing of game and the curing of fish showed, decomposition did not inevitably bring putridity; that was an aberrant outcome.[42]

According to Willis's etymology, *fever* had evolved from *fervor*, suggesting the "effervescency" of "a working must": "It plainly appears . . . that the Blood doth hugely boil up and rage in a Feaver." Heat came from the pressure of effervescing spirits in fervid blood or nervous fluid.[43] Willis would be largely responsible for Anglophone physicians' fixation on the qualities of the blood they removed from patients. Just as the multiple forms of coagulated milk signaled different fermentations, well into the nineteenth century one measured the speed of clotting, the "buffiness" of the blood's surface, or the separation of components as signs of fever's fermentation.

But while the term *ferment* linked Willis to the chemists, his rising spirits and farmstead analogies recalled Hippocratic kitchen concepts, and his focus on altered blood and scheme of fever classification were adaptations of Galen. His chief innovation—a theory of heat that was largely kinetic (with pressure from ebullition), generated in part by the release of sulfur—can be seen as an ulterior explanation of Galen's theory of fevers.[44]

Bellini

Where Willis was concerned with blood's instability, the Pisan anatomist Lorenzo Bellini (1643–1704), a student of Galileo's associate G. B. Borelli, emphasized blood physics. Fever was a hydraulic phenomenon. In contrast to Santorio's experiments and Willis's bucolic analogies, Bellini drew on the axiomatic approach of contemporary mathematical philosophy. His 1683 *De Febribus* was a veritable *Principia* of fever. From a few first principles, he worked out the mechanics of blood flow and derived the phenomena of fever as necessary truths about fluid dynamics. The result is a tome as turgid as Santorio's was sprightly. The English translator felt the need to apologize for the complexity both of Bellini's Latin and of his ideas, offering a gloss that made it unnecessary for readers to actually read the book. When that translation finally appeared in 1720 as *A Mechanical Account of Fevers,* a so-called Bellini-inspired "Newtonian" medicine, chiefly developed by Archibald Pitcairne in Edinburgh, had already begun to decline in popularity.[45]

Like Willis, Bellini emphasized separation, but it was gravimetric: reflux within a closed system of pipes would cause a settling out. The thinnest blood would be in the smallest capillaries, while larger vessels would be clogged with big particles, labeled *lentor* for their slowness. Bellini worried not only about their size but also about their shape and viscidity. Intravascular forces would reshape these lentor particles, Bellini found, elongating them and creating a tangled web that coated the vessels' walls. Impeded flow would disrupt the circulatory equilibrium, producing further separation and more clogging. In vessels just below the skin, such clogging might block Santorio's invisible perspiration. Such phenomena constituted fever. The friction from propelling sticky masses through narrow channels manifested as elevated heat. Medicines that altered the chemical properties of the blood could affect its physical properties and allow the system to re-equilibrate.

Bellini's elegant system, far removed from the realities of the bedside, suggests how important high science had become in maintaining learned medical authority. Lentor particles could not be extracted, nor could their stickiness be measured, but well into the nineteenth century, the term *lentor,* like *buffy,* to describe blood (and occasionally humors and spirits) remained part of the explanatory repertoire. Bellini's approach probably did help to revive interest in the pulse, a mode of assessment that had been neglected in early modern Galenism notwithstanding its importance for Galen. Until its displacement by thermometry around 1870, aberrant pulse would be the most important designator of fever. Still, we should not write these schemes off as speculation run amok. While *lentor* was primarily a hydrodynamic deduction, the phenomenon of plasmodia-caused blood thickening would become an important element of malaria pathology in the twentieth century.

Hoffmann

The post-Galilean hydrodynamic tradition on which Bellini drew was, if sometimes at a distance, grounded in experience. Not so the metaphysical mechanics of Friedrich Hoffmann (1660–1742), appointed as the first professor of medicine at the new Pietist University of Halle in 1693.[46] Hoffmann's fever theory was grounded in the first principles of the greatest metaphysician of the age, René Descartes (1596–1650). Descartes was not a doctor, but he could not avoid medical matters in exploring how plenum and extension could account for all phenomena of the human body (except its mind). He acquired medical followers, notably Henricus Regius of Utrecht, who developed and taught an (unauthorized) Cartesian medicine in the 1640s.[47] Hoffmann did not belong to that group but would be far more influential.

In the Cartesian cosmos, vortices accounted for change. As some particles were flung to the periphery, others fell into the middle. Fever, in Hoffmann's view, was simply one manifestation of that process. The chill felt at a fever's onset was from blood particles concentrating in the body's core. There, under compression, they popped; being incompressible, the finest ether "rupture[s] the pores of the blood, whose particles it drives into circular motion." The results were a pounding pulse and perceived heat. The jetting forth of ether from a squashed blood particle also explained the separation of the blood and Willis's effervescence. All these phenomena followed logically from axioms of plenum and extension, Hoffmann held.[48]

Hoffmann and his colleague (and rival in matters of theory) Georg Ernst Stahl would make Halle the leading medical school in Protestant Germany. There, as in modern medical schools, arcane science provided grounding and authority. Medical students took a long course in anatomy and physiology before studying the pathological domain. But Hoffmann recognized the gap between high science and practical doctoring. Most topics did not receive the Cartesian gloss; students still learned traditional aphorisms and standard remedies couched in familiar terms.

Sydenham

To some, the aloofness of theory was problematic. It was hard to see how theories developed far from the bedside could be applied there. But the adequacy of theories was of course itself a theoretical issue, and the foremost analyst of that issue was Willis's contemporary Thomas Sydenham (1624–1689), known as the "new Hippocrates." The English Civil War, in the 1640s, had shaped the careers of both men.[49] They had fought on opposite sides, and at the restoration of Charles II in 1660 Willis became an Oxford professor, while Sydenham, passed over for a fellowship in the Royal College of Physicians, plied his trade, amassed experience, and became critical of theories in general, instead espousing a simple Baconianism: with "Nature for my guide, I should swerve not a nail's breadth from the true way." And as he later wrote, "I directed my attention to the close observation of Fevers" and "at length, hit upon a mode of curing them."[50]

Sydenham's first book, *Methodus curandi febres* (1666), would be expanded into the famous *Medical Observations on the History and Cure of Acute Diseases* (1675). Where other theorists sought unity, Sydenham acknowledged variability. He sought patterns: not only did fever vary seasonally but the prevailing character of fevers shifted every few years. Thus five successive "epidemic constitutions" held sway in London between 1660 and 1675. In each, febrile diseases of various types also exhibited characteristics of the ascendant constitution, and each constitution had the features of a familiar disease. The first constitution was of an "ill-conditioned tertian." Plague, cholera, smallpox, and dysenteric constitutions followed. During the reign of each constitution its signature disease was uncommonly prevalent. In Sydenham's view, these patterns had implications for therapeutics and nomenclature. He would replace vague and unsystematic adjectival labels, like "*putrid, petechial, malignant,*" with constitution-based clinical descriptors. Thus, he wrote, "I may, perhaps, be allowed to designate the present fever as the *variolous fever,* from its likeness to the smallpox."[51]

In placing clinical generalization above theory, Sydenham has been seen as inaugurating the modern concept of specific diseases and the familiar trajectory on which clinical induction of fever species allows investigation of their unique causes, culminating in the discovery of their microbe agents. While Sydenham recognized the first steps in that trajectory, he challenged its metaphysical foundation and questioned its practicality and utility. Designating a disease species presumed access to its essence, but there was no such access. Indeed, "Parent Nature" moved things from "the abyss of cause into the open daylight of effect" by "immutable" laws, but "their essences, their constituent differentiae, their quiddities" were "veil[ed] . . . in . . . deepest darkness."[52] Nor would Sydenham touch the issue of remote causes. Perhaps, as with the inexplicable waxing and waning of plague, geophysical and/or meteorological factors, generating some mysterious "miasma," were responsible, but the empiricist Sydenham repudiated the call to explain: "Etiology is a difficult, and, perhaps, an inexplicable affair; . . . I choose to keep my hands clear of it."[53]

Insofar as disease species could be recognized, it was by observation, leading to induction, and not by "Reason." But even that was tricky. The "constitutions" were not fever species. During each there were still "symptoms common to all fevers alike," such as "heat, thirst, restlessness, &c," as well as complications, idiosyncrasies, and intercurrent disease.[54] Sydenham's self-critical sobriety is refreshing. In later centuries successors would pursue this mirage of species, lumping, splitting, and redescribing as they struggled with the eternal problem of induction of how to group. Often partisans would present themselves as Sydenham's true legatees.

Sydenham also rejected the claim, later used to defend the pursuit of disease species, that rational therapeutics required knowledge of cause. "The question . . . how we can cure diseases, whilst we know nothing of their causes, gives us no trouble," he insisted, for "it is not by the knowledge of causes, but by that of methods at once suitable and approved by experience, that the cure of the majority of diseases is accomplished." As with pathological theories, preoccupation with cause violated his mantra, which was that "the art of medicine was to be properly learned only from its practice."[55] Regardless of the prevailing constitution, treatment was still a matter of adapting available tools to individuals according to their age and condition. Sydenham exploited the full range of pharmaceutical options, including the new and dubious mineral medicines.

Sydenham's vehement rejection of essences and causes is understandable as a reaction to the excesses of contemporary fever theory. But he could not escape theoretical language or theory itself. As much as possible, he tried to avoid the question of what was going on within the fevered body; if that was impossible, he tried to describe such matters in neutral terms. Substituting the atheoretical term *commotion* for Willis's terms *fermentation* and *ebullition* would avoid dis-analogies, for fevered blood was really nothing like vinegar. Still, one had to communicate with one's peers: "I shall, now and then, use the words *fermentation* and *ebullition,* inasmuch as they have been . . . adopted by medical men of late years," Sydenham apologized, but readers should know that "the terms are used . . . solely for the sake of illustration."[56] And occasionally he too speculates, writing, for example, of the persistent cough in continued fever: "No wonder; the volume of the blood is in vehement commotion; it has grown tumultuous; there is a faction, and sedition, and rebellion amongst its elements; its humours are let loose; they take leave of the general mass; they ooze through the vessels of the lungs; they spring out upon the membranes of the trachea, [which] are very delicate, and exquisitely sensitive."[57]

Sydenham's own fever theory was a generic Hippocratean coction-expulsion model, resembling the views of Santorio and Willis. It assumed a "Nature" that stabilized the body in health and restabilized it through fever when it went awry. Thus he explained the initial shivers of intermittent fevers: "The febrile matter which . . . has been imperfectly assimilated with the volume of the blood, has become, not only useless, but inimical to Nature whom it frets and vexes. She, . . . stirred up by what we may call a natural sense, and planning, as it were, an escape, creates a shivering and a shaking throughout the body, as the evidence and the measure of her aversion."[58] This "Nature," he admitted, was itself a metaphor. Yet it was not merely a collective noun but signified coordination by "the highest reason."[59]

Differences in theory need not generate polemics. Where they did, these were usually rooted in issues other than the nature of fever—struggles for appointments, personal or professional animosity, even religious or political agendas.[60] Often the theorists were also false pretenders to originality. In fact, notes Bates, they were merely rebottling ancient wine.[61] They did not have significantly greater access to the body's interior than Galen had. Both Willis and Bellini focused on blood (Galen's premiere humor). That it now circulated was a fact to be accommodated, but both understood Harvey within a broadly Galenic tradition. Earlier writers had also worried about separating fluids and clogged vessels. Synochus, John Jones had explained, was due to "the thicknes and binding of the meats or ways of the pores in the body & flesh," which "detyneth many hymors and slimye excrementes, wythin the body."[62] By themselves, new concepts and terms did not threaten the traditional framework.

Marginal Men: Thompson and Talbor

However varied their views of fever, Santorio, Willis, Bellini, and even Sydenham were respected medical insiders. For others, such as the Flemish alchemist-physician-philosopher Johan Baptista van Helmont (1579–1644) and his followers, fever offered another occasion to attack not only the official medicine of the "schooles" but also the sorry world they represented, even including states and their churches.[63] Reading exhaustively on his own (including more than four hundred works on fever, he claimed), the reclusive van Helmont found the bastard Galenism of his day to be rife with anomalies and arbitrariness. So complete was his self-alienation, however, that his views became widely known only after his death as disciples put his iconoclasm to work.

One such was James Thompson, an otherwise obscure "student of physick," author of the 1657 dialogue *Helmont disguised, or, The vulgar errours of impericall and unskillfull practisers of physick confuted more especially as they concern the cures of the feavers, stone, plague and other diseases.*[64] In it, Pyrosophilus, a Helmontian philosopher, defends his new book, which proves that "fever is unknown to the Physick Schools . . . in its essence, root, and properties, [and] the remedies thereof." Surely, his naive friend Philiatrus observes, such a work will "offend Physitians, and especially anger such."[65] Indeed it will, Pyrosophilus replies, but there is a divine duty to seek truth and state it.

Van Helmont had challenged the coherence of conventional theory on many fronts. First, he had denied Galen's claim that heat was the essence of fever. It was not, said van Helmont; in some so-called fevers there was neither unusual heat nor its proxy, thirst. Many fevers started with chills; in intermittent fevers heat and cold alternated. If the center of the body produced the heat, moreover, there should be a gradient between center and periphery. Inflammation from a splinter should be impossible, since nearby blood would not be unduly hot and its presence would prevent the arrival of hotter blood from the heart. Nor could putrid humors account for fever. Except in cases of gangrene, living substances did not putrefy, nor did putrefaction produce heat; if it did, corpses would be warm, not cold. And, he claimed, the Galenists did not even agree on whether blood was a composite of other humors or itself a humor, or on how a single humor, black bile, could cause both quartan fever and melancholia. The individual criticisms were not new, but they gained power as a single concentrated critique.[66] And van Helmont's iconoclasm would be influential. In the succeeding century the development of inflammatory and nerve-based fever theories would divert attention from heat and the heart and provide a new basis for the study of both local and whole-body aspects of fever.

Yet that was in the future. The Helmontian alternative, at least as presented by Thompson, is striking in its utter arbitrariness. The human body was under the control of a master archeus, or "way-making spirit" and "worker of all alteration." This archeus was responsible for fever as well as for health: "Whatever bringeth forth sound actions . . . the very same is it, which uttereth, faulty, or unhealthfull actions in diseases. For this very spirit heats, a man naturally in health, which in fevers is inflamed." Thompson does not explain what has "displeased the Archeus" and made it inflame the body, but he agrees with Hippocrates that recovery will be a coction- and crisis-based process and that this restorative process will be undermined by strength-sapping bloodletting and most other forms of physic.[67]

And yet it is not obvious that we will or should recover. Thompson (and van Helmont) pointed out a fact that most contemporary fever authors adroitly avoided, namely, that fevers killed. They decimated armies and filled graveyards. And so they should, declares Thompson, for fevers were providential instruments for ending lives. If God had decreed one's death, it would be both impious and futile to indulge the pretensions of human healers. To rely on bleeding, purging, and the other "novelties, impossibilities, or aery toys" of physic to preserve one's life was to flirt with "the most dangerous enemy of Mankinde [Sathan]." The pieties were familiar, but Thompson is unusually rigid in their application.[68]

In fact Thompson was departing from van Helmont and from Paracelsus before him. They had posited a central controller but had also focused on the subordinate archei, which carried out the work of each organ, and on the invading hostile archei, which subverted that work and caused diseases. They had held out hope that the arcana of nature might yield an antidote to each of these hostile archei. But Thompson's single archeus was as arbitrary and implacable a ruler as Calvin's God.[69] In Thompson's radical approach, health (or fever) was grace.

In stark contrast to Thompson's angry self-righteousness is the tone of Baconian openness struck by Sir Robert Talbor

(c. 1642–1681) in his 1672 *Pyretologia, A Rational Account of the cause & cure of agues with their signes Diagnostick & Prognostick*. Talbor's motives, however, had more to do with social and economic than with religious marginality.[70] On the grounds that it was rife with intractable quartan fever, Talbor had gone to live in Essex in order to study it by "that good old way, observation and experiment," or so he claimed. His book would bring to medicine the benefits of the new "Experimental Philosophy" in conjunction with the heritage of Hippocrates, allowing a "golden mean" of analysis and experience. In particular, the arts of chemistry would allow a new kind of "anatomizing" by which the modern savant might learn "the effects and constitutions of the whole body or parts." After detailing the anatomy (and presumed physiology) of the spleen and the stomach, the organs chiefly implicated in quartans, Talbor turned ethnographer, considering folk and occult remedies. If gathered under the influence of Jupiter, cinquefoil and verbascum were (said to be) powerful specifics; they lacked potency at other times. Noting the faith of the country folk in anti-ague charms, he related sad tales of those who relied on them.[71]

Notwithstanding the goal of integration, the book is a motley combination of disconnected elements—until, that is, Talbor gets round to the true cure of quartans: a "specific Emetocathartick Powder" of "three *Herculean* Medicines, each . . . requiring twelve . . . labours in their preparations," together with a fourth ingredient, called "Athleta" because "like a powerful Champion, it dissipates, and expels all Natures enemies." There are even more potent formulae, including a "specifical splenetick medicine" of five or six vegetable ingredients (including two that are foreign, Talbor notes). After a course of these, the patient is then to walk or ride. Talbor never lists the ingredients of these medicines or explains how his knowledge of them derived from his Essex fieldwork or laboratory dabblings. He does, however, caution readers against mere "palliative Cures, and *especially* . . . Jesuits Powder, as it is given by unskilful hands; for I have seen most dangerous effects follow the taking of that Medicine . . . Convulsions, Epileptick Fits, Phrensie." But, he observes as an afterthought, "this

Powder [is] not altogether to be condemned; for it is a noble and safe medicine, if rightly prepared and corrected, and administered by a skilful hand."[72]

This Jesuits' powder, or "bark," from the Andean cinchona tree, was the raw material of quinine. It has had a singular place in the history of fever, regarded as the sole effective specific. It cured one kind of fever, malaria, and hence vindicates one rationale for disease specificity, that with accurate diagnosis of real diseases there can be true cure. It illustrates as well other themes: the importance of medicinal motives in global exploration and conquest and the revolutionary impact on medicine of organic analysis and then synthesis. Quinine, the active ingredient, would be isolated in 1820. Efforts to manufacture it and its structural analogues would lead to the invention of general antipyretics, including commonplace drugs like acetaminophen, by the end of the nineteenth century.[73]

During the rest of the 1670s, Talbor was parlaying his secret remedy into big money. He cured Charles II, then moved to the court of the Sun King, Louis XIV, and ended up selling the secret to the French Crown for a small fortune. It would turn out that the central ingredient of what would be known as the "English remedy" (presumably Talbor's "Athleta") was none other than a heavy dose of bark.[74] But had Talbor simply advocated cinchona, he would have been invisible in the medical marketplace, for European doctors had been using it since the 1630s. They disagreed, however, about its utility, dosage, and indicating conditions. Prudent practitioners such as Willis and Sydenham shared Talbor's caution: here was a good medicine if rightly used. But Talbor, a struggling apothecary's apprentice unable to afford a Cambridge medical education, probably living in Essex because he could not compete in London, lacked access to the lucrative world in which respectable physicians sold their sober thoughts. Hence his deception. To foster the illusion that he had found something new and marvelous, Talbor employed all the chief contemporary tropes of progress: the Baconian ideals of gathering and sifting facts and of accrediting lay knowledge while stripping it of superstition, the allure of the chemist's laboratory, even the exciting prospect of powerful exotic ingredients. Sydenham and others were peeved at having been duped. A con man Talbor may have been, but without the deception he would not have had the opportunity to cure kings.[75]

The Consolidator: Boerhaave

Talbor reminds us of the wide gap between theory and practice. Facts kept pouring in—from anatomy, chemistry, and natural history, from voyages, laboratories, and microscopes. New cures such as bark challenged old and new explanatory schemes. The theorizing did not quickly stop. Eighteenth-century fever theorists emphasized mysterious vital and nervous processes rather than bubbly or coursing blood, but many appreciated the need to check speculation when it came to practice. At Hoffmann's Halle and at many other universities the teaching of medical practice was separated from instruction in pathophysiological theory, known as the "Institutes of Medicine." Cartesian philosophy had limited utility at the bedside.

The most singular consolidator, both between theory and practice and of the multiple theories of fever, was Hermann Boerhaave (1668–1738). He was for a time virtually a one-man medical faculty at Leiden, where he taught both institutes and practice and sometimes chemistry and botany as well.[76] Precluded from a career in the ministry, Boerhaave taught himself medicine by reading ancients and moderns, synthesizing as he read.[77] He became fluent in humors, the alchemists' three elements, and the mechanists' particles and structures. His agenda was one of translating, reconciling, and simplifying. Mechanical explanations were more ulterior than chemical ones, he believed, but chemistry was often the better stopping point.[78]

While medical novelties lengthened the list of what could go wrong, Boerhaave emulated Galen in inventorying possible malfunctions. He tended to regard any so-called disease as a composite or series of discrete malfunctions—eyes that could not see, ears that could not hear, stomachs that vomited, places that swelled and hurt.[79] Ideally, the approach would make the concept "fever" a superfluous abstraction. No longer a primary malfunction of either solids or humors, much less the expression of some pathological essence, it would become merely the suite of symptoms that constituted febrility—shivers, heat, rapid pulse, and weakness.

In three respects, however, Boerhaave foreshadowed distinct approaches that would dominate the understanding of fever in the centuries to come. First was his focus on nerves. Boerhaave and many other eighteenth-century theorists began to view the nervous system as the ulterior locus of explanation. Fever might be felt as heat and traced to sticky particles or to the friction of coursing blood, but what made that blood rush was a nervous force.[80] Yet how did that force act? Boerhaave's intermediary between nervous force and felt fever was a hypothetical anatomical unit, the fiber. As the body's ultimate structural unit, fibers were forerunners of tissues and cells. Woven into membranes, fibers formed the body's fabric. Membranes rounded into hollow cylinders became vessels. Fibers were the recipients of nervous force, and their state determined health. They might be too strong, rigid, and cohesive or weak and lax, even soluble.[81] It was spasms in the fibers of the capillaries that initiated the symptoms of fever. In a loose sense, this was ancient Methodism in the garb of modernity, tinged with anatomy, microscopy, and corpuscularian philosophy. Fibers replaced humors as a commonplace of medical vocabulary. Another medical fossil, they have lingered on into the present as a quack's cash cow. High strung? Tightly wound? Take a tonic, one would be told. It would restore the *tone* of one's fibers.

Boerhaave's second innovation was to recognize toxins as remote causes. From the seventeenth-century standpoint, toxins—for example, arsenic, venom, or contagia—were substances that in minute doses inexplicably produced a feverish response. Rarely could one explain how they did this, Boerhaave admitted, on "account of the Minuteness of their Particles; [yet] as when a Viper has bit a Person, all the Symptoms of an acute Fever follow."[82] Intelligibility had been a hallmark of classical medical systems. One expected to understand how one's interior was being harmed, because the dynamics of the body's interior were presumed to be analogous to those in the exterior world. Appeal to arbitrarily acting poisons was thus a last resort in explanation, though writers on the plague had often adopted it.[83] Over the next century, Boerhaave's declaration that inexplicability was not an obstacle to accepting a pathological cause would help make people comfortable with the idea of specific disease entities, whose differences might be ascribed to an unknown agent with disease-generating properties that were effectively arbitrary.

A key proving ground for this enticing analogy between toxins like arsenic and contagia would be the great armies of the eighteenth century. Soldiers on campaign seemed to suffer fevers more as populations than as individuals. Their diseases often seemed to reflect exposure to a place. Toxins did not immediately displace airs, waters, or the other traditional non-naturals, but they certainly supplemented them. Boerhaave was no camp doctor, but his eminent students John Pringle and Johan Baptista Bassand were, and Bassand sought his advice.[84] By the early nineteenth century epidemiologists were commonly referring to the "morbid poison" of each febrile disease. Later, early bacteriologists would make *toxin* a placeholder for the unique but often poorly understood damage a pathogenic microbe inflicted on its host. Remarkably, despite regular appeal to the toxin concept, the science of toxicology itself would evolve in isolation from pathology, bacteriology, and clinical medicine.

Boerhaave's third innovation was his concern with temperature, particularly with regard to the question of a body's equilibrium with its surroundings. Part I began with a Hippocratic workout. It ends with Boerhaave's skater, who, on a Brueghelesque winter's day will come down with fever, literally at the drop of a hat, while skating along canals or *zees* into a cold wind that has swept down from the Baltic. "There is no Cause capable of destroying the human Body so soon," declares Boerhaave, "as when the severe external Cold is overcome by Heat generated by muscular Motion of the Body. . . . When men are scating [*sic*] upon the Ice, more especially if at the same time they move contrary to

Study of Three Figures Skating, an eighteenth-century drawing by an anonymous Dutch artist. Having evidently not read Boerhaave, these happy skaters do not know that they risk severe fever.

Reproduced with permission. Image copyright © The Metropolitan Museum of Art. Art Resource, NY.

the Wind . . . the Solids act violently upon the Humours, so as to generate Heat by Attrition, which overcomes the external Cold, whence the Blood becomes heated to a degree more than usual." A positive feedback then kicks in. As the finer particles fly off, the density and viscosity of the residual humors increases. Their circulation generates ever more friction, more heat, and more rapid perspiration of finer particles. A physiological disaster ensues: "*The most violent local or muscular Motion may in one Hour produce the same Effects with a severe Plague.*" This happens to horses, Boerhaave claims: "An infinite number . . . die suddenly, when having been educated to Idleness, they are upon some occasion urged to violent running."[85]

Boerhaave, friend and patron to one Daniel Fahrenheit, recognized that this heat might be quantified: "The Winter's Cold, when Water begins to freeze, is the thirty-second degree of the Thermometer, but the Heat of the human Blood is the ninety-second degree." This difference of sixty degrees is generated by what we call metabolism, but what Boerhaave understood as "the motion of our Solids and Fluids." Yet the temperature gap was a misleading measure, he recognized, for the body must compensate both for the absolute difference in temperatures and for the rapidity with which the wind removed some entity—call it heat—that was more than a mark on a thermometer.

In today's terms, Boerhaave was worrying both about the effect of wind chill on a sweaty, underdressed teenager and the absolute temperature. As we know, and he knew, merely measuring temperature alone does not show how much work the body must do to maintain its degree of heat, much less the damage resulting from that work. But to us, for whom fever is merely elevated temperature, Boerhaave has missed an opportunity. He has conflated two distinct questions—how the body regulates its temperature and how much energy the body is using—and has mistakenly taken the former to be a product of the latter.[86] In the nineteenth century these questions would again be separated. It would become clear that in a fever a thermoregulating animal set its thermostat—the metaphor would be inescapable by 1950—slightly higher, regardless of the ambient conditions in which it finds itself. The pattern of temperature maintained by the animal would come to have diagnostic and prognostic significance. At the same time it would also become clear that a body's thermoregulatory capacity was finite: bodies that could not adequately cool (or warm) themselves experienced febrile symptoms. But these two pathological processes, however similar, were understood not to be the same thing.

The matters about which these philosophers debated are now well understood by biomedical scientists. But I suspect that few people experiencing fever think about the mysterious processes within them that are causing their shivers and sweats, aches, heats, and hallucinations. Putrid humors, lentor, popping corpuscles, acrimonious bile, or spasmodic capillaries no longer register. But they once did, and not exclusively for learned audiences. Revisiting these ancient vocabularies of the body should raise questions about what concepts and terms rule our own interrogation of our interior. Simply to accept that one's body is hot because one has a passing "infection" explains all by explaining nothing.

PART II

Fever as Social

❖ ❖ ❖

It is easy to understand what made plague social. Not only did it provoke a powerful public reaction as authorities took steps to cleanse their towns, seal off affected areas, and hospitalize (or bury) victims but it invited individuals to be mindful of the people around them, who might be spreaders of plague, and of the adequacy of the institutions that were to meet the crisis.

Fevers—epidemic, seasonal, or sporadic—continued during the centuries of plague, but they are much less visible. Historians of the early modern period have been struck by chroniclers' and diarists' relative lack of interest in epidemics that were not plague. Generally, even when many were stricken and many died, fevers remained a matter of individual bodies, as they had been for the authors of the Hippocratic *Epidemics* and for Galen.

It is tempting to ascribe the differing responses to epidemics to rival theories of cause and associated possibilities for prevention. Most saw plague as imported and somehow communicable. Fever had been the hallmark of a medicine of the non-naturals, a matter to be dealt with by victim, family, and medical adviser. But as plague waned during the eighteenth century, plague-based institutions and sensibilities, including a concern with communicability, would be slowly and incompletely transferred to epidemic fevers.

Yet while fever did become a social entity during the eighteenth century, worry about catching fever was not the leading edge of its socialization. Instead, at least in western Europe and its colonial dominions, fever's status changed as part of broader changes, particularly the rising sense that communities, nations, and even humanity were bound together by common experience.[1]

For my neighbor's fever was becoming *my* business. Sometimes, interests were at stake. For slave owners, fever threatened invested capital; for military commanders, it threatened the "strength" of the forces; for promoters of health resorts, it threatened reputation; and for governments, it represented social disorder that could become political disorder. By the early nineteenth century fever had become an element of identity. Certain fevers were being associated with certain groups—prisoners, soldiers, women, Irish beggars, indigenous peoples, and later, with world-historical consequences, eastern European Jews. While their fever made such groups problematic for others, it did so in various ways. In some cases such groups would be seen to pose a contagious threat. But fever signified in other ways too: as the nervous sensitivity of some groups, the pathologies of a homeland, or a risk associated with an occupation. Precisely because all humans were subject to fevers of some sort, the characteristic forms of fever provided ideal signifiers of difference.

As humanity came to be seen as a community, at least in the romantic imagination of western Europe, fever began to evoke sympathy, that curious projection of the imagination onto the presumed experiences of others that so fascinated Adam Smith.[2] A growing interest in the fevers of others was accompanied by a growing and changing interest in one's own fevers. An older form of introspection (why hast Thou stricken me so?) was giving way to fascination with febrile experience (what is this wild ride, and where does it fit into my life's plot?).[3]

For increasingly fevers were part of plots, and not random events or inexplicable divine judgments. Novelists and reformist doctors alike sought to discern the grand lesson of each fever. In this golden age, fever writing broke free from medical theory. Fever would figure importantly in letters, novels, and autobiography, as well as in geographic inquiries.[4] Febrile delirium again became central as the most common and dramatic form of alienation from the normal self. Fever carried with it both the prospect of deadly danger and the frisson of fear—not only of contagious others but of deadly exotic places and of activities, experiences, or states of being that might induce it. Fever would be seen as both the product and the cause of heightened emotion.

By the end of the eighteenth century, fever had become gendered, usually female. Galen's paradigmatic fever victim had been a plethoric youth who had drunk too much wine; by 1800 the classic victim would be a nervy anemic young woman whose shivers and faints progressed to fits, delirium, and then invalidism or death.

The result was an increasing sense of obligation to care for fever and prevent it. Ultimately, as Margaret Pelling has noted, fever, and not cholera, would become the nucleus for modern institutions of public health in the mid-nineteenth century.[5] Recognition of that obligation had democratizing implications, although these played out differently in different polities. Because each person had the capacity to affect and be affected by the biotic states of others, fever became a currency of exchange that bound people together, willingly or not, into the dictatorship of the fevered and the potentially fevered. If that transformation was clearest in bourgeois and liberal Britain, an analogous orientation was evident among the medical police of central Europe and in the work of the Royal Society of Medicine in France. There, the "citizen patient" of the revolutionary era would enjoy both rights and obligations.[6]

How much of this change, one may wonder, was an artifact of new modes of expression? Surely people had cared; they simply hadn't sent a card. But consider Thomas Sydenham. While he was fascinated by the chronology and character of London's epidemic fevers, he was not especially interested in London's people—in who was getting the fever or how they were to be cared for. Fevers did not bring a focus on community. The contrast with plague in Milan a century earlier could not be sharper. There all aspects of life were the public's business, and doctors guided the authorities in responding to the "social causes of plague," including access to care and poor living conditions.[7] And like John Jones, Sydenham saw no obligation to remain to fight a deadly epidemic. Flight was both rational and morally unproblematic.[8] A century and a half later many medical men would see staying as a clear duty to community and society at large.

If culture was changing, so, too, were the predominant diseases. Plague was waning; syphilis was no longer new. On the other hand, there were new fevers. A "nervous" or "low" or "slow" fever struck soldiers on campaign and raged in military hospitals, on ships, and in jails. It also gained the reputation for generating spontaneously among the urban poor and spreading to the rich. A "putrid," quasi-feverish "scorbutic" disease struck sailors and soldiers. The postpartum febrile illnesses later known as puerperal fever became epidemic in the new lying-in hospitals. A "yellow fever" arose to menace the slave-based industries of the Caribbean, the slave-supplying areas of West Africa, and later North American coastal cities and southwestern European ports. But certain mild and lingering fevers also became fashionable illnesses, marks of heightened sensibility.[9]

Chapter 4 focuses on the emerging sociality of fever; chapter 5, on its individuality and gendering.

CHAPTER FOUR

Communities

❖ ❖ ❖

Fever became a social and then a public matter only gradually over the long eighteenth century. It became an issue for friends, countrymen, townsfolk, soldiers, and sailors.

FEVERED LIVES: MADAME AND HER FRIENDS

Friends first. Of fevers "I am always apprehensive . . . they say they consummate, but it is life itself which receives a consummation from them."[1] So wrote the gossipy, introspective, well-read and well-educated Marie de Rabutin-Chantal (1627–1696), better known to readers of her thousand-plus letters to her distant daughter Françoise as the Marquise de Sévigné. A minor planetoid in the orbit of Louis XIV, Mme de Sévigné became a literary figure only in the mid-eighteenth century. Readers admired her independence, insight, wit, and compassion. In her letters, which focus on family and friends, fever is a frequent topic. She writes about many aspects of fever, providing a well-connected lay person's reaction to the theories and practices of the day. Her own life would end in a fever brought on by caring for her beloved and fevered daughter.[2]

Fevers were simply a fact of life in Madame's world; she writes of Mme de Coetquen, for example, that she is "ill here of a fever."

Marie de Rabutin-Chantal, marquise de Sévigné (1627–1696).
Frontispiece of *Lettres de Mme. de Sévigné* (Paris: Firmin Didot, 1853).

Rarely are the fevers named. Many were evidently Galen's ephem-erals, symptomatic affectations of the spirits. "My son has had an attack of a fever," she writes, adding that he hopes that "like that of last year" it will resolve within twenty-four hours. But one never knew at the outset. In a letter of 1671 she writes of Monsieur de Mans: "Dead he is, and of a slight fever, without having had time to think of heaven and earth. He passed the time of his ill-ness in a stupidity; it was a tertian that carried him off."[3]

Sensitive to the manifold possibilities of mismanaged non-naturals, aristocratic bodies seem exceedingly susceptible to fe-vers. Hazards included chocolate, which "occasions the vapours and palpitation of the heart. It flatters you for a time indeed, but presently lights up a continual fever, which at length carries you to the grave." Functionaries and courtiers worked themselves into fevers, sometimes from anxiety over inadequate obescience. She writes that Madame La Fayette has isolated herself, "resolved neither to think, speak, answer, or hear: She is quite wearied with saying good-night and good-morning, and has almost every day a touch of fever, which a little rest always carries off." Cardinal de Retz is "never free from a low fever" from the exhaustion of his philosophical labors. Mme de Sévigné cautions the pregnant Françoise to keep "as quiet as possible, and not hazard a fever for the sake of talking a little." Even writing long letters may overstrain one. Fever also came from immoderation. The duc de Rohan was "at the point of death, of a violent fever he got with swallowing two glasses of brandy upon a debauch of wine."[4]

Nature too, in the guise of season and place, threatened these aristocrats. Mme de Sévigné writes that the court will move to Fontainebleau: "It is [thus] imagined, that . . . they will overset the Dauphin's fever, which seized him this season at St. Germain; for this year it will be cheated; it will not catch him there again." In danger zones fever could strike suddenly, as one was entering a carriage, for example, or crossing Pont Neuf.[5] It is best, she writes, to follow Pascal, who teaches that "all our ills come from our not being able to endure keeping within doors." She believed that epidemics were borne on winds, though occasionally she hints at concern with contagion. She tells her daughter that she avoids being continually in the chamber of a fevered friend, going regularly into the garden.[6]

Even slight fevers warrant comment. But rather than disrupting social life, they were part of it; one visited fevered friends and enjoyed their company as far as the fever permitted: "I alighted at the house of the good d'Escars, whom I found in a bilious fever, but replete with the kindest and most hearty good will imaginable."[7]

These were visits of commiseration. Friendship was feeling, its depth measured in the pity a fevered person evoked. Shuttling between her fevered friend Mme La Fayette and her dropsical and sometimes fevered aunt, Mme de Sévigné clarified these bonds: "It is no purpose to take care of one's self, while there are persons who are dear to one, and stand in need of our assistance: we sympathize in all their feelings; their troubles and uneasinesses become our own, and, in short, all the pleasures of our own lives are buried in their want of them." Yet "that wicked fever has made another attack upon poor la Fayette, and my aunt grows worse and worse."[8] (Both mother and daughter worried that the other's excessive compassion would itself engender fever.)

Some solicited that pity. In March 1671 Mme de Sévigné found the duc de la Rochefoucauld, himself an eminent moralist, "roaring out violently . . . light-headed, and with a violent fever from the excess of pain" (jointly fever and gout). He wanted the distant Françoise to know this: "So dreadful a fit of the gout, . . . with such a violent fever, . . . worse than ever you saw him yet: He intreats you to pity him; and I would defy you to see him without having the greatest compassion."[9]

Confident in her own expertise, Mme de Sévigné judged theories and therapies alike. Fever vulnerability was a matter of blood, which might be too thick or too thin. She writes in 1677 of Mme La Fayette: "I pity all those whose blood is so extremely subtle, that the least trifle is sufficient to set the whole machine in a flame. My dear child, when one has a great regard for a person it is not ridiculous to wish that a blood, for which one is particularly concerned, should grow sedate and cool; but you, as I apprehend, should endeavour to render yours thick." Later that year she counsels her daughter, who is suffering from a sore throat: "Will you never think of moderating that prodigious heat of your blood?" For once "one's blood boils . . . all is soon over with one."[10]

But rarely was bloodletting the way to moderate hot blood. Mme de Sévigné well knew (and shared) Molière's sardonic views of contemporary medicine. She would have known too of the raucous midcentury antimony war, the attack by heterodox proponents of mineral medicines on the establishment neo-Galenists of the University of Paris, whom they condemned as inveterate bleeders.[11] Persons in her circle did worry about bloodletting, especially of their children and by certain practitioners.[12] Hence, royal enthusiasm for it was a concern, for courtiers must follow fashion. Suffering a "violent fever" (caught at Versailles), Mme Coulanges "must certainly lose every drop in her body, so great is the esteem they hold for her at that place." Mme de Sévigné relates, for its shock value, the number of times a dying or, occasionally, recovering fever victim has been bled;—in the case of Bassompierre, bishop of Saintes, it was thirteen times in the course of a fatal twenty-five-day fever.[13]

But Mme de Sévigné was no friend of antimony either. Her preferred remedies were Hippocratic gruels, although she was intrigued by the magic bullets of the day: bark, Talbor's mysterious fever specific (which would beggar the doctors of France, she delightfully declared), and the skills (and violent remedies) of a group of Capuchin monks, whose undescribed rigorous regimen did wonders. She would prescribe too: "Ask the Chevalier de Grignan [her daughter's husband's brother] whether I took not a great deal of care of him, whether I did not procure him a good physician, and whether I am not an excellent one myself? . . . I treat the Chevalier's health in a very serious manner; I see how his medicines begin to operate on him, and see him always in a good way before I quit him."[14]

Fever was also, as it would be increasingly in the eighteenth century, a matter of states of mind. Strong feeling or other mental intensity might generate fever, which in turn affected the mind, producing unsettledness, loss of reason, change of mood, as well as delirium. Mme de Sévigné would not accredit a letter from her friend Mme La Fayette, because it was "from a person just recovered of a fever."[15] Fevers, some of them diseases of spirits, are also tied to dispiritedness or depression. Her son-in-law's fever has brought on a "lowness of spirits and falling away [that] alarm every one." Not surprisingly, a favored medicine is identified as a "purge for the brain."[16]

"Ravings" feature regularly in Mme de Sévigné's descriptions. Delirium, not heat, was the measure of a fever's severity. While acknowledging the unreliability of her own fevered mind, she downplays the severity of her delirium in a 1676 fever: "As to my ravings, they only proceeded from want of proper nourishment, for I could swallow nothing but a little broth; besides there are some people light-headed all the time of a fever." She and her son, who has attended her throughout, joke about "the nonsense I used to talk, [it] makes me ready to die with laughing at the repetition." And yet an earlier letter suggests uneasiness: "I have refreshed my memory of every thing that a week's fever had made me forget, for you know that I have been subject to so many wanderings, that my poor brain has made a mere hotch-potch of truths and falsehoods."[17]

But fevers were also a matter of catharsis and redemption: "I return [from a walk in the woods at her Brittany estate] . . . changed as one just out of a fever," she notes in an early letter. She thinks she must make "proper use of this affliction" (her 1676 fever) but feels that she has not.[18] At the same time, fevers were a subject of grim humor. One popular theme was that fever and bloodletting physicians exist to help people die. "They were resolved to kill him in form," she writes of a smallpox sufferer bled eleven times. Boasting to her daughter that her son's acuity in selecting mistresses is superior to that of her son-in-law, she declares that none of Charles de Sévigné's mistresses will "die of a malignant fever, he has chosen them so well of late."[19] Others boast of the seriousness of their fevers. That another's fever has warranted extreme unction makes an acquaintance jealous: "Eh! Why do they not perform it for me? I am sure I deserve it full as well." (Both recovered.) Having married a younger woman, an aguish codger finds that his periodic fevers have gone: "We have got a remedy for the ague, but who can tell us the dose?" asks a wit. More disturbingly, a niece welcomes her husband's death by a "dreadful fever": she "has long wished to be a widow" and will inherit a great estate.[20]

That fever was a grim reality, an assault on the weakened self, a condition equally of mind and of body, a commodity in the commerce of sentiment, would all become familiar themes in the next century and a half. Care and sympathy during fever would be an important element of friendship in Mme de Sévigné's world, but there is little sense of any brotherhood of the fevered. She is unconcerned with epidemic fevers that decimate the general population, which signify mainly as estate-management problems.[21] Her own rheumatic fever was from such an epidemic, she explains, apparently to avoid accusations that she had brought it on herself.

THE FEVERED NATION AND DR. JAMES'S PANACEA

Rule Britannia, and take Dr. James's powder!

By the 1770s whoever and wherever one was in Britain's wide realm, one's fever could be cured by Dr. James's fever powder (sold exclusively by Newbery's, St. Paul's Churchyard). Dr. Robert James and his partners, the publishing firm of John Newbery, did create a brotherhood of the fevered, albeit as a marketing ploy. Their vision was of a global British society (not quite an empire) bound together by the administration, often as an act of charity, of this most British panacea. Notably, their vision of the connections people acquire through fever was not a matter of communicability. James was not interested in what caused fevers; he simply wanted to cure them. And since anyone's fever might be cured with his powder, morality commanded that it must be. James well reflects the new consumerism of the mid-eighteenth century, but in addition to creating a communal demand for a unique product, he sought to inspire in his countrymen a sense of obligation to ensure universal access to that product.

Trained at Oxford and London, with a Cambridge MD, Robert James (1703–1776) was no garden-variety quack.[22] After a decade of practice in the Midlands, he moved to London. There he turned to medical authorship to supplement his small practice, only to find that patients considered him too bookish to be a good doctor.[23] Within a few years of the deaths of Boerhaave (1738) and Hoffmann (1742) he had united their views in *The modern practice of physic, as improv'd by the celebrated professors, H. Boerhaave and F. Hoffmann* (1746). That work followed his three-volume *Medicinal Dictionary* (1743–45), published a few years before the famous general dictionary of his friend and Litchfield schoolfellow, Samuel Johnson.

Meanwhile, James was testing an antimonial fever powder that he had invented (or expropriated) around 1740 and would patent in 1747. Antimonials were not new, but James managed to distance himself from the controversies of the previous century. He would represent this powerful and poisonous mineral medicine as a gentle, natural alternative to the sweat-inducing hot treatments that were still popular.[24] As his claims for the powder inflated steadily, the phenomenon of fever grew ever simpler. There was really only one fever, James thought. Following Boerhaave, he held that all fevers shared three features—shivers, heat, and a quickened pulse—but as only the quickened pulse persisted throughout a fever, it became the essence of fever, and its causes became fever's causes.[25] Those causes were "numberless and infinitely varying," James wrote. Among them were mismanaged non-naturals (including too much exercise), obstructed vessels, poisons or toxic bodily fluids, and, at the end of a long list, "intense Study" and "excessive venery," both of which, if we believe his biographers, were applicable to James himself. His list, which drew from virtually every species of contemporary causal theory, included humors, poisons, epidemic constitutions, acridity, corrosion, causticity, friction, contagia, states of mind, surfeit, cold, hunger. But the full cause would rarely be known and didn't matter: "Let the variety of causes . . . be ever so great, . . . the effect is nearly one and the same, a fever." Behind the endless quibbling were a few common clinical challenges: languor, cold (and its alternation with heat), tremors, "anxiety" (primarily a cardiac rather than an emotional condition), thirst, nausea, vomiting, debility, heat, delirium, coma, sleeplessness, convulsions, sweats (bad and good), diarrhea, and rashes or eruptions.[26]

This unitary fever required no theory. James, who was consolidating the views of the consolidator Boerhaave with those of Hoffmann and all others to "*render the whole practice of physic compleat*," declared that theories were fictions: "Galenists, . . . Chymists, . . . Mechanics, and an infinite variety of other theorists, [were] all equally uncertain, abstruse, and deceitful. Every man practices according to his favourite . . . and . . . is led astray by that enchantress." Mere "amusement," theory lacked both "physical" and "moral" utility. As is often the case, the empirics' argument democratized: the powder's effectiveness obviated all recondite private knowledge, and verbosity could make no reply to the one hundred thousand cures claimed for James's powder. At the same time, it is worth keeping in mind that James's dismissal of theory was a *learned* one.[27]

In fact, James had a theory. He saw fever in Hippocratic terms as a concoctive and expulsive process. The powder was an expulsive aid; it might act as a vomit, purge, or sweat. Even when it seemed not to be doing anything, it might be enhancing insensible Santorian perspiration. It was what Sydenham had *sought:* a specific for all fevers and for many nonfebrile conditions as well.[28] It cured the new yellow fever (important for the West Indies), smallpox, rheumatism, worms, stroke, mad dog bite, poisoning by arsenic and nux vomica, and the inability to urinate. It also worked (in today's terms) as an antidepressant for "universal Languor or Dejectedness," which James regarded as a kind of fever.[29]

In 1750 James sold a half-interest in the powder to the publisher and publicist John Newbery (d. 1767). In the next decades, the powder would be successfully marketed by Newbery's son and his nephew, both named Francis.[30] The firm sold other patent medicines too. The Newberys were also pioneers in children's literature and publishers and/or friends of Oliver Goldsmith, Tobias Smollet, and Samuel Johnson.

James's success, coupled with the dubious practice of patenting a medicine (whose manufacture nevertheless seemed inimitable), fed professional jealousy. Many questioned his claim to originality; others attacked the powder's safety or complained of its price, 2s. 6d. for two twenty-grain doses. (Since, claimed James, it actually cured, the cost would be far less than the expense of weeks of doctor's visits and futile doses.)[31] Practitioners did gradually adopt it, probably in response to public demand.[32]

A key tool in marketing James's powder was his *Dissertation on Fevers and Inflammatory Distempers* (1748), a real book, if a promotional one, and at sixpence far cheaper than the medicine it

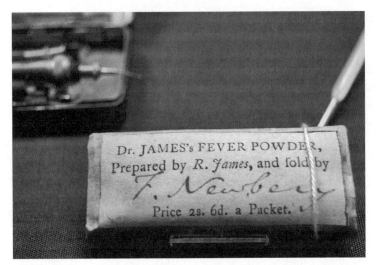

A sachet of Dr. James's fever powder. Counterfeiting of the powder was an ongoing problem. Newbery's signature and the centralization of sales were means to combat that problem.

Oxford Science Museum. Photograph by Mo Castandi.

advertised. Seven editions appeared during James's lifetime, but it is in the posthumously published eighth edition (1778), edited by Francis Newbery the nephew, that we see most clearly the novel vision of a society made happy, healthy, and productive by the universal availability of fever powder.

Like Walt Disney two centuries later, the Newberys imagined and marketed a moral universe—English society as a harmonious system of distinct stations, high and low, in which loyalty and charity, experienced as sympathy and sentiment, mediated relations. They broadcast that ideal on all channels, through entertaining and edificatory digests for all ages on everything from natural philosophy to theology but most famously through illustrated children's books. Best known is *The History of Little Goody Two-Shoes,* which opens with the death of little Margery's father from a "violent fever, in a place where Dr. James's powder was not to be had" (the Newberys were pioneers in product placement).

The story is about virtuous landownership. In the beginning Margery's parish is ruled by greedy Sir Timothy Gripe and his minion Graspall, who withhold charity and drive tenants off the land. At the end, Margery acquires the estate and transforms it into a yeomanly utopia. Along the way, convenient fever deaths allow "large" fortunes to accrue to heiresses, notably Margery, who can then become ladies bountiful.[33]

Benevolence and charity unite that society; its most important divisions are between benefactors and recipients. A particular concern of James's were servants and laborers dismissed when they became ill. Bankrupted by fever, "scarcely any thing remains for the Women but the Brothels, and for the Men but the High-way." The artisan too might recover from a few weeks' fever only to end up in a debtors' prison. James declared that he would get "no small Satisfaction, if I should contribute to prevent these Calamities, by pointing out . . . a Method of retrieving the Health in a few Days, and at a Trifling Expense."[34] Cost was never an issue; indeed, if the powder didn't work, one was simply to take more.

Claiming the high ground of magnanimity was no less important than establishing the powder's efficacy. Charity would be central in the James-Newbery business plan, as was evident in their striking division of roles. Where the Newberys sold the powder, James gave it away. He had, the Newberys would boast, "given away to the indigent, near as many [as have been sold]; . . . it has been refused to no poor person who has asked for it at his House."[35] Moreover, successive editions of the *Dissertation* were full of reminders that others gave it away too. Given that disease was often the occasion of need, and fever the most common form of disease, it made sense that the powder should be a mainstay of charity. "Blessed with Leisure, Affluence and Dignity," genteel neighbors of the fever-stricken rural poor would surely help were there "a Prospect of Success." But there was: in simply stocking the powder local worthies were purchasing virtue, "a Satisfaction of which the Proud, the Thoughtless, and the Cruel, can form no Idea."[36] (The Newberys appealed to self-interest too. In direct-mail marketing to parish overseers they maintained that stocking the powder would lower poor rates. There were discounts for bulk purchases.)[37]

Often the givers of powder were laypersons, frequently strangers with incidental ties to the stricken. For the society of benevolence had no need for professions. Not only would the powder save the patient from the deadly and mercenary ministrations of unenlightened practitioners but its use required no skill beyond what the good burgher, clergyman, squire, or goodwife possessed. Having used the powder on themselves, they were competent to administer it to others. In 1761 "persons of quality" visited the fevered taverner Howell, who was being attended by two other doctors. Respecting the conventions of medical ethics, James himself would not interfere, but Howell's family wanted the powder given, hence a "Mr. Collyer, upon the principles of the good Samaritan, and without the least pretence to any medical knowledge, except what his own experience upon himself had furnished . . . gave Mr. Howell ten grains" (and later another 20). Other lay prescribers included colonials who had brought back plenty of powder from visits to London, as well as overseers of slaves. We may understand physicians' ultimate acceptance of James's powder (and of analogous unpatented antimonials, such as tartar emetic) as the profession's effort to regain authority over prescribing.[38]

But perhaps the most powerful testimony of a nation united by fever is the cases themselves. Jettisoning theory, marketing what was effectively a secret remedy, James appealed to experience. While the *Dissertation* includes a brief defense of Hippocratic empiricism and a section on dosing for various conditions, case histories make up its core. In early editions, James's chief concern was mere credibility. Since his practice was not an elite one, the testimony of persons of "quality" to the cures achieved was important. The recovery of the first case he lists, that of a Mrs. Morton, might be "fairly ascrib'd to nothing but the Medicine," he declares. "Many Gentlewomen were present during her whole Disorder, saw her take the Medicine, observed the Effects, and are ready to give Testimony to the Truth of what I have asserted."[39] But as James and the Newberys realized, the mere enumeration of recoveries, even from varied and dire conditions, was less effective than portraying the universal applicability of the powder. As testimonials accumulated, they dropped from later editions some of the cases of London servants, clerks, and shopkeepers who had, in effect, been James's experimental subjects in the early 1740s. In the 1778 edition, Newbery artfully arranged cases to suggest that nowhere on earth was there a British subject who could not benefit.

Near the head of the parade is the octogenarian actor Colley Cibber, who declares it his "duty . . . to communicate, for the general advantage of mankind" his recovery from the brink of death. Among victims of the Old Bailey epidemic of May 1750, the most recent of the celebrated "Assize" epidemics, in which prisoners brought from their fever-generating styes into the genteel courtroom communicated their deadly fevers to worthy administrators of justice, "not one" of the infected attendees who had taken James's powder had died. North America is represented by letters from Boston, where the writer tells of using the powder to save a neighbor's "negro girl, which he had a great value for," and from Savannah, where the importer of fifty shillings' worth of the powder has offered it to a nearby Carolina town, where "they died of pleuritic fevers like rotten sheep." "I do not know any medicine . . . fitter for the distempers to which this climate is subject," the correspondent declares.[40]

Representing rural England is the Lincolnshire shepherd Isaac. A clergyman has offered the powder to Isaac's wife on the promise that she will dose Isaac according to his directions (here, however, the "fever" was cured by the passing of worms, noteworthy in their size and number). Rural England is also represented by a 1755 epidemic in Kilpeck, Herefordshire, chronicled in the *Gloucester Journal.* There beneficiaries include a 20-year-old woman suffering from "violent fever . . . intense heat, excessive thirst, . . . watchfulness, and deliriums," a man of similar age, an 8-year-old girl, her mother, the mother's sister, and "a robust strong man" on whom the powder at first has no effect. A 25-year-old woman has given herself up to die, yet after reluctantly taking four doses over two weeks, she discharges a large yellow mass and then sleeps soundly for the first time in three weeks. Within a week she is nursing a neighbor afflicted late in pregnancy with a pleuritic fever arising from her own nursing of her fevered husband. Three seven-grain doses break that woman's fever, "an happy event for herself and family!" She then goes into labor and quickly delivers, only to be stricken with a postpartum delirious fever, which is quickly relieved by another dose. "Philanthropos," the chronicler, enlarges on the implications: "If such have been the consequences of these Powders in one small country parish, in . . . only a few months, it is easy to conclude how many individuals, how many families, may owe their lives, and every comfort of their lives, to this [James's] publication."[41]

Newbery turns then to cases of infants at opposite poles of society. Her benefactress, Lady Vere, reports the May 1756 case of the foundling Mary Yatesby, who suffers from vomiting and purging and could not suck or pass urine. Two four-grain doses (ordered by Lady Vere) yielded "lumps, very foul and discoloured." Later in the year, the declining twenty-month-old Marquis of Lindsey, having been repeatedly blistered and bled for complications from measles, is finally given the powder at the request of his father, the duke. The four-grain dose produces, after just thirty-nine minutes, an hour of vomiting and two stools. Additional doses over the next few days bring full recovery.[42]

Also included are reports from places where extremes of climate might be presumed to challenge this most English remedy (in fact, hot climates seemed to require lower doses).[43] A functionary in the East India Company writes from Persia, which he insists has an unhealthier climate than Guinea or Madagascar. Fevered for two weeks, he finally tries the powder "sold at Mr. Newbery's in St. Paul's Church-yard London. . . . It removed my disorder: . . . I soon recovered, and continued in health in all Heats." Lord Barrington, the secretary of war, forwards a letter from the surgeon Browne in Madras, who has used the powder on fevered soldiers during the long voyage out, with only one death in forty cases. James replies that were all regimental surgeons to follow Browne, "no doubt but some thousands of brave fellows might have survived to serve their king and country . . . a consideration of no small moment in this tedious and destructive war."[44]

No less important was the powder's contribution to Britain's slave economy. James gets double duty from the tale of a Mrs. Payne of Guadeloupe, who saved seventy-three of her seventy-four fevered slaves with the powder but died when her own doctors refused to administer it. On St. Kitts, the agents of Augustus Boyd had paid a yearly salary of sixty pounds to a medical attendant prior to employing the powder, but for ten years they have relied almost exclusively on the powder, administered by an overseer. "It has not cost him [Boyd] above seven pounds a year; . . . before, at least six of his negroes died for one that he loses at present." The point is clear, yet James hammers it: "The value of the West-India estates should seem to depend principally on the preservation of the Negroes. . . . I therefore judged, that if my Powder proved as effectual in the cure of acute distempers in the West-Indies as in England, it would contribute to the enrichment of that very valuable part of the British dominions."[45]

A key theme for this commercial empire was rapidity of recovery. To the servant Sarah Francis, Mr. Pringle the cabinet maker, Joseph Needham of the Strand, and William Parry the attorney, the powder brings sleep and, within "a few Days," competency "to pursue my Business."[46] Rapid recovery is paramount for Gilbert Douglas, Esq., "a Gentleman very well known in both Houses of Parliament." James treats him at the command of the Countess of Anglesea, whom he assures that Douglas will be well enough by "*Wednesday*, to transact some Business of very great Consequence." In fact, Douglas is "so well by *Tuesday*, as to get up and write for several Hours, and to settle some Proceedings in a Cause of great Moment."[47]

Buying James's fever powder would strengthen the bonds of nation. Political economy might teach that there must be sufficient specie, but James's powder was itself a key currency. Once fever struck, it would be too late to summon this social capital; it must be kept on hand. And it was sold only at Newbery's, St. Paul's Churchyard.

FEVER AS SOCIAL CONTRACT: PLACE AND CLASS

For James and the Newberys the altruistic engagement with the fevers of others was absolute, uncontaminated by any hint that in combatting the fevers of others one might be protecting oneself from a deadly contagion. They were, after all, peddling pills, not running public-health programs. James knew that fevers came in bunches, but he pretended that each person succumbed as an individual. He even understood the prominence of smallpox as an index of insufficient access to his powder. Likewise, the "Assize" fevers, like the Old Bailey fevers of May 1750, were triumphs of therapeutics rather than the icons of communicability they would become.

But many contemporaries were focusing on cause, responsibility, and, with it, blame. To them, the most important thing about anyone's fever was what had led to it—perhaps the distribution and movement of contagious bodies, or unremoved filth, unrelieved indigence, or even a poorly built environment, all often acting in conjunction with states of nature. For these contemporaries, such public causes and the preventive actions they might imply were the high road to universality. If fever was communicable (in a broad sense), they were the community's business. Sufferers could not be allowed to experience their diseases individually even if they had benevolent, powder-dispensing pastors. Achieving the private good of gaining freedom from fever required investing in the public good of preventing *and* curing fever in all others.

Especially in liberalizing England, publicizing the private was a delicate matter. The first fever hospitals, begun in progressive northern towns in the 1780s, would be designated "houses of recovery." Not all recovered, but recovery was the message to would-be patients. Donors were to take comfort that in these "houses" contagious "inmates" were off the streets. Their confinement lowered transmission, mitigated epidemics, and hence saved the lives of donors, who were as much the clients of the hospitals as the fevered inmates.

But in sharp contrast to James's theme of universality, for those concerned with communicability the key issue with fever was almost always place. That one was fevered was not inherently problematic; that one was fevered *and here* was. Often, as with plague, the chief unit of concern was the town. Plague had been only an occasional visitor to any town. Usually, its likely route of importation had been clear, and quarantine had been a flexible and often an effective response. Fevers, however, were more often endemic, with seasonal fluctuations in character and severity, which could not always be tied to importation. Similar problems arose in mobile communities, armies and ships. While the cause might not be clear in those cases, responsibility often was. I focus on exemplars of each: Joseph Rogers's 1734 analysis of Cork's "popular" fever and the midcentury treatises by John Pringle on camp and hospital fevers in the British army and by James Lind on shipboard health.

Carnivorous Townsmen

Why care about Cork? Joseph Rogers declares that he wrote his *Essay on Epidemic Diseases, more particularly on the Endemial Epidemics of the City of Cork . . .* reluctantly, after fellow townsmen, faced with a new and deadly fever, pressed him to organize a quarter-century of case notes for "the Benefit of Mankind: and more particularly for the use of the Inhabitants of this City."[48]

A modern reader who understands *epidemic* and *endemic* as exclusive alternatives for describing the relation of disease to place may well balk at Rogers's phrase "Endemial Epidemics." We might expect him to be writing about some universal fever that happened to come to Cork or, alternatively, to be addressing the motley of individual fevers that would be unique in Cork as in any city. In fact he is doing neither, nor in the main is he doing what local fever writers would be doing at the end of the century, that is, soliciting local worthies to take public measures to fight and prevent fever in their communities. Instead, he exemplifies what has been recognized as a neo-Hippocratic approach, in which the most important determiner of the characteristics of the prevailing disease is place itself, here a combination of climate, terrain, built environment, human activities, and perhaps even race.

Rogers insisted that Cork was facing a new kind of fever, one so alien to received theory that by Boerhaave's definition it was not fever at all: the pulse did not race. Debilitory rather than inflammatory, this fever seemed more likely to be seated in the solids than in the humors; dissection usually revealed "some Organ, Bowel, or larger part, injured or somehow altered from its original State of Perfection."[49] These slow or low fevers had occurred sporadically since about 1710. Always "it hath been the same Contagion," Rogers asserted.[50]

Rogers knew that similar diseases were arising elsewhere, but he maintained that Cork's was unique. The claim does not reflect any compilation of comparative clinical data from multiple towns, but is instead axiomatic: since disease specificity was a function of geographic specificity, Cork's slow fever would differ from Derry's, and even more from those in England. Their fevers would "of Course require another Sort of Treatment," while Cork's fever must be treated by those who, like Rogers, were uniquely experienced in the diseases of Corkonians.[51] In appealing to ineffable differences, Rogers may seem solipsistic, but he was an insistent empiricist, relying on both bedside observation and dissection. With "so many incontestable Data," all "Object(s) of our Senses," he and his colleagues will reach sound conclusions, he asserted.[52]

To a modern reader, the disease Rogers describes is very important as a universal disease, louse-borne typhus. It was a fever of rapid onset with "slight Horrors," followed by a sense of weight, pain, and intense headache, "particularly in the Forepart and just over the Eye-brows," but without a change in pulse. Following these initial symptoms was "an universal Petechial Efflorescence, not unlike the Meazles, [which] paints the whole Surface of the Body, Limbs, and sometimes the very Face." The headache might lead to a "Coma or Stupor, or . . . Delirium." The heart and lungs were also involved, and the urine was "pale, crude, and limpid."[53]

This new fever was to be treated according to principles opposite to those followed in ordinary "accidental" fevers. The usual evacuants would exhaust a patient with a low nervous fever. Still, many cases did not correspond to type, and intercurrency was the norm, not an aberration.[54]

Rogers was indeed a pioneer in recognizing what has been seen as a major shift in northern Europe's dominant fever, from inflammatory to debilitating or "nervous" fevers, chiefly typhus. The eminent historian of epidemics Charles Creighton lauded him as among the best eighteenth-century epidemiologists, but again, to see Rogers as mistakenly claiming that a fever must be unique to Cork because it is new, misrepresents him. Nervous fevers were arising in other places, but aspects specific to Cork were more important than any general features.[55]

In Rogers's view, clinical uniqueness was grounded in causal uniqueness. The usual generic causes of fevers were not responsible in this case; those stricken had not committed "any remarkable Errors . . . in the Non-naturals." Rogers's long list of causes is eclectic. It violates any neat division between universal causes and factors particular to person and place and includes entities that resemble the sorts of specific agents that would become synonymous with *cause* two centuries hence. "All Epidemics," Rogers asserts, "arise from some peculiar Deleterious Particles, conveyed into our Blood and Juices."[56] These were "contagia" or "Morbid Miasmata." These terms, often seen as antitheses, were here complementary, even interchangeable. Rogers goes on to say that the source of Cork's common disease must be its "common Food," the atmosphere, in which "float Particles of all kinds, detached from the Animal, Vegetable and Mineral Kingdoms. From . . . these variously combin'd, result Mixtures of a certain determin'd Nature, which affect sometimes Mankind." These were the "noxious Miasmata" or "Morbid Effluvia." The severity of outbreaks would depend on the number of particles and on atmospheric states, like heat and humidity, which enhanced their action. Cork's epidemics came in warm, wet seasons, which facilitated putrefaction.

Yet these aerial entities are not quite our modern disease

agents. The contagia were not passed from sick persons to well; the particles were neither the triggers of individual cases nor the essences of some disease species. In imagining states of Cork's atmosphere that might account for its prevailing disease, Rogers was mainly trying to translate Sydenham's constitutions into mechanico-chemical terms. But he would not "misspend Time in pretending to determine, what are the particular Size and Figures of these Morbid Particles."[57]

Cork's fever was equally a product of the bodies with which these particles interacted. And those bodies, in turn, were affected by activities and milieu. Other determinants of Cork's special fever were filth, particularly the "Ordure and Animal Offals that croud our Streets, . . . Allies and Lanes"; seasonal slaughtering, a mainstay of Cork's economy; and the combination of meat eating and water drinking, especially among "the better Sort of People," who "imagine," notes Rogers, "that they may Riot with Safety, in the Delicacies of the Animal Diet, by doing Penance in the insipid Plainness of their Drink." But it was the meat eating that stood out. In "no Part of the Earth," declares Rogers, "is a greater Quantity of Flesh-Meat . . . consumed, than in this Place, by all Sorts and Conditions of People during the slaughtering Season."[58]

If none of these were unique to Cork, still their combined effect might be unique. When Rogers turns to the question of how these factors produce Cork's fever, the effects (and the entities themselves) blur. Thus, it was the effluvia of Cork's dung-filled alleys and offal pits, not the atomic contagia or miasmata of the general atmosphere, that seemed to act as ingested specific agents: these, "arising from the Corruption of Animal Bodies, are of a most active and subtil Nature; they are the Salts and Sulphurs of such Bodies, [produced] by a Fermentative Putrefaction, wrought up to the highest Pitch of Exaltation." Tiny and active, "they most readily insinuate themselves into us, . . . we every moment swallow 'em with our Spittle; with our very Food they pass into the Blood and Juices of our Body; and our Lungs receive them with every Breath we draw." Rogers then appeals to plausibility: "'Tis easy then to conceive, that many and great Alterations

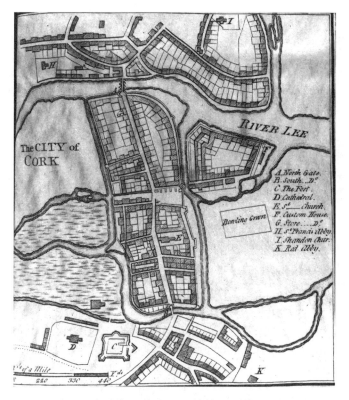

This 1690 map of Cork suggests the compact, medieval walled city, surrounded by marshes, that was rapidly growing during Joseph Rogers's lifetime.

City of Cork, c. 1690, cartographer unknown. From Cork: Past and Present, Cork City Library, www .corkpastandpresent.ie/mapsimages/corkinoldmaps/c1690thecityofcorkmap/1690.pdf.

must necessarily be produced, by the Admission of such acute and destructive Instruments, into the inmost Recesses of an Animal Machine." There they acted as irritants. They could convert "the whole Juices of a sound Body, into an equal State of Corruption," especially when that body was predisposed by a meaty alkaline diet (the autumnal slaughtering season coincided with the fever season).[59]

Thus, to a certain extent Cork had brought this fever on itself.

Community action—cleansed alleys, altered diet, even sanitary slaughtering—might help. As dietary preventives, Rogers advises, along with bread and sea salt, acid vegetables, including those James Lind would famously enlist against scurvy, "Vinegar, Lemons, Oranges or Sorel," as well as cider and wine.[60] But Rogers was not a strident reformer. He believed he had explained the communal origins of the Cork fever by treating the city as an integrated whole and distinct from other towns. He did not problematize single aspects of the city. Nor did Cork's fever seem to suggest conflicts among sectors or groups within the city. Rogers's "popular" fever might not be likeable, but he assumed that Corkonians enjoyed febrile equality: all ate the same meat and breathed that "common Food," the contagious air.

In the same decades, the 1730s and 1740s, inquiries like Rogers's were being carried out in provincial towns across Britain and Ireland. There too, practitioners, thinking in Hippocratic and Sydenhamian terms, sought to attribute uncommon fevers to some idiosyncratic relation of place and people. Only slowly would the medical literature transform such reports into the paradigmatic European fever, typhus.[61] A century later that disease would be regarded as peculiarly Irish, but more as an unwelcome attribute of Irish persons than as an aspect of place. That recognition too would have democratic implications, but the forms of equality it would engender had much more to do with cross-class bargains—cures and alms in exchange for a reduced risk.

Wet Soldiers

Cork's fever reflects the democracy of place; other authors would focus on the democracy of common circumstance. Among them was Sir John Pringle (1707–1782), author of the 1752 assessment of diseases encountered in the British army's recent campaigns in the Low Countries and against the Stuart pretender in Scotland. In the mid-1740s, Pringle, a student of Boerhaave's, had migrated from teaching metaphysics and moral philosophy at Edinburgh University to become physician-general of the army; he happened, fortuitously, to be the personal physician of successive commanders of that army. Although his military career lasted only six years, it had implications far beyond military hygiene: it would be the departure point for much urban public health. Pringle himself would become a medical experimenter in London and serve as president of the Royal Society in the 1770s.[62]

Where Rogers had to adapt therapeutics to Cork's climate, society, and economy, Pringle could both measure the effects of pathogenic variables and, in some cases, lower exposure to them. If the army was sometimes a fever factory, it could also be an epidemiologist's laboratory and a health-maintenance machine. The effects of place could be traced as the army moved. The effects of conditions could be traced by correlating illnesses with men's rank and branch of service. Why did the cavalry suffer less fever than the infantry? Because their cloaks protected them not only from cold and damp but also from the "putrid exhalations" of soils. Why not, then, provide "a blanket for every tent of the infantry," along with an oilcloth groundsheet?[63] Command had its limits, Pringle admitted. Generals could not control the weather or other things that "a soldier shall have in his power to neglect" (e.g., food, drink, sex, and thoughts). But there was much they could fix, including accommodation (ventilation and density), sanitation, and the water supply.[64]

All these means of prevention might be mobilized against a suite of increasingly distinct febrile diseases that threatened Pringle's troops. First were the sporadic inflammatory fevers, pleurisy and pneumonia. Generally these arose from exposure, such as sleeping on the cold ground in wet clothes. These diseases dominated in the spring but resolved quickly. Deadlier were the autumnal diseases. The tertian of the Holland marshes might turn into continuous fever. It was often accompanied by other diseases, including scurvy, a communicable dysentery, a bilious remittent (linked at least conceptually to the West Indian yellow fever), a cholera, a relapsing fever associated with jaundice and dropsy, and an ardent fever accompanied by great thirst, delirium, and biliousness. Complicating all the diseases were infestations of worms, as well as colds and "influenza."[65]

Though all these diseases concurred and could evolve one into another, Pringle saw his task much as a later epidemiologist might. Disease species could be understood both in terms of courses of symptoms and in terms of circumstantial or causal variables. Those distinctions could be used to rationalize therapy and plan prevention.[66] Bleeding, required in inflammatories and intermittents, would be dangerous in a low nervous fever. But since it was difficult to predict how a fever would evolve (and treatment could affect that), it was important to know the fevered person's history of circumstances and exposures. Those with inflammatory fevers should not be moved, while those with autumnal fevers could be; in fact, in the latter a change in locale might improve their condition.[67]

The deadliest disease of the army was a "hospital fever," an "inseparable attendant of foul air from crowds and animal corruption." Pringle equated this fever with the famous gaol fevers, not just the Old Bailey fever of 1750 but also the notorious Oxford Assize fever of 1577.[68] It might also be linked to the sixteenth-century *morbus Hungaricus* or to the "camp fever" that had killed Thomas Willis's parents when it broke out in the army besieging Oxford in 1644. In the next century such linkages would be important in forging "typhus" as a modern disease species.[69]

For Pringle, hospitals and jails were more than sites where specific diseases frequently occurred; they were veritable pathogenic machines. He understood much hospital fever as a product of the hospitalization of soldiers suffering from autumnal bilious remittent or dysentery. In a confined space, the dysenteric discharges "vitiated [the air] to such a degree" that they not only infected other patients with dysentery but also generated that "still more formidable disease, namely, the *hospital* or *jail-fever*."[70] This process was one of amplification: diseases generated in the field from insufficient protection from a malign climate became more deadly, and contagious, through concentration. *Contagious* here need not imply transmission of a particulate agent specific to the new disease. Rather, through concentration, emanations, perhaps generic, acquired sufficient power to generate the new disease in others.

Tracking exposed persons would become important, however. Nairn hospital surgeons need to know if a disembarking regiment has been exposed to this low fever, Pringle insists. Leaving the fetid decks of troop ships would increase sick soldiers' chance at recovery, but it would put those on shore at risk. For in Nairn and at Newcastle the epidemic jumped the hospital firewall, killing in Newcastle three apothecaries, four of their apprentices, and two journeyman. Even worse were the effects on residents of an island in northern Holland where such a debarkation led within a few weeks to fifty deaths, "a sixth part of the inhabitants."[71]

The process Pringle describes would be the most important *medical* foundation for recognition of the sociality of fever. The causes of fevers in individuals varied, and each might be specific to a unique history of misery, but those fevers that were not originally contagious might *become* so. Thus, the ephemeral fever of a single person was potentially the deadly typhus of all. In distinguishing generators of contagion from recipients (and retransmitters of contagion), this model raised questions about blame and responsibility. Were recipients (and retransmitters) less or more blameworthy than generators? In one moral narrative that would become popular in the succeeding century, recipients—persons in comfortable circumstances—had only themselves to blame if they did not act to stop the generation of contagion in those who were powerless to avoid fever-generating circumstances. Such persons might be displaced, impoverished, overworked, or ill-housed civilians, or prisoners, or Pringle's private soldiers marched hither and yon. If they did not take preventive and charitable actions, the powerful would receive their just deserts; their fevers would be the mark and measure of unmet obligation.[72] Later writers—conspicuously in Ireland—noted that the fever one caught was deadlier than the spontaneously generated fever, as if the payback for neglect came with interest. The moral is plain: fever bridged social division, put the rich at the mercy of the poor, and cre-

ated a common interest in the health of all. Medical staff mediated this leveling. They were, on the one hand, relatively innocent
runners of risks and, on the other, brokers, uniquely empowered
to demand charity on grounds of fever prevention. Visitations
of plague had put doctors in such a role in earlier centuries; the
ever-present threat of an endemic fever becoming epidemic would
regularize it.

Pringle made clear that much could be done. Hospitals were
not uniformly bad. Battalions in Zeeland were struck with Malignant fever, but the "spacious and well aired" wards in Ravenstein
hospital prevented it from spreading further ("tho' several of the
sick were brought in with petechial spots"). For Pringle, as for
Florence Nightingale a century later, ventilation was the key, so
important that he would urge breaking the windows of the cells
of the fevered. Moreover, Pringle stated the manifesto of the sanitarianism of the next century. For towns were like army hospitals:

> From this view of the causes of malignant fevers and fluxes, it
> is easy to conceive how incident they must be, not only to all
> marshy countries after hot seasons, but to all populous cities,
> low and ill aired; unprovided with common sewers; or where
> the streets are narrow and foul; or the houses dirty; where water
> is scarce; where jails or hospitals are crouded, and not ventilated
> and kept clean; when in sickly times the burials are within the
> town, and the bodies not laid deep; when slaughter-houses are
> also within the walls; or when dead animals and offals are left to
> rot in the kennels or dunghills; when drains are not provided to
> carry off any large body of stagnating or corrupted water.[73]

While these elements of insanitation had often been included in
the long lists of disease causes, it was Pringle who asserted their
supreme importance. Armies and army hospitals could be better
managed, towns too.

Putrid Sailors

As well as setting an agenda for the future, Pringle was connecting
to the pathology of the past. He had been moved to write, Pringle

explained, by the "great number of putrid cases, that were under my care in the hospitals abroad."[74] He would share this interest in putridity with two other authors of midcentury medical best-sellers, his naval colleague James Lind (1716–1794), a physician at the Haslar naval hospital at Plymouth and author of the famous *Treatise of the Scurvy* (1753), as well as the 1757 *Essay on the Most Effectual Means of Preserving the Health of Seamen in the Royal Navy* and the 1768 *Essay on Diseases incidental to Europeans in hot Climates,* and the civilian doctor John Huxham (1698–1768), also of Plymouth, author of the celebrated *Essay on Fevers* (1750) and probably the foremost English fever theorist of the day.[75] For these three and for others, putridity would become the most worrisome pathological mechanism in febrile diseases.

Putridity had been central in Galenic fever theory. Referring to a hypothetical pathological process of humoral corruption, the term had designated the large class of fevers caused by that pu-trefaction. Criticism from many quarters, ancient and modern, had exposed ambiguities, to the point of incoherence: without undue flippancy James could claim that while he did not know what "putrid fever" was, his powder would cure it. Yet rather than disappearing, "putrid fever," in an altered version, became the centerpiece of the eclectic fever theory of the mid-eighteenth cen-tury. In Creighton's view, that concern probably reflected a new form of febrile disease that came to prominence around 1740 and declined after 1775.[76]

For, increasingly, *putrid* was becoming an observational rather than a theoretical term. It referred to a distinct clinical profile. Symptoms included blackness of lips and gums, foul breath, foul wine-colored sweat, great livid or purple ecchymosal blotches that might suddenly blossom or evolve out of petechiae, and swol-len legs and joints. Blotches might turn into ulcerous sores; hem-orrhages were frequent. The black spots, often swollen, in turn recalled the buboes of plague. Thus at Feckenheim "petechial spots, blotches, parotids [e.g., swellings in the parotid glands, as-sociated with plague], frequent mortifications, contagiousness, and the great mortality set forth its pestilential nature. . . . It was worse than the true plague, [in] that there was no security against a relapse; but on the contrary, almost a certainty of it, if the person continued in the infectious air."[77] The other major analogue was scurvy. Though Lind had struggled mightily to narrow the term's use, even for him the generic *scorbutic* implied both a disposition to scurvy and a suite of symptoms that might culminate in it: darkened and roughened skin, large spots, hemorrhage from various sites, swelling and joint pain, and stinking breath.[78] The classic pathognomic, putrid gums, did not occur in all cases (some presumably died before it developed), nor was it always distinct from the putridity around and in the mouth that accompanied other fevers. As was the case with the slow nervous fevers, scorbutic symptoms were associated with exhaustion—"listlessness . . . [and] aversion to any sort of exercise." Scurvy too was linked to dampness and to cellars and prisons.[79]

For the next century, putridity, along with nervous exhaustion, would reign over fever discourse. As they had been for the Hippocratic writers long before, putrid symptoms were exceptionally grave. The body seemed to be decomposing while still alive; postmortem decomposition too was often said to be unusually rapid. While explicitly humoral allusions were waning, there remained strong interest in the chemistry of putrid blood. Withdrawn blood failed to coagulate in a healthy way and quickly putrefied.[80]

Coupled with greater confidence in the empirical distinctiveness of the suite of putrid symptoms was renewed interest in the underlying process of putrefaction. Following the lead of Francis Bacon, Pringle would devote much of his later research career to the chemistry and physiological effects of putrefaction, as would Huxham. They were seeking to make *putrid* more than metaphor, a placeholder for similarity between obscure pathological processes and decomposition of dead animal matter. They wanted to understand a common process that not only was responsible for both inner and outer putrefaction but might reflect a dynamic reciprocity in which external putrefaction begat internal, which in turn might be recommunicated to environmental media through discharges from the victim's putrefying body and thence communicated on to other vulnerable persons. At several points Pringle gropes toward such a concept. He writes: "As the exhalations from marshes [contain] putrid *effluvia*, . . . it is not surprising, that the distempers incident to such as breathe that air, should be of so putrid and malignant a nature." He notes that putrefaction-corrupted air was the "most fatal" of the causes of army diseases; that dysenteric or fevered soldiers acquired their disease from "the corrupted water of marshes; . . . from human excrements lying about the camp in hot weather, [or] from straw rotting in the tents." Then, in the hospital, "crouded with men ill of putrid distempers," new states of corruption might be inhaled, though ideally they would pass right through the body if the pores were working properly.[81]

While there was nothing new in recognizing environmental rot as a cause of disease, internal, pathological putrefaction had not usually been equated with external putrefaction. It would be. In Justus Liebig's zymotic theory of disease, popular in the half-century after 1840, specific fevers would be explained as the communication of particular modes of putrefaction from external filth to vulnerable parts of the body. Putridity would be both the mechanism of communicability and the primary mode of pathological activity. That zymotic theory would be the analogical bridge that allowed Pasteur to associate microbes, responsible for particular types of fermentation or putrefaction, with specific diseases. It was one of the main roads to modern germ theory.

Pringle and Lind were not thinking of putridity in terms of disease specificity, however. They viewed it more as a complication or variation than as a species. In Pringle's practice, putrid symptoms often followed hospitalized cases of low nervous fever or of some intermittent, supervening so frequently that they might seem to be a late stage of the disease itself.[82] In Lind's, where putridity was primarily scorbutic, febrility preceded (or occasionally followed) scurvy. On the HMS *Salisbury* it was those recovering from respiratory inflammatory fevers who became scorbutic; indeed, conversion of fever to scurvy might demarcate a "crisis."[83] Was scurvy itself a fever? Not by Boerhaave's definition (rapid pulse, shivers, and excess heat), but nor were many nervous fevers. While writers occasionally referred to scorbutic fevers, it was more important that scurvy and fever overlapped epidemiologically and clinically.[84]

In serving as a bridge between the acute epidemic diseases, such as plague, and scurvy, the exemplar of chronic, environmental-constitutional, and even *cachexic* diseases of weakness and decline, the putridity paradigm had not only effaced a central categorical division of ancient medicine, that between acute and chronic, it had also refocused attention on the grand question of the relation between episodic fever and the ordinary life. The boundary between acute and chronic had long been problematic. Though they might only burst forth once some presumed tipping point of predisposition had been reached, fevers exemplified acute disease: they had a clear course and a crisis. One succumbed or recovered according to a recognizable schedule. Usually that had distinguished them from chronic, or wasting, diseases, though not in the case of Galen's two fevers of the "solids": marasmus, for us a condition of malnutrition, and "hectic fever," most familiar as a symptom of pulmonary tuberculosis. The alternative was a continuum view, which obscured the division between normal and pathological: one might envision an infinite series of states between full health and serious febrility.

The weakening of that division had profound implications for the democratization of fever. It clarified the mechanisms of pathogenic interconnectedness, both of humans with one another and of humans with the environment. It also vastly enlarged the domain of preventive action, both qualitatively and temporally. Broader possibilities for taking action to prevent fever in turn raised questions of public responsibility for taking those actions.

Pringle and Lind were particularly interested in the implications of putridity for military and naval command. Commanders had the power to mitigate exposure both to environmental putridity directly and to conditions that contributed to the generation of internal putridity. A key factor, both for scurvy and for fevers that became putrid, was exposure to wet and cold, which

Lind and Pringle, following Santorio, saw as blocking necessary perspiration.[85] Pringle counsels senior officers, and the governments and citizens they serve, to keep in mind the burden they are exacting: "What a change a soldier undergoes, from warm beds and the landlord's fire, in England, to cold barracks, scanty fuel and sharp winters in the Netherlands; and that without any addition of cloaths." And Lind writes that "the inconveniences which persons suffer in a ship during a damp wet season, are infinitely greater than people who live at land are exposed to; these latter having many ways of guarding against its pernicious effects, by warm dry cloaths, fires, good lodging, &c. whereas the sailors are obliged not only to breathe in this air all day, but sleep in it all night, and frequently in wet bed-cloaths." Uninterested in differentiating the "acute epidemics" from the "scurvy" of marshy Holland, Pringle emphasizes instead that it is "the richer sort" who "keep freest from the diseases of the marshes. For such climates require dry houses, . . . proper exercise, without labour in the sun or in the evening damps, [as well as] a just quantity of vinous liquors, and victuals of good nourishment."[86]

In the case of scurvy, a famous feature of that prophylactic or therapeutic regimen was diet. We view Lind's famous discovery of the antiscorbutic potency of lemon and lime juice in terms of a vitamin-centered concept of deficiency disease. But for all, the potential for putridity was tied more generally to internal alkalinity. Lind would call for "seeds, such as the garden-cresses; which in a few days, . . . will supply . . . a fresh antiscorbutic salad." He would also stock onions (and make sure that there was plenty of Dr. James's powder aboard).[87] Pringle, like Rogers, counseled against a diet containing too much meat, "without a proper mixture of bread, greens, wine or other fermented liquors," and thought that especially in summer, "vegetables, . . . ought to make a great part of the diet."[88] Both Pringle and Huxham considered lemon juice and orange slices effective in treating putrid fevers.[89]

Behind concern with dietary balance were broader, quasi-metabolic concerns. Fevers had long been associated, not with obscure hidden hungers, but with the overt hunger of famine. Some-

how an imbalance between intake and metabolic activity, which
might vary with activity and exposure, could induce putridity. In
what would become an exemplar of famine-induced febrility a
century later in famine-era Ireland, Huxham reported the case of
a mad patient who would not eat. He fell first into a feverish state:
"flushed in his face, and very hot in his head; . . . pulse . . . small
but very quick." But putridity quickly followed. "In four or five
days his breath became exceedingly offensive, his lips dry, black,
and parched, his teeth and mouth foul, black and bloody, his
urine vastly highly-coloured, and stinking . . . ; the sweat . . . very
dark yellow . . . , [and] of a most nauseous stench." Starvation,
Huxham explains, brings "acrimony" to the blood, "which begets
Fever, Frenzy, and such a Degree of Putrefaction, as is utterly destructive of the vital Principles." Deprive a wet nurse of food for
twenty-four hours, he asserts, and her milk will become foul, "yellow, nauseous, and even stinking." The effects will be even worse
if she already has fever.[90] The message was clear: failure to meet
basic needs induced a deadly fever that was communicable as well.

In all these vignettes, fever is a focus of community, sometimes
even a force of communality. Mme de Sévigné belongs to a community of commiseration. Gifts of fever powder bind James's
British subjects. A common and unique fever unites Corkonians.
Putridity in the soldiers and sailors who serve their country can
and should be combated, Pringle and Lind declare. In different
ways, each points toward a sensibility that would evolve into a
commitment to comprehensive public health. Camps, hospitals,
and ships translated easily into towns, dwellings, and workplaces;
soldiers and sailors into town dwellers.

CHAPTER FIVE

Selves

❖ ❖ ❖

The flip side of growing interest in the fevers of others was fascination with one's own. The terrible rotting fevers of soldiers and sailors might tell us of the perilous life they led, and Cork's unique fever might warn us of the dangers of the beefy Irish diet, but what was the message of one's own fever? Usually, those who indulged in what may be seen as the autobiographical turn in fever writing found that *their* fevers arose from their sensitivity, a state of their nerves. Partly owing to Boerhaave's influence, fever was coming to be mainly about nerves. In time past, "people of fashion had not the least idea that they had nerves," recalled the Scottish physician J. M. Adair in the mid-1780s in a snooty essay on "Fashionable Diseases."[1] But nerves connected fever to feelings, minds, and selves.

To ask what triggers the nerves to initiate fever is usually to ask the wrong question, for more often, the state of the nerves was the sum of all the impacts upon them. Accordingly, the cause of one's "fever" was effectively the course of one's life. The fever itself, no mere random life-punctuating event, thus became an occasion to assess the mix of events that had brought one to such an unfortunate juncture. Almost always, a biographical approach to fever was a moral one too: *victims had been victimized*—by their own errors or by others', by ill-conceived or unjust institutions, or even by merciless fate. Often the fever tales are tragic; we can see fever arising inexorably from the workings of circumstance.[2]

This new sensibility had far-reaching effects. The mandate for public action to prevent fever that matured in the nineteenth century was predicated on the view that fever indicated error, that there was an implicit entitlement to be free from dangerous fevers. Nerves would also be the vehicle for the gendering of fever, though the effects were neither quick nor clear. In 1730 the stereotypical overwrought fever victim might well be male, but after 1800 that victim would likely be female.[3] Such gendering would be a product of sensibility rather than epidemiology.

One's fevers were not merely occasions for reflection; they were central to the plot of one's life. Just as deathbed performance displayed important truths of character, one's enactment of fever revealed otherwise inaccessible elements of self, if in a more ambiguous way. After all, in fever all aspects of being are simultaneously under attack. Reason and control no longer rule; the will succumbs to an abject sensate body; the senses fail or plumb new, barely expressible realities. One is on a wild ride, but more dangerous, dramatic, and unpredictable (and much longer) than any roller-coaster ride. Delirium may obliterate the self altogether, usually temporarily but sometimes permanently, or it may reveal a self normally hidden: "*In delirio veritas,* as *in vino veritas,*" noted one authority.[4] If, in the romantic imagination, living is an intensity of experience, fever is intensity squared. Among literate and reflective persons, it was considered the most meaning-laden disease before yielding that status to cholera and tuberculosis in the nineteenth century and cancer and HIV/AIDS in the twentieth.

Hence it is appropriate to recognize a romantic era of fever as more than just a matter of periodization. I examine theory first, asking how a disease of bubbling blood or impeded circulation suddenly became one of nervous tremors. I then focus on six aspects of the romantic culture of fever: (1) the emergence of a phenomenological concept of fever emphasizing the subjective experience of change; (2) a renewed interest in felt heat as fever's defining feature; (3) an interest in the ambiguous beginnings and endings of fever and in its integration into the story of a life; (4) an interest in febrile mental states, particularly delirium; (5) fever's feminization; and (6) the centrality of fever in a literary exploration of intersubjectivity in which one apprehended the fevered other as the self one might as well be.

ROMANTIC FEVER THEORY

Not one but three roads led from blood to nerves—one primarily pathological, one epistemic and metaphysical, and one rooted in practice. We may wonder at the need for any road, for surely systemic states of the body will reflect the rule of the brain. But recognition of brain dominance is relatively recent. Early modern theorists still imagined the body as a federation. Though usually the heart was king, in particular conditions other organs might rise to power.

In part, the emergence of nerve-centeredness in the eighteenth century was a product of the search for an ulterior explanation, in particular of fever's oscillatory character. In the classic intermittents, intense heat succeeded intense chill. In the new low fevers, exhaustion might alternate with frenzy. If normality was equilibrium with small perturbation, fever was equilibrium with large perturbation. Attempts at a dynamic explanation followed a common template, already evident in the Hippocratic picture of the fevered interior. Usually one looked for an initial event that had triggered a cascade of overcompensations, manifesting as the disturbances of fever. In vascular theories like Bellini's, the oscillation was primarily hydraulic, arising from back pressure in the capillaries caused by an obstructing lentor. Boerhaave, seeking to reconcile the new mechanists and chemists with the ancients, retained Galen's heart-centered view of fever. But he saw the cardiac effects as secondary consequences of the condition of the vessels, though he stressed their hypothetical mechanical properties less than their independent contractility.[5]

But contractility was mediated by nerves. In what would become a paradigmatic romantic example, Boerhaave invited readers to imagine a person "suddenly affrighted." There was pallor. "The heart immediately palpitates very swiftly; because the veins being contracted urge the blood more swiftly towards the heart, and the arteries being contracted or lessened make a greater resistance to the blood's expulsion out of the heart."[6] Mechanism was not logically dead—Boerhaave might still imagine nerves to be channels filled with a fluid or force that innervated the minute fibers from which the capillaries were formed—but practically the focus was shifting to whatever external or internal circumstances were evidently affecting capillary performance.[7]

In Anglophone academic medicine, the most prominent nervicentric theorist well into the nineteenth century was the great Edinburgh medical teacher William Cullen (1710–1790).[8] In Cullen's view, fever began with a capillary "spasm." Contracting capillaries prevented blood from reaching the body's extremities, creating chills and shivers or "horror, . . . universally the prelude to all . . . fever." After some hours, the circulatory system would compensate, leading to a hot stage of *reaction* as the heart pushed blood back into the extremities. This overexcitement continued until the wise body managed finally to purge itself via a classic Hippocratic crisis of copious sweat or sediment-laden urine. Exhausted but at peace, the suffering body would then require a long convalescence.[9]

But what made capillaries spasm? Particularly for nervous or slow fever, in which the reaction was never strong, the usual generator was *debility.* But the term was ambiguous. Theoretical debility did not correspond to the feeling of being debilitated or to anything else. It presumably registered the sum of debilitating forces, but as these could not be independently measured, *debility* remained as much a placeholder as *lentor* had become. Some fevers, Cullen recognized, were triggered by contagia, but contagia (and miasms, cold, and fear) were "of a sedative nature," that is, debilitating. Thus the concept was circular: one only measured debility in the fever it presumably had caused.[10]

If one asked where in the body this debility was located, the answer was in the nerves. Debility was sometimes coupled with depression, as both its cause and a chief symptom, and with anxiety. But these too were considered physical states more than states of mind. *Depression* was a lowering of "vital power"; *anxiety* an objective state of cardiac irregularity.[11]

On the Continent, the chief lines of critique arose as an appeal to a new vitalism, associated in German-speaking lands with the ascendency of Georg Ernst Stahl (1659–1734) over his Halle colleague Friedrich Hoffmann. The mechanist Hoffmann would derive febrile phenomena from Cartesian necessary truths, but Stahl was skeptical: glib mechanical hypotheses were far too simplistic to account for febrile phenomena or much else. Van Helmont's archei too had reflected recognition of irreducible properties, both in the action of particular parts and in the coordination of the whole.

If, as the Helmontians suggested, a master archeus was in charge both in a healthy and in a fevered body, it followed that fever might be the manifestation of that vital controller adopting, under stress, a new mode of operation involving the squeezing of the capillaries. Health might return, but only after an arduous healing process; the idiom *laboring under a fever* would become common in the period. Having drunk deep of such theories of vitality at the University of Ingolstadt, young Dr. Frankenstein succumbs to the very processes he is trying to comprehend: he cannot control his own nerves. Exhausting himself in making his new creature, he becomes pale, emaciated, and anxious, unable to escape the destructive maelstrom: "Every night I was oppressed by a slow fever, and I became nervous to a most painful degree; the fall of a leaf startled me, . . . Sometimes I grew alarmed at the wreck I perceived that I had become. . . ." Here fever has subverted the systems responsible for judgment, will, and action. Alienated from his usual self, Frankenstein can perceive but not intervene.[12]

Proponents of such wise-body notions of fever, like J. C. Reil (1759–1813) and Christian Hufeland (1762–1836), were closely linked with the German romantic movement as physicians and as comrades of Goethe, Herder, and Schiller.[13] But such metaphysi-

This chart, titled "Table of Fever," from the lecture notes of an American pupil of William Cullen's famous student Benjamin Rush, illustrates themes of Cullenian fever medicine. The 50 mark indicates good health, with overstimulation indicated above and understimulation below. During a case of fever, a typical patient would move both above and below that midline.

From Notes of Robert Maxwell, 1807–8, in Benjamin Rush, Collection of Lecture Notes, 1783–1810, MS Coll. 225/14, Kislak Center for Special Collections, Rare Books and Manuscripts, University of Pennsylvania Libraries. Photograph by Sara Naramore.

cal vitalism would also be championed by French protomaterial-
ists impressed by its empiricism: ultimately one might know the
laws of life, and terms like *vital* (or *mechanical*) would become
superfluous.[14]

In addition to academic medicine and philosophical vitalism,
a third context of nerve centrism was society doctoring (and perhaps epidemiology too). Around 1710, notes Creighton, English
doctors were struck by the emergence of vague feverish diseases.
Generically these were known as *febricula* ("little" fever), but there
were other terms: *hysterick fever, hypochondriasis, spleen,* or simply *low spirits.* Besides common febrile symptoms of weakness
and sweats, trembling was prominent, as were states of mind—
"apprehensions, . . . unaccountable fears of death," and dejection,
also sometimes irritability and unsettledness. These illnesses rarely
had a clear course and might come and go indefinitely. They were
certainly nervous diseases, and they were linked theoretically to
fevers. Some saw them as incipient fever; others, like the Bath
spa doctor George Cheyne (1671/72–1743), who considered them
at length in his celebrated work on nervous diseases, *The English
Malady* (1733), hinted that they were the would-be fevers of bodies
too spineless to mount a proper fever. Referring to them as "vapours," a neo-Galenic term for products of the evaporation of putrid humors in the vena cava (and which would become a familiar
diagnosis in cases of invalidism), he regarded them as merely a
"slow and languid" presentation of what would be "*hot, brisk,* and
eager" (i.e., fever) in those of firmer fiber. Associated both with
low stress tolerance and accentuated sensitivity in both men and
women, these diagnoses—febricula and its associates—would un-
derwrite the new concept of a *valetudinarian,* one appropriately
preoccupied with one's delicate health.[15]

However tempting it may be to pass over this hypochrondriac
illness as the indulgence of the peevish rich, it represents the forging of some important linkages. First, through Cheyne, who had
begun his career as an advocate of a Newtonian mathematical approach to fever (similar to Bellini's), it carried the authority of the
most advanced science. It carried that authority to a very familiar

form of medical practice, however, for the implications of nerves differed little from those of humors. Cheyne's hygienic advice was but a more moralized version of Galenic hygiene, based on the non-naturals.

What Cheyne (and the sardonic physician and essayist Bernard Mandeville, author of another book on these new illnesses) added, however, was an autobiographical element. Both had suffered from these vague and fleeting fevers and had written about their own experiences, addressing their work to general readers, not other doctors. That these physicians explored their own vulnerability not only allowed others to do so but effaced divisions between the medical and the literary and between objectivity and subjectivity.[16] Last, through the common element of assaults on nerves, was the linkage of these fashionable febriculae with the dangerous low and slow fevers marked by prostration, including those we call *typhus* and *typhoid,* which were often regarded as the fevers of poverty or crisis. "Nerves," then, were a means of democratizing experience.

While Cheyne set the course, the late-century consolidator of such medicine would be Cullen's notorious protégé John Brown (c. 1735–1788), who would gain pan-European celebrity with another simple nervicentric model.[17] Brown transformed Cullen's multiple qualitative distinctions into a single continuum: diseases were excesses or, more commonly, deficits of some entity akin to energy or excitability. Like Cullen, he emphasized understimulation. Not only were the slow and low fevers manifestation of an "asthenic" or debilitated state, even fevers of pounding pulse and the hot, swelling redness of inflammation might be manifestations of the body's overcompensation for its debilitation.[18] Certainly in low fevers bleeding was inappropriate. The chief Brunonian restoratives were opium and wine; the latter and sometimes the former would become a therapeutic mainstay in the first half of the nineteenth century.

Brown's success is an index of how widely shared was this image of the fragile nervous body. The nerves integrated all incoming pathological and hygienic forces, physical, emotional, or both. Some bodies might possess significant inertial resistance to disruption of their equilibrium, but one could never know how much. Better to see health not as a default but as a fortuitous and contingent state that might fail from accident or from the briefest lapse in discipline. Any adequate hygienic security system must acknowledge the body's frailty and be on guard against every conceivable assault, singly or in combination with others.[19]

Hence relocating fever from the heart to the nerves went beyond pathological explanation, theory, or therapeutics to involve a profound reorientation of the way people imagined the relation between disease and self. It involved greater appreciation of the subjective state (since feelings were both a consequence and a cause of one's nervous state), and with it fascination with mind-body reciprocity. Earlier hopes of uncovering a fever's essence in humors (i.e., blood) or solids (some inflamed part) came to seem naive.

WHAT IS FEVER? WHO GETS TO SAY?

Among the casualties of the theoretical tumult of the seventeenth century had been the old understanding of fever as excess heat. What, then, was fever? Cullen's pathology was clear enough, but explaining fever was not the same as defining it. For new fevers were challenging old definitions. Could *fever* still be an observational term? And for what states of illness? Boerhaave had addressed the question inductively: If one took all the conditions that were called *fever* and found their common elements, these elements (rapid pulse, with heat and shakes secondary) must be fever's essence. But in the new "slow" nervous fever, the pulse did not race, and there might be no unusual heat or shakes. Accordingly there was, at the end of the fever-fixated eighteenth century, *no* widely accepted objective definition of fever. One need not doubt the reality of the entity, however. Bartholomew Parr coyly begins the entry on fever in his 1819 *London Medical Dictionary* with a diagnostic enigma. An 82-year-old woman with cool skin and a moderate pulse was nevertheless "generally allowed to be in a fever." Parr confirms this diagnosis, citing a "peculiar feel" of the pulse," tension in the palms and wrists, and an undescribed countenance.[20] Good doctors knew. They could not say how.

One way to define fever was simply to describe it. In 1834, James Copland admitted the impossibility of "a *definition* of fever altogether applicable to the various forms and states it . . . exhibits. . . . Description instead of . . . definition . . . is necessary."[21] The mainstay of description would be sequence. Tertian malaria, the template of fever from Galen to Cullen, followed a regular schedule. Perhaps all fevers exhibited the same regular sequence, differing only in the length and character of the several stages, with changes occurring on critical days. Thus one might speak of the rhythm of a disease.[22]

Cullen himself had insisted that his "theory" was simply such a sequence: "Now this is what I call the theory of fever, but take notice what kind of a theory it is. . . . It is no more than saying that there are certain states of the body [debility, chills, and heat] which are combined together in a certain order of succession, and . . . from this constant combination . . . are to be considered as a series of causes and effects." To Cullen this was "a matter of fact." He saw no problem in being unable "to explain in what manner, or by what mechanical means these states severally produce each other."[23] To his friend David Hume, that regular succession might be cause enough, but others would criticize Cullen for failing to explain *how* the body shifted from stage to stage, for example, from spasm to reaction.

Copland's own expanded model had six stages rather than the usual three or four. He put much stress on a prodromal stage, days or weeks in which the patient sensed something amiss, perhaps an "oppression in the chest," "a sense of uneasy depression," or changes in appetite or appearance. Second was a sudden "invasion" of chills and rigors. Third was "excitement," hot flushed skin, possibly delirium—Cullen's "reaction." Then came crisis, "decline," and "convalescence." Copland's model, particularly its emphasis on gradual beginnings, was intended to include the new nervous fevers, but all too often these seemed to drift along in defiance of any particular stadial model, resolving without overt crisis. One option, attractive in Germany among followers of Stahl, was to acknowledge the protean character of each case. In Germany, case histories were undertaken less for typing fevers and more for confirming their irreducible individuality.

Such a "historical-biographical approach" threatened any assumptions of disease specificity.[24] Not only the species of fever but the genus itself was in question. Sydenham had hoped that sharper differentiation of fevers would lead to better therapies. But increasingly, therapeutics was itself being seen as a key factor in creating variability. Did bleeding, purging, or diet shape the fever, for good or for ill? Or would (and should) a fever follow its course to the end? Galen had argued both sides. Since he thought he knew how a fever would evolve, he could act quickly to preempt it. But his lesser successors struggled with the fact that often conformity to some known type could be confirmed only well into an illness, and even then it might be an artifact of the measures already taken. Increasingly, however, there would be more focus on the timing, extent, and indications of routine therapeutic practices rather than on the elements of the pharmacopeia (and especially the exotics) that one proposed to use.[25]

That fevers did have a course but might sometimes be diverted from it could be accepted, but only as a general truth. The skilled physician must know "the nature of all mankind in general" but also of "every one in particular," van Swieten sagely observed. Like their Hippocratic forebears, eighteenth- and nineteenth-century fever physicians would make much of their irreducible tactical experience with fever. But biography made a difference. The ancients (Hippocrates and Celsus) well knew, noted van Swieten, that those who "practice only by generals" were at a disadvantage: "in two people of equal knowledge, a friend may be a better physician than a stranger."[26]

The retreat of theory, the admission of fever as protean and biographical, left more room to bring subjective elements into medical description. For Copland, fever was neither a theoretical entity nor exclusively an observational one. Among its most general elements were what patients *felt*—"*Painful lassitude, with debility of the corporeal and mental faculties,"* also *"increased thirst, and abolition of the appetites."* Copland also notes the changes in circulation and secretions, but he uses *debility* not in the sense of Cullen's theoretical construct but to refer to the sufferer's own feeling of functional impairment. He noted other strange sensations that arose during the cold stage in many fevers, including "a sensation resembling cold running down the back" and "formication" (the feeling that ants are crawling on one's skin).[27]

A key aspect of accrediting patients' statements as medical evidence was who got to speak for heat. Particularly for the stage of chills, later writers, following Boerhaave's student Anton de Haen, would often note that patients' perceptions did not correspond to thermometer readings. Often fevered persons continued to feel chilly even when they were demonstrably hot. Some contested the finding or disregarded it, continuing to focus treatment on what the patient felt. Still others saw the malfunctioning of the human thermometer as itself an important aspect of fever pathology. To Cullen, "imperfect sensations" were markers of "the diminished energy of the brain." By 1900 all that would change. The aura of objectivity and the practice of thermometric reductionism would convert patients into dummies, unaware of their own bodies, or liars, asserting what was not real.[28]

Ironically, the road to objectivity led through heightened subjectivity. Among the first to return to a heat-centered conception of fever was the Liverpool medic James Currie (1756–1805), reformer, poet, political philosopher, an introspective soul prone to a "romantic, melancholy humour." Currie began medical study at age 21 following an adventurous career as a loyalist merchant in Virginia at the outbreak of the American Revolution, where he suffered shipwreck, dysentery, and at least three serious fevers.[29] He would gain fame for the cold-water fever cure, in which one doused the heated head with a bucket of cold water. Currie, however, credited another Liverpool medic, William Wright. Struck by fever on his return from Jamaica in 1777, Wright had decided that if bracing sea air helped, sea water might also help; three buckets of water gave "immediate relief."[30] Intrigued by the effects of hot and cold, Currie found an opportunity to test the remedy systematically when fever broke out in a regiment garrisoned in Liverpool in 1792. This led, five years later, to his *Medical Reports on the Effects of Water.*

Currie understood his discovery in terms of submission to wise Nature. He had cast off medical sophistry to hear the voice of the pained body. Heat was pain; it was also a concomitant or cause of disability and a dangerous state in its own right. While Currie included Galen among the few precursors of the cold-water cure, he made plain that neglect of so obvious a response showed how far theory had ignored the experience of suffering.[31]

Currie went far in his redefining. If cold stopped fever, heat *was* fever. He was jettisoning not only particular pathologies but the primacy of pathology itself. Boerhaave had attributed fever's heat to a perspirational anomaly. The skin was hot and dry because perspiration was suppressed. To Currie all this missed the point: the heat itself was problematic. "Can we suppose," he asked, "that a heat, six or seven degrees greater than that of the blood in health, however generated, will not have the most important effects on the system . . . ?" Someone should have noticed, he pointed out, that when temperature returned to normal, *the other symptoms vanished.*[32] Whether internal or external factors generated that heat was moot:

> If a person is confined in a hot room, or in the hot bath, till his heat rises three or four degrees above the natural standard, his pulse will be found of a feverish rapidity; wandering pains will soon be felt over the body; languor, lassitude, and at length great debility will take place, with most of the symptoms of regular fever. . . . These symptoms cannot be expected to go off till the inordinate heat is removed, and if the person remain some time in the heated medium, he will find, that the inordinate action of the heart and arteries continues after leaving it, and even after his own heat has subsided to its natural standard.[33]

That redefinition had implications for causation and responsibility. Some conditions in which soldiers, sailors, and workers ordinarily existed not only were risky but produced a physiological state that *was,* at least generically, fever. As with the hot bath, their bodies might not always reequilibrate; exposure might generate a full course of serious febrile illness.

In representing the body as struggling and sometimes failing to govern its temperature, Currie saw himself as applying Cullenian principles, but more consistently. For Cullen, the hot stage of fevers arose as a reaction to a critical level of debility, but he posited also an inherent restorative power that the wise body used to return itself to health. However, Cullen seemed to forget that people died from fevers, Currie complained. There was no innate tendency toward recovery: people survived because of timely interventions to beat the heat. To sit and wait for Nature was metaphysical claptrap masquerading as medicine; fever needed to be cured:

> We are not therefore to wait for the sanative process by which nature is supposed to separate this *virus,* and to throw it off, watching her motions, and assisting her purposes; but to oppose the fever in every stage of its progress with all our skill, and to bring it to as speedy a termination as is in our power. When we dispel the morbid heat, and reduce morbid re-action in the hot stage of the original paroxysm, by the powerful means of the cold bath effusion, the whole of the febrile symptoms vanish; a sufficient proof, that . . . these symptoms arose from inordinate heat and inordinate action, and not from a poison circulating with the blood.[34]

However impressed he was by the power of hot and cold to transform frail bodies, Currie realized that fever was more complicated. Cold water did not cure all fevers; often people did acclimate to extreme temperatures. He would admit that contagion was a key element in some fevers. Yet always it was the sufferer's sense of feverishness, and particularly fevered heat, that should be preeminent. Individuals were the most important thermometers. Feverable bodies might not be precise, but they accurately registered both illness and danger.

The most enthusiastic reception of Currie's work would be in that homeland of agonized introspection, romantic Germany. Christoph Wilhelm Hufeland would be his translator and disciple.[35] Currie's thermometric research would be foundational in the thermometrization of fever in the Leipzig clinics of Carl Wunderlich a half-century later, a change that has been seen as signaling the triumph of objectivity. But Currie would not have approved. Hotness, unlike other fever symptoms, can seemingly bridge the gap between the sufferer's feeling and the physician's assessment. Both, ideally, sense the same entity from opposite sides of a wall, the skin. But whose reality should rule? For Currie, what patients or subjects felt might correlate with measured temperature, but there is no sense that persons should calibrate their feelings to a mark on a glass.

Elsewhere interest lapsed. Cold-water therapy might remain on the list of fever therapies, but Currie's medical philosophy would be forgotten or tainted with the brush of "fringe" medicine. Yet declining faith both in theory and in traditional therapeutics after 1850 would open the way for a resurgence: by the 1880s, at the cusp of the aspirin age, the giving of cool baths was a key skill to be mastered in the new specialty of fever nursing, especially useful in typhoid cases. Yet Currie, if remembered at all, was just another resident of the benighted prescientific medical past who had stumbled on an important truth.

THE BORDERS OF FEVER: BEGINNINGS AND ENDINGS

Any description of a fever's course required determining its beginning and end. Determining a fever's onset was an ancient and important clinical problem, for one could not know which were the critical days unless one knew when the fever had begun. The problem was even more difficult in the case of the new nervous fevers. Unlike the intermittents of the ancient writers, they sometimes came on gradually. An indeterminate period of disorientation might occur before the rigors and headache hit—Copland's prodromal stage.[36] A common approach was to discount all prefebrile uneasiness and measure from the *invasion,* the technical des-

ignation for the transition from any prefebrile stages to febrility. Here the readiest index was the day the person had taken to
his or her bed. But of course people responded differently based
on strength, habits, or situation. Some had to work, while others "take to their bed immediately upon every slight occasion."
Those who insisted that they had known the exact moment fever
hit were committing the error of assuming that a big effect must
have a big cause—something akin to an electric shock, noted the
Austrian theorist Johann Valentin von Hildenbrand.[37] But if fever
was a continuum phenomenon, the question itself might not be
a valid one. As with Victor Frankenstein, there might be no sharp
distinction between the long train of physical and moral events
that exhausted the nerves and the ensuing state of febrility.[38]

Ideally, a person would recognize when he or she was falling
into fever and then quickly secure medical aid. Yet not only were
the early stages often subtle but the fever itself might impair one's
ability to assess it. In the week before his fever, Rev. Cyril Pearl,
author of an 1845 delirium memoir considered below, recalled
"frequent sensations unusual with me. . . . capricious morbid appetite; . . . an indescribable heaviness, and an indecision, which I
could not shake off." In retrospect, he admitted that his refusal to
accept aid from those who saw his coming illness had itself been a
mark of the "insanity" of his incipient disease.[39]

Prodromal dis-ease came in many forms. The developing fever
might be signaled by "fatigue upon the least corporeal or mental
exertion . . . stupidity, loss of nervous and mental energy . . . irritability, moroseness, or impatience . . . uneasy sensations." Or
it might be indicated by "frequent sighing, gaping, forced and
lengthened inspirations, and . . . a sense of uneasy depression,
or nausea." Or one might experience "confusion of ideas; [feel]
morose, low-spirited, sluggish, indolent, or incapable of exertion,
and of directing his attention long to any object. . . ."[40]

But for hypochondriacal persons like the spa denizens who
were Cheyne's clients, such symptoms were more or less normal.
Besides, there was the worrisome prospect that in inventorying oneself for such signs one might create or exacerbate them.

Worry itself could bring on a fever. As one of six categories of non-naturals, passions of the mind had long been recognized as causing fevers, chiefly ephemerals. Usually other factors, like the ambient environment, overexertion, or surfeit, were more important, though as van Swieten, following Santorio, noted, sorrow was particularly dangerous, since "nothing more promotes a free perspiration than chearfulness." Earlier writers had often seemed to imply that the passions were as manageable as any of the other non-naturals. Van Swieten told of the Dutch physician Isbrand van Diemerbroeck, who, during an epidemic, "carefully avoided all passions of the mind as much as possible, so that he lived free and without fear in the midst of a thousand who died: But if at any time he became sorrowful, which was hardly to be avoided in so desperate a calamity, he washed away his cares with wine till his spirits were revived."[41]

For romantics, who found strong emotion both attractive and uncontrollable, this was not an option. The mind-body feedback that van Swieten described was inescapable: not only did "grief and fear obstruct the perspiration, but . . . the perspiration, when once obstructed . . . is followed with grief and fearfulness." While "cold, fatigue, general privations, or mental depression" at a fever's outset all worsened the prognosis, noted the Edinburgh professor Robert Christison, the "worst" prognostic signs were . . . mental emotions of the depressing kind."[42]

A key concern was sleeplessness. Again van Swieten: "To be long watchful . . . consumes the spirits; dissipates the most subtle humours, and dries up the rest; breeds melancholy, . . . settles it in the hypochondria, and disturbs the functions of the brain: Hence follow delirium, ravings, madness, &c." Helpfully, he notes that the cure "of fevers arising from over-watchings, requires sleep in the patient." But might it be the "violent passions of the mind" that kept one awake?[43] Hence the double bind of morbid introspection: one harmed health (self) by worrying about health (self). This situation led to paradoxes in responsible victimization. Since states of feeling both caused fevers and were produced by them, how was cause to be separated from effect? When was one being appropriately solicitous of one's health, and when was one's worry both cause and product of one's disease? In such a climate the beginnings of fever became the subject of an existential angst fest. To a sufferer, "When did your symptoms begin?" might be heard as "What has brought you to this juncture?" and taken as an invitation to indulge in sorrow, blame, and hope.

That openness to introspection would disappear in the next century. But even before it became common to attribute fevers to a microorganism taking root within, some were recognizing an "incubation" period that separated an infectious event from the onset of symptoms. In such a scheme patients' testimonies no longer mattered. Knowing the disease and its incubation period, one could backtrack to identify the date of infection.

Not only was fever's beginning unclear, so too was its end. In theory, a successful crisis marked the transition from *status*, the height of the fever, to Copland's "decline" or "defervesence" (literally, defevering).

For the pill promoter Robert James, the marks of the end of fever were mental. A patient might seem physically well, but until cognition and affect were normal, the fever was not over. Thus, where today we might declare ourselves recovered by saying "My temp's back to normal," James and his patients saw recovery as a "return to [one's] senses." Writing in 1746, Elizabeth Doddimead says that she has "recovered her senses" by taking James's powder after a three-week "violent fever" left her so "light-headed and in- sensible" that she did not "know her own child." She had also lost her sight.[44] James finds one recovering patient "without fever, and without delirium, except that sometimes he made some little mis- takes, such as are not unusual when a person in health just awakes from a good sound sleep." Beyond avoiding actual error, a fully recovered patient would exhibit reason, prudential judgment in understanding and complying with his or her regimen, and "good spirits," a phrase that referred to mood even if it also retained overtones of Galenic pathology.[45] (The return of sexual desire was also a common marker of the end of fever, noted Copland.)[46]

After defervescence came convalescence, which might last indefinitely. If the continued presence of late-stage fever might sometimes be a contested matter between sufferer and practitioner, convalescence was to be the expert's preemptive strike against the patient's presumed willfulness. Patients thought themselves well, but they weren't: the pulse might still race, there might be brief daily bouts of febrility (a return to meat eating might trigger both), and there was real danger of a serious or even fatal relapse (though those who died in convalescence never died from their fevers, Galen had insisted).[47] There was great potential for conflict: did the relapse come from failing to follow a wise regimen or from adhering too strictly to an unwise one?

Not only did patients often fail to appreciate that they were still in danger but the feelings on which they based their assessments, a postfebrile euphoria, were themselves sequelae of fever. The early nineteenth-century essayist and traveler Rev. Timothy Flint described his early convalescence following a fever of fifty-five days. First there was debility without anxiety, "the weakness of an infant: . . . [with] all its freedom from cares and desires." This was followed by "tranquility and repose, which excludes all uneasiness and all vexation," and which approached "the serenity and satisfaction which are supposed to be the portion of the blessed." Last was a sense of magnificence: "I shall not forget . . . the sensations excited by the first view of the earth, the trees, the river, and the heavens. . . . Every object had a new aspect and a new colouring; . . . I beheld the beauty of nature, as if for the first time." This bliss was more than relief at recovery, Flint thought. He saw it in conversion terms as a spiritual turning point of his life. The "vexing and bad passions" of his prefevered self had gone.[48] Indeed, some convalescing patients did display "extraordinary activity of the mind, clearness of thought, facility of expression, and brightness of memory," while others exhibited "feebleness of intellect" or "indisposition to mental exertion, sluggishness in conversation, and defective memory."[49] Yet both constituted abnormality. Convalescents were advised not to sign contracts or make major life decisions, just as is now the case with persons recovering from anesthetics.

The end of convalescence might be understood as the state of again being truly oneself. But that might require months or years. And for some it might never occur. At the end of the nineteenth century there was great interest in psychoses of febrile convalescence. In 1881 Emil Krapelin, the great systematizer of mental diseases, had called attention to the fact that "with very slight external cause a psychosis may develop after every evidence of the typhoid process has disappeared." Henry Berkley's *Treatise on Mental Diseases,* based on his 1899 lecture course at Johns Hopkins, included a long chapter on the psychiatric sequelae of all the major forms of fever, which he claimed accounted for a small but significant proportion of cases in state asylums.[50] There might be epidemics of postfebrile psychoses. In Switzerland in 1890 a young man convalescing from influenza unknowingly murdered his mother and his sister. During that year, allusions to *la nonna,* a postfebrile "asthenic psychosis," made the rounds of medical journals. Associated in the Italian Alps with the pointing finger of the crone, the condition was understood to arise in persons who had been exhausted before their fevers and received poor care during them. There would also be much interest in postmalarial psychoses, with characteristic symptoms of short-term amnesia, severe depression, and paranoia.[51]

But there was also interest in fever's power to strengthen resistance to future disease—the references may be to acquired immunity, though this is not always clear—and to cure chronic diseases. The same period saw the first of the "shock" treatments, the therapeutic induction of malaria to cure neurosyphilis by the Austrian psychiatrist Julius Wagner-Jauregg.

MIND FEVER

Postfebrile psychoses were infrequent; much more important was the delirium of fever itself. Fever is now mainly a matter of body, not mind. Classical medicine was holistic, but it did not follow that delirium was always equally important or important in the same ways. During the romantic era, however, it was both a key indicator of a fever's severity and character, and an intrinsically fascinating matter to many in this introspective age.

In making phrenitis and lethargy acute diseases, ancient writers had paid attention not only to the behavior of the fevered but also to the link between febrile experience and other transitory states of being. To some, fever, with its "unconquerable" feelings and paroxysms of "excitement," was of a piece with ecstasy and catharsis, Sibylline and/or sexual. Fever, like sex or transcendence, was a sequence of uneasy feelings rising to climax or crisis, followed by peaceful exhaustion. That state of relief, understood as a cathartic purging or cleansing, was precisely what the Hippocratic doctor sought to achieve in treatment.[52]

Galen had not ignored such matters, but many of his successors had. Though surely the fevered still raved, many early modern authors passed that by to focus on perspiration, pulses, or putridity. Febrile aberrations of mind had no diagnostic or therapeutic significance, though as with Mme Sévigné, sharing a chuckle about her fevered ravings with her nurse-son, they might become subjects of conversation.[53] In 1819 Bartholomew Parr puzzled over that relegation: in any list of fever symptoms, "mental alienation" should be included along with horrors, heat, and pulse.[54] Boerhaave, who had learned directly from the ancients, took an interest, attributing delirium to overstimulation, which was manifest in the racing blood. He located it in the hypochondrium (the site of the ancient phrenes). It was important to try to cure it, perhaps by evacuants. (To Cullen, by contrast, delirium often implied understimulation. Wine was the cure; one patient needed eight bottles a day to remain free of delirium.)[55]

The resurgence of interest in delirium arose in multiple contexts. First was the Sydenhamian project: for purposes of description and differentiation, it was important to include the severity and character of delirium. Second, delirium was coming to be seen as a management problem: patients might harm themselves or others or resist treatment in their ravings. Last and most important was a fascination with the bizarre that was both scientific and sensationalist: what was going on in the fevered mind?

One sees changes over the course of James's career. In the 1740s, delirium was an incidental detail. "Raving Night and Day" in her "violent fever," the teenage Miss Sudwel went "without . . . a Moment's sleep, or . . . any Sustenance for near three weeks."[56] By the 1770s it had become graphic embellishment; James includes accounts of unending restlessness or of awful delusions and direful actions based on them, as well as major disorders of mood. Representing loss of customary civility, "raving" had become the measure of the power of fever to disrupt the self. In Bath a man recovering from smallpox relapsed into "delirium or frenzy . . . [he] foamed at the mouth, refused all liquor, and endeavoured to bite the attendants." His "violent ravings" persisted even after a double dose of James's powder, retreating only after a second dose a few hours later. Retiring on a July night in 1774, a London draper was suddenly "seized with delirium." Early the next morning he descended the stairs, "naked, and behaved in every respect like a person light-headed. . . . He was very outrageous; and [it] was with great difficulty that two or more persons . . . could keep him in bed." On a slow boat to China in 1772, a fevered crewman went to the forecastle, ax in hand, "swearing, that he would cleave any man that came near him." Forty shipmates would succumb to that fever, "most . . . also delirious." (Two doses of powder always cured the fever.)[57]

Seeing similar symptoms among fevered troops, John Pringle reverted to the ancient diagnosis of phrenitis. He saw it as a brain inflammation resulting mainly from sleeping drunk in the sun. On one occasion, soldiers exposed to morning fog became feverish and delirious; "a few . . . were so suddenly taken with a phrenzy, as to throw themselves . . . into the water, imagining they were to swim to their quarters." Pringle recognized manic behavior (agitated patients trying to leave their beds) and mood ("great dejection of mind") as key features and major problems of fever management. Like James, he associated them with a disturbed cognition. Such "confusion," in his view, might be induced or exacerbated by inappropriate treatment.[58]

Many sought to combat delirium by moral and environmental means, deflection and reassurance. Van Swieten suggested that coma vigil, in which terrifying visions set in whenever the patient tried to sleep, might be alleviated by a soothing bedtime concert or a creative engagement with the patient's *idée fixe*.[59] Cullen worried that darkness and silence allowed "thoughts . . . to wander." Light, along with familiar voices or objects, might restore patients but might also upset them. However, others held that darkness and quiet were precisely what was needed for relief from the "commotions of the mind." Always one must remember that the fevered brain was weakened: were it possible to "suspend . . . thought," Cullen would.[60]

By the mid-nineteenth century, most accepted that fevered minds would rave; one must simply manage. Still, some writers, such as Charles Murchison (London), Robert Christison (Edinburgh), and Robert Lyon (Dublin), were classifying varieties of febrile delirium. The robust empiricism of the "numerical" method of the Paris hospitals required cataloging symptoms, and some hoped that patterns of delirium would aid diagnosis, particularly in the vexed distinction between typhoid and typhus. Thus Murchison asserted that almost all cases of typhus involved mental impairment, while some typhoid patients had no delirium, and among those who did, it was milder, usually came at night, and followed a different course, noisy and ferocious and then lapsing into prostration. By contrast, in typhus almost all fatal cases showed "great delirium."[61]

Correlation required standardization. Agitation was the usual measure. At one end of Murchison's scale, patients muttered incoherently but could be roused into brief periods of cogency. Next was *delirium tremens,* a condition not exclusive to drinkers, which involved fidgeting, sleeplessness, tremors, and efforts to rise. The furtiveness of patients suffering from this type of delirium would be evident in their "ferrety" eyes, yet they were "easily cowed by a gesture . . . and a firm look or word." In the rare *delirium ferox,* however, a "bold and fierce" expression replaced the "ferrety" eyes; "the patient . . . rolls his head from side to side, shouts and screams incessantly, and makes constant attempts to leave his bed." Such delirium was accompanied by uncommon strength: "it may require several strong attendants to keep him in bed." (But there was also a "pleasing delirium with quickness of expression, and a tendency to smiling," noted Christison.) Just as delirium varied in severity, it also varied in duration. Lyon warned students that the raving and sleeplessness might last for days. An attendant could do little but could not safely leave: "Hour by hour he sees his patient expanding the last remnant of his vital powers." Reluctantly, he recommended straitjackets.[62]

Suicide attempts and attacks on others reflected the compulsion of feverish hallucinations, these writers recognized. Christison noted the frequency of attempts "by young people in fever in Edinburgh to leap from their windows." He saw these attempts as imitative or as paranoid reactions to steps taken to prevent them. One patient killed himself trying to sever his penis; another tried to climb a chimney; many attacked attendants. But febrile suicidality was not always tied to hallucination or paranoia. It might reflect an *idée fixe* in a mind that retained "its acuteness of perception and vigour of reasoning" or even exhibited "an elevation of the mental faculties" but lacked complementary restraints.[63]

If, as seemed plausible, fevered mentation bore some relation to physical and mental normality, the type of delirium might conceivably be linked to class or occupation. Again Murchison: "In the poor and badly nourished, and likewise in the aged, . . . the delirium is almost always low and muttering . . . in the young and robust, and still more in persons in the upper class, it is often acute." Lyon hypothesized that "organs which are in the highest state of physiological activity in . . . health, are the most prone to disease, . . . as the cerebral organs are more exercised amongst the better classes, they are more readily seized upon by the typhous [*sic*] influence."[64]

DELIRIUM AS DISCOVERY

What do "ferrety" eyes see? By the mid-nineteenth century, doctors thought they knew what went awry in the fevered mind. Murchison saw delirium as fixation combined with confusion of memory and perception. Hence reason, while sound, was operating on faulty premises. He and others could appeal to their own fevers. In his first fever Murchison had fixated on rare plants gathered on a field trip, but in his second paranoia had predominated: he had believed that a nurse and a friend were trying to kill or imprison him and had had to be tied down. Hildenbrand had fixated on removing an ornament from a stove. One of his students had taken on the identity of a catcher and swallower of snakes. Doctors also recounted patients' bizarre delusions: a cowherd tried to herd the other patients, and some patients imagined themselves to be two persons, or to be dead, or to be generals, kings, or high churchmen.[65]

Plainly the fevered mental state differed from the normal mental state, but some wondered why the unfevered "me" should be preferred to the fevered—a characteristically romantic conceit. In his 1848 poem "For Annie" Poe represented normality itself as the fevered state, complaining of ". . . the fever called 'Living' / That burned in my brain."[66] Both body and brain worked harder than normal during fever. Lyon repeatedly used a steam-engine analogy: not only temperature but the nervous and circulatory systems and metabolism were "kept *above par* continuously. . . . [During fever] the human machine is, as it were, working at a high pressure, with an unremitting strain on its various parts."[67]

While a body might not be able to sustain that rate for long, some held that delirium often produced uncommon genius as well as manic strength. Recalling the brilliant operatic duets and choruses he had composed in his imagination during a 1742 fever, Jean-Jacques Rousseau lamented: "Could but the dreams of the fever-wrought brain be preserved, what great and sublime things might not . . . its high-scaling flights bring home!"[68] Timothy Flint's friends told him that during his long fever he had "repeated whole passages, in the different languages which I knew, with entire accuracy. I recited, without losing or misplacing a word, a passage of poetry which I could not so repeat after I had recovered my health."[69] Cyril Pearl claimed to have composed, during a half hour of delirium one evening, the equivalent of a six-hundred-page "closely printed octavo, written with more clearness and strength, and beauty of language, than I could ever command in health."[70]

Alfred Russel Wallace's fevered insight had more lasting significance. In Malaya in early 1858, in the midst of a fit of malaria, he came up with the concept of evolution by natural selection: "It suddenly flashed upon me that this self-acting process would necessarily *improve the race* . . . . That is, the *fittest would survive.* . . . At once I seemed to see the whole effect of this." Wallace "waited anxiously" for his fever to pass "so that I might at once make notes for a paper on the subject." Two days later he would send the results of his fevered dream off to Charles Darwin.[71]

While most fever doctors saw delirium as amusing when it was not dangerous, long before Aldous Huxley a few wondered whether fever might open doors to levels of reality otherwise inaccessible. New England transcendentalism provided a congenial home for such explorations. Pearl, a Maine Congregationalist and mason, was the author of *Spectral Visitants, or a Journal of a Fever, by a Convalescent* (1845), a series of reassuring letters to his wife about the course of the feverish pleurisy he contracted on a trip to Portland in January 1845.[72] In fact, they were actually written during his convalescence and were intended to teach the moral "lessons in sickness" and to reassure children, who might be frightened unless they knew what fevered brains were apt to do.[73]

But Pearl also felt that he had glimpsed something special. In *Spectral Visitants* he details hallucinations on six successive nights (it was at night, writers agreed, that delirium peaked and fever was hardest to handle). Each night brings a different entertainment. Many are variously disturbing or irritating: a doctor and nurses who disagree on his treatment; teenagers who are supposedly visiting him but are more interested in one another; a thief; a group of androgynous dwarves who dance on his bed. Thursday's apparition is Disneyesque, with miniature animals acting as humans: a toad tries to inspect him through opera glasses, a mouse tries to put on neckwear. Pearl quickly realizes that none of this is real. The delusions disappear when he opens his eyes; his psychological and philosophical readings have prepared him for them.[74]

Sunday night's vision, coincident with the fever's crisis, seems different, however. A stranger who is both unknown and yet familiar ("an earnest and deeply experienced christian") leads him to a two-seater open carriage "moved by electricity" (which the Yankee Pearl is pleased to see being finally used). Moving faster than lightning, they soar over Casco Bay and off into space: "Jupiter presented a disc apparently about seven feet in diameter. Saturn, too, appeared in similar proportions. . . . I could distinctly see living beings in human form." Pearl believes that his experience will be an important contribution to astronomy and that it has prophetic importance as well. Looking back, he sees earth "at peace. As the evening advanced, the voice of prayer and praise seemed to ascend from every dwelling.—There was no distressing poverty—no grasping avarice. Plenty and prosperity everywhere abounded." The fever has offered Pearl access to profound truths, which he records (within the vision), recognizing all the while that these insights are workings of a disordered mind. But by journey's end, he has appreciated the impossibility of expression: "My discourses . . . could never be published in the language of sounds."[75]

To the missionary Flint, febrile delirium brought a different form of spiritual transcendence: "flutes playing harmonies in the most exquisite and delightful airs. . . . I remember well that every person who came into my room seemed to come with an insufferable glare of light about his head, like a dazzling glory; and that every one about me seemed to walk in the air." This was not heaven, he acknowledged, but it was a preview.[76]

If Pearl strayed into the deep end of the transcendentalist pool, Edward Hitchcock, a geologist and president of Amherst College, upheld canons of empiricism in exploring the same matters.

Hitchcock made himself an experimental subject, combining his own recollections with those recorded at his bedside by family members and students. These data he published as an open letter to his psychologist colleague Nathan Fiske. Hitchcock's ravings are more mundane than Pearl's. Many involve travel, often to places he has never seen. But he too leaves the planet: "A party of us in a barouche seemed to come in sudden proximity with a barouche of ladies dressed in white, whom I understood to be from Saturn, and my impression is that we met somewhere near the orbit of Jupiter."[77]

GENDER AND THE MORAL REGISTER

Also occurring in the late eighteenth century was a gendering of fever as it was linked to hysteria through a common pathology of nervi-centrism. That gendering in turn would augment the emerging moral significance fever was acquiring. Increasingly, fevers arose not merely from error (much less chance) but from wrongs, even though contagion of some sort might be the mechanism.

The changes are striking. For the Hippocratic authors and for Galen, the typical victims of plethoric fevers had been shortsighted males suffering the consequences of their own excesses—Hippocrates' athletes or Galen's hard-partying youths. Pringle and Lind likewise were concerned with men—soldiers on campaign or cold and wet sailors unable to perspire. In James's vision, fever's universality obscured both gendering and finger pointing. Cheyne had addressed his best-selling *Essay on Health and Long Life* (1724) to hypersensitive gentlemen, but as a Bath spa doctor and physician to the novelist Samuel Richardson, he would have a central role in creating the stereotype of the overwrought young woman who descends into fever.[78]

The device of heroines falling into "brain" fevers would persist through the nineteenth century in works of both major and minor novelists. But almost always, fever was a currency of moral exchange. It was a manifestation of the state of the nerves, and in turn of all the "feelings" that affected that state. The enormity of being jilted, for example, might be measured in the severity of the disappointment-based fever it caused. For succinctness it is hard to match Laura Jean Libbey's *Parted by Fate* (1890), in which on learning of "handsome Rutledge Chester's approaching marriage," Uldene Dean swoons:

> "Miss Dean, surely you are ill! You *are* going to faint."
>
> "No!" muttered Uldene, piteously. But, despite her denial, she suddenly threw up her white hands and fell face downward in a death-like swoon to the floor at Emily's feet.
>
> Poor, tortured soul! she had borne all she could. Had her heart broken with one awful throb in her bosom at last? . . .
>
> The usual remedies which they applied failed to bring back the fleeting breath to those pale lips, and, in alarm, the nearest physician was summoned.
>
> "It's a bad case of brain fever, induced by some great and sudden shock," was the doctor's verdict, as he bent over beautiful, hapless Uldene. "I fear the young lady is destined to be confined to her bed for many a weary week."
>
> "Is it dangerous? I mean, do you think she will die, sir?"
>
> "It is a pretty severe case," replied the doctor, dubiously. "But while there is life there is hope. The chances are evenly balanced as to her recovery or—"
>
> "Her death!" breathed Emily, in a low voice.
>
> The doctor nodded. If Emily could have read the future, she would have prayed Heaven to take Uldene then and there.[79]

Readers knew about such pathology not only from Richardson but also but from Fanny Burney, Emily Brontë, Mrs. Gaskell, and others.

Cecilia

In a long section in the middle of Burney's 1782 *Cecilia* the eponymous heroine works herself into a dangerous fever by running through London streets hoping to forestall bloodshed resulting from a misunderstanding that only she can correct. Found shrieking by a pawnbroker's wife, she is recognized as genteel (and hence remunerative) and locked up as a presumed escapee from an asylum. Still shrieking, she becomes "bereft of sense and recollection" and unable to identify herself.

In these passages, Burney uses *fever* both in a loose sense—"the fever into which terror and immoderate exercise had thrown her"—and to refer to a dangerous acute disease. Bad treatment exacerbates Cecilia's case; thwarted in her efforts to right the wrong, she gets further out of control: "Every pulse was throbbing, every vein seemed bursting; her reason . . . could not bear the repetition of such a shock." This vindicates the verdict of madness, as the landlady notes that "though she was in a high fever, refused all sustenance, and had every symptom of an alarming and dangerous malady, . . . [Cecilia's] case was that of decided insanity." Before long Cecilia becomes delirious. A friend finds her and summons aid, but Cecilia, with the "temporary strength of delirium giving her a hardiness that combated fever, illness, fatigue, and feebleness," seeks to go out, only to collapse. Finally, multiple grave physicians arrive, each pronouncing not madness but high fever and delirium, "so loved a voice uttering nothing but the incoherent ravings of lightheadedness." Then begin many days of open-eyed stupor that seem likely to end in death. But crisis finally comes. Cecilia sleeps, and on her awakening her doctor pronounces her "quite rational," her "intellects sound."[80]

Cecilia's condition *becomes* a serious continuous nervous fever, but it is difficult to say exactly when that fever begins. Madness intervenes between her resolute peacemaking and her fever, but here too there seems no clear boundary: perseverance begets frustration and in turn madness. Burney does not object to a mode of medical sensibility incapable of disentangling appropriate solicitude for one's friends from madness and deathly fever. Instead, she suggests that Cecilia's fate is a risk the sensitive incur. Even though Cecilia's friends assail the pawnbroker's wife, Burney goes out of her way to explain if not defend. Mad is as mad does; too much madness is fever.

Cathy

Likewise for Catherine Linton of *Wuthering Heights,* married to the good but boring squire Edgar of Thrushcroft Grange. Forbidden by Edgar to meet her friend-lover Heathcliff, Catherine fevers herself: hers is a hold-my-breath-till-I-turn-blue, you'll-be-sorry gesture that, in keeping with contemporary designations of fevers, might be called a "petulant malignant." Following a confrontation, Cathy takes to her room, with "A thousand smiths' hammers . . . beating in my head!" She instructs Nelly, her housekeeper-maid and confidante, to tell Edgar that she is "in danger of being seriously ill. . . . I want to frighten him." Nelly is to remind Edgar of Cathy's "passionate temper, verging, when kindled, on frenzy." Nelly's unwillingness to join in this emotional extortion leads Cathy to escalate: Nelly and Edgar find her "dashing her head against the arm of the sofa, and grinding her teeth, so that you might fancy she would crash them to splinters!" Cathy fasts, or pretends to. When this too fails to sway Edgar, "she increased her feverish bewilderment to madness, and tore the pillow with her teeth; then raising herself up all burning, desired that I would open the window." This worries Nelly. During an earlier feverish fit the doctor ordered that Cathy "not be crossed." Cathy begins to pluck feathers from the pillow (a version of floccillation, a delirious picking at the bedclothes, which Hippocratic physicians had recognized as a dangerous sign). Paranoid hallucinations set in: Nelly transforms into a hag who will bewitch the cattle, and Cathy's reflection, which she cannot recognize as her own, moves by itself.

Cathy ultimately acknowledges her own alienation—"Why am I so changed?"—in conjunction with the sequence of moral causes, beginning with "utter blackness" following Edgar's declaration, and a sense of "going raging mad" and losing "command of tongue, or brain." "Why," she asks "does my blood rush into a hell of tumult at a few words?" But she cannot stop the "burning." When Nelly refuses to open the window, Cathy, with "delirious strength," does, and then she threatens to throw herself from it.

4/30 Set 4   Cathy in Delirium   clare Leighton 1938

Clare Leighton, *Cathy in Delirium*, 1938.

Allen Memorial Art Museum, Oberlin College, Ohio. Gift of Mrs. Malcolm L. Mcbride.

Accepting the fever as self-induced, the local general practitioner, Dr. Kenneth, asks about its moral causes, for "a stout hearty lass like Catherine does not fall ill for a trifle." She will survive, he decides, but "with permanent alienation of intellect." Newly solicitous, Edgar nurses Cathy's "brain fever," which lasts for two months, "patiently enduring all the annoyances that irritable nerves and a shaken reason could inflict." He risks his own health, mistakenly believing that physical recovery will bring Cathy's mind "to its right balance also." Wise Dr. Kenneth knows better, recognizing that what survives is "a mere ruin of humanity."[81]

Ruth

Last, Mrs. Gaskell's *Ruth* (1853) is a novel framed by fever. From one standpoint, its message is simple: as ye sow, so shall ye reap. The fallen woman Ruth dies of a fever caught when, many years later, she nurses her seducer, Bellingham. Her sacrifice redeems her but cannot save her life.

But there are wrinkles. Ruth and Bellingham meet over a case of fever. Bellingham's original desertion follows his (first) fever, when, as he later says, "in my weak, convalescent state I was almost passive in the hands of others," that is, his mother, who extricates him from unsuitable lower-class Ruth. That desertion in turn triggers Ruth's first fever, which is diagnosed as "a thorough prostration of strength, occasioned by some great shock on the nerves."[82] Her fatal nursing of Bellingham at the end of the book comes after she has reluctantly decided to forgo care of their son to work as a fever-hospital nurse during an epidemic. She should avoid such duty, she is told, until she can conquer her anxiety about her son's welfare, since her "tremulous passion will predispose [her] . . . to . . . the fever." Success in the fever hospital brings the opportunity to nurse the wealthy Bellingham. Her virtue (or nursing skill) immediately calms his raving: "She had gone up to the wild, raging figure, and with soft authority had made him lie down: and then, placing a basin of cold water by the bedside, she had dipped in it her pretty hands, and was laying their cold dampness on his hot brow, speaking in a low soothing voice all the time, in a way that acted like a charm in hushing his mad talk."[83]

It is the exhaustion of awaiting his crisis (not his contagiousness) that generates her own fatal fever: "Every sense had been strained in watching—every power of thought or judgment . . . kept on the full stretch." When finally her relief comes, she is "oppressed with heaviness," and, quickly losing all sense of time, place, identity, and will, is overcome by a "whirling stupor of sense and feeling." Ruth's fatal delirium is peaceful, and Gaskell feels the need to explain why: "It might be that, utterly exhausted by watching and nursing, first in the hospital, and then by the bedside of her former lover, the power of her constitution was worn out; or, it might be her gentle, pliant sweetness, but she displayed no outrage or discord even in her delirium." Ruth dies singing, her watchers duly edified.[84] Ruth's son's grief at her death then nearly puts him into a fever, and his convalescent father, Bellingham, nearly suffers a relapse from the strain of confronting his son.[85]

A MORAL ECONOMY OF FEVER

By the end of the eighteenth century the stage had been set for the public-health movement of the nineteenth century. Its priorities and practices were not new. Plague writers had pointed to the need for better sanitation, and plague-management institutions were adept at tracking the movement of disease to and within towns. Lacking was an imperative to act. The profound change that brought that imperative has been located in philosophy and ethics (humanitarianism), political theory (expanding notions of rights), psychology (Adam Smith's concept of "sympathy"), economic theory (utilitarianism), and lived religion (evangelical activism). But a concept of fever as a function of feelings served as a binding agent, providing plot, meaning, immediacy, and a moral map.[86]

The fever-struck heroine was a literary device; compilers of fever statistics did not find that women were consistently more fevered. But the feminization of fever was nevertheless central to the emerging public-health movement because both rested on a strong link between injuries, feelings, and fever. That strong feelings could bring on fever was not new. Any Galenist could have explained how each of these heroines had mismanaged the non-naturals. But while each of these novelists considers the possibility that her heroine's high-strung nature contributes to her fever, none comes close to advocating Diemerbroeck's even-keel regimen of wholesale, alcohol-induced apathy. The feelings themselves have acquired moral standing. To feel deeply, to acknowledge the power of sympathy, and to cultivate sensibility were cardinal virtues. Translated from the personal to the social, such sensibilities brought fever into the sphere of humanitarianism: Like the owning and trading of slaves or the maltreatment of animals, a deadly fever resulting from injured feelings was a wrong in need of correction. Hence the novelists' concern was not so much with the sequence of physical events that led to fever but with the sequence of wrongs. In making Bellingham's fever "the proximate cause of Ruth's death," Ruth's benefactor, the Unitarian minister Theophilus Benson, is referring not to contagion but to blame.[87]

By giving emotional injury concrete manifestation, fever kept score: it backed the currency of feelings. It was particularly important because it integrated so wide a spectrum of wrongs, stretching from the personal to the social to the physical. In the medical-welfare literature of the early nineteenth century, *misery* and *miserable* would be applied to both depressed mood and squalid living conditions. Both debilitated; the nerves weighted them appropriately.

By the late eighteenth century, at least in some places, fever had begun to be a nucleus for a kind of citizenship. People did not expect to escape fevers entirely, but at least in some places they were expecting accountability, for fevers were public problems, not mere accidents of fate. Some responsible "public" should be acting to prevent fever and care for the fevered. The unwritten expectation fell particularly on medical professionals and on those who, like Ruth, filled the void of caregiving. Meeting those obligations was dangerous. In Edinburgh "not a single fever nurse escapes who remains long enough at her post," noted Christison.[88] Many, like the pioneering prison reformer and jail-fever investigator John Howard (1726–1790), died in the process. By the 1840s it was easy for professional leaders in Dublin and in London to exploit the idea of fever martyrdom. The groundwork had already been laid; the stricken doctors (and the priests and nurses, their comartyrs) were simply walking obvious paths of duty. In London in 1847, those raising funds for the widow of the fallen Dr. Jordan Lynch recalled that Lynch had recognized the "danger, but said cheerfully there was no avoiding it—it must be undergone."[89]

Victimization, whether of caregivers like Ruth or of the multitudes whose deadly fevers society had failed to prevent, was to incite change. But what should change? The moral entanglements that brought about fever in the novels were complicated indeed. If, in the broadest sense, fever was an injury to the feelings, a more sensitive and humane society would prevent it. But often, the moral of the fever tales was that intervention must happen far back in the train of cause and effect. And the main threat to the community was not the isolated fevers of bourgeois young women but epidemics that arose among the abject and transient poor. Mrs. Gaskell paints a scene common in the social novels of the day. To Ruth's town of Eccleston

> there came creeping, creeping, in hidden, slimy courses, the terrible fever—that fever which is never utterly banished from the sad haunts of vice and misery, but lives in such darkness, like a wild beast in the recesses of his den. It had begun in the low Irish lodging-house; but there it was so common that it excited little attention. The poor creatures died almost without the attendance of the unwarned medical men, who received their first notice of the spreading plague from the Roman Catholic Priests.

As the doctors dither, the fever "bursts forth in many places at once—not merely among the loose-living and vicious, but among the decently poor—nay, even among the well-to-do and respectable."[90] The moral is plain. Race- and religion-based prejudice had blocked accurate assessment of incipient fever in a part of the community. Such assessment, accompanied by positive action, would have prevented the epidemic.

Once a framework for confident management of health, the non-naturals had become a motif of pervasive vulnerability when applied within the context of a nervi-centric pathology and to the population of an industrial town. However carefully the wealthy might manage their non-naturals, their efforts would be overwhelmed by their failure to attend to the non-naturals of the many. Remarkably, as the passages above suggest, nervous-exhaustion explanations flourished during an age of rising concern with contagion. Usually the frameworks were complementary. Acting as sedatives, Cullen's contagia were just one more debilitator of the sensorium. Not all those exposed to a contagion fell ill; those who did must somehow be vulnerable, their nervous strength already so weakened as to be unable to resist its assault. Additionally, the ancient belief in the malleability of fevers persisted into the mid-nineteenth century. Fevers not originally contagious might become so.[91]

Such moral tales proliferated in the early nineteenth century. Often, besides mapping sin and injustice, fever was the mechanism of natural redress. In a case from the Scottish physician William Pulteney Alison's 1840 campaign for a Scottish poor law that was publicized by Thomas Carlyle in *Past and Present,* a destitute elderly woman arrives in a large town where, in keeping with the practice of discouraging displaced persons from relying on urban services, she is refused alms. Her fever spreads, and seventeen die.[92] In *Yeast,* Charles Kingsley's 1848 novel of agricultural injustice, the lady bountiful Argemone Lavington dies from a fever caught as a result of visiting cottagers on her family's estate. Kingsley makes clear that her death is payback for generations of aristocratic neglect, low wages, and poor housing.[93] Here, fever is karma, retribution. Alison would make fever incidence a proxy for injustice: "The existence of epidemic fever in any great community, particularly if there be neither war nor famine to explain

it, . . . is a most important test to the legislator of the destitute
condition of the poor, . . . of the deficiency of the funds which, in
a better regulated state, are applied to their support."[94]

A "better regulated state"? If fever registered the competence
of governments, might it also measure their legitimacy?

The beginnings of a coherent institutional response appear in
the wake of a series of fever epidemics in the 1780s and 1790s that
overspread Europe, where they were often related to hunger or
to war, and North America, where they followed a series of yellow fever epidemics in coastal cities.[95] The precedent had been
set by John Howard. If fevers prevailed under certain conditions
and in certain populations—newcomers, the hospitalized, jailed,
or encamped; in slums or among the starving—that was not accidental but the consequence of preventable causes. Charitable
dispensaries, which often provided home visits as well as clinics,
were one response. Another was the fever hospital, including the
"houses of recovery" introduced in chapter 4. To proponents like
John Ferriar of Manchester and Currie in Liverpool, fever hospitals could break the cycle of disease and poverty. They were promoted with long lists of collateral benefits, symbolic and practical,
which reduced to the congruent interests of two groups, victims
hoping to recover and benefactors hoping to prevent the spread
of contagion and thus to survive. Fever hospitals were a firewall: if
a sufficient number were built, fever might be eradicated, at least
if the fevered could be convinced to immure themselves for long
enough.[96]

And, of course, if fever were propagated exclusively by contagion. But that was not the case, insisted the Dublin dispensary
doctor Henry Kennedy, protesting a planned fever hospital in
1801: "The fevers common to the poor . . . are those called nervous, and may be traced to such causes, as cold, moisture, fatigue,
and want. . . . I will not say this fever does not ever arise from
contagion, . . . but . . . contagion appeared oftner [sic] the effect,
than a primary cause . . . in other words, when one person in a
family got a fever from cold, the disorder would, from the want of
early attention, and of common necessaries, acquire . . . a degree

of malignity, fully capable of communicating infection to others."
Kennedy accepted the exanthems as contagious, but he worried,
as had Pringle, that a fever hospital would merely convert mild
fevers into dangerous ones. The cure was food and rest, and it was
better supplied in the home than in the hospital.[97]

If few were as hostile to fever hospitals as Kennedy, others too
acknowledged that the welfare of the rich depended on that of
the poor. In the community of sympathy that would develop, the
poor—biological land mines—got the best deal when the rich
felt most threatened. But by the mid-nineteenth century that
framework had largely been displaced by its antithesis. The non-
naturals might still cause trivial febricula, but serious fever would
be demarcated by its intrinsic *independence* of socioeconomic context. Some of the non-naturals might still be implicated in its
transmission, but fever ceased to imply the redistributive mandate
recognized by Kennedy.

The age of narration as the key to understanding fever would
be brief. It would be superseded by the middle of the next century
with the arrival of the lesion-based approaches of the pathological anatomists, the data-digesting projects of epidemiologists, the
culture and antigen tests of microbiologists and immunologists,
and, among clinicians, the ascendency of the temperature chart,
which would reduce a unique illness to the dots and lines that
constituted each disease's signature.

PART III

Fever Becomes Modern

❖ ❖ ❖

Shortly after 1800, fever changed in three ways that made it recognizably modern.

First, it became plural. Centuries of ambiguity about how to characterize fever ended as broad and loose adjectival classifications gave way to well-defended nouns—the modern array of many "fevers," each essentially distinct. This approach was dominant after 1840.

Second, the distinctiveness of these fevers came to be attributed to the distinctiveness of their agents. Interest in the transmission of fevers was still sporadic in the 1790s; by the 1860s it had become central. By then, the practice of considering cases singly, in terms of mismanagement of the non-naturals, had largely given way to exposure-based epidemiology. Catching a disease would be an accident, having little to do with one's previous health. Fever might still unite a population, but it did so in a different way: each infectee became a potential infector, directly or through an intermediary.

By 1900 it was clear that the agents in many of the worst fevers invaded from insect or arthropod landing craft—mosquitoes, flies, or ticks. By 1920 most of the responsible microbes and vectors had been identified, their lifestyles worked out, and their

habitats mapped. It had often been hoped that the deadliness of
fever zones, particularly to newcomers, might be mitigated by
well-managed non-naturals. These stealthy enemies—tiny, silent,
plentiful, and persistent—doused such hopes. With the old worry
about catching a contagious fever from contact with the sick came
the new worry about being caught by one. The bedside of the
fevered was far easier to avoid than the nocturnal mosquito bite.

Third, fever would be redefined to refer to internal temperatures higher than 98.6°F. Galen had defined fever as preternatural
heat, but many of his successors had seen elevated temperature
only as an incidental and transitory aspect of some fevers. What
set in between 1860 and 1890 was a preoccupation not merely with
heat per se but with the diagnostic and pathological significance
of a temperature that was too high. To the medical soothsayer, the
temperature chart would both define the disease and chronicle the
challenges it posed, those overcome as well as those to come. To
"get the temp down" became a key therapeutic goal.

These were profound changes. They transformed how physicians interacted with patients, how individuals perceived their
illnesses and held themselves accountable, and how they viewed
others and related to them.[1] In fact, the three transformations occurred independently.

CHAPTER SIX

Facts

❖ ❖ ❖

In 1842 the most doctrinaire of the Paris-trained American doctors, Elisha Bartlett, then on the medical faculty of Transylvania University in Lexington, Kentucky, pronounced the death of fever. "There is no such disease as that . . . expressed . . . by the term *fever.*"[1]

What Bartlett was arguing for is less clear than what he was arguing against. He was jettisoning wholesale the theories proposed to explain fever. But not merely were the theories wrong or impossibly speculative, there was no disease to have a theory about. Even before Galen the learned had been seduced by a word with "no intelligible signification. . . . [Fever was] a creature of the fancy; the offspring of a false generalization and of a spurious philosophy. What, then, can its *theory* be, but the shadow of a shade?"[2]

We have seen the problem coming: bile, heart, blood, vessels, and nerves had all been nominated as the substratum of fever; heat, tremors, mental alienation, and rapid pulse had all been its defining clinical features. None had stuck. The very bounding of the domain, evident in the confused relation of symptomatic to idiopathic fevers and reflected in the nouns *fever* and *febricula* and the adjectives *feverish* and *febrile,* was fraught with arbitrari-

ness. Cullen's approach—treating fever as process, not thing—
had merely invited the romantics' introspective and idiosyncratic
indulgences in what one might call their "fever journeys." Bartlett
was not alone. The Scot Robert Christison noted that by around
1800 "the sentiments of physicians as to the nature of fever [had
become] so divided, that it is scarcely possible to say what were
the prevailing principles of any of the great medical schools of
Europe."[3] Bartlett's problem is the focus of this chapter—not just
the distinguishing and naming of fevers but the concepts underlying those activities. I examine France, America, Ireland, and England but leave the principal German response for chapter 8.

Bartlett declared that in the new medicine there would be no
theories, only facts. Ironically, however, pervading Bartlett's own
medicine would be one enormous and overarching theory. Fever as abstraction might be fiction, but fevers as diseases (or at
least a few of them) would be very real. Most important were a
newly recognized "typhoid" and a redefined "typhus." The former
would be defined principally in terms of a characteristic lesion of
the Peyer's patches along the lower part of the small bowel, the
latter largely by its absence.

When it comes to infectious diseases, modern medicine is
Bartlettian. There are distinct, singular diseases. Diagnosis can be
secured in principle—less often in practice—by isolating a specific agent in a culture medium or by tracing in the blood serum
the body's unique response to that agent. Familiar with typhus
and typhoid as real diseases, we forget how hard it was to make
them real. To us, their distinctiveness is grounded in a series of
confirmations—epidemiological, environmental, microbiological, immunological. But these confirmations, each problematic in
its own right, took place only decades after a clinical distinction
whose oddity and enormity increase the closer we get to it.[4]

Bartlett saw himself as ideologue in chief for "Paris" medicine.
Often, early nineteenth-century Paris has been seen as the site
of the beginning of modern medicine, with all that went before
regarded as a wasteland of speculation. The Parisian doctors did
many things right. They repudiated the past and sought to limit

themselves to fact. While they did not wholly disregard the patient's own sense of illness and the superficial signs, they knew enough to look within the body for traces of the hidden events ultimately responsible. Paris medicine is celebrated for its reliance on clinical-anatomical correlation, for the linking of symptoms and signs with the underlying structural "lesion," or "seat," of disease, which could be found by dissection.[5]

The typhus-typhoid distinction has been rightly seen to exemplify that approach, but much less clear has been the enormity of the conceptual change required to conclude that anatomical differences registered essentially different diseases. Such a conclusion came only after a bitter contest within Paris medicine, between "essentialists" (Bartlett's position) and those who held a "physiological" view in which illness was a continuum of pathological conditions rather than a set of discrete states. To the essentialists, the distinct exanthems, diseases like smallpox, would become the model for fevers. Followers of the physiological view would highlight the inflammation-based fevers that might arise in any part of the body.

What Charles Rosenberg has called a "revolution in specificity" was occurring.[6] By 1860 not only did most practitioners of European medicine see typhus and typhoid as discrete entities but they were on the hunt for other distinct diseases in what once had been a great sea of fever. Henceforth, diseases, presumed fundamental and permanent entities, would be the most important units. Diagnosis would be definitive and exclusive; a patient had one disease or another, not some mongrel mix. Ultimately that view would sanction the search for the unique external agents of diseases, but by no means did it require that agenda; clinical specificity might merely reflect the body's limited repertoire of ways of being sick. However raucous their squabbles, earlier theorists had not usually divided on that particular issue. For both Sydenham and Boerhaave, there was no essential tension between recognition of the "genius" of a prevailing fever (like smallpox) and its transformation into another disease. In his concept of "constitutions," considered in chapter 3, Sydenham had recognized the shifting character of the dominant fever. And, while it is more common to view modern medicine as the victory over discrete disease, it is important to remember that the physiologists and the biochemists won too.[7] Modern biomedicine is a composite; genetic and epidemiological ways of knowing are important too.

PARIS CLEANS HOUSE

The doctrine Bartlett promulgated as Parisian was an unintended and ironic development. The Parisian medical revolutionaries in the 1790s were not committed to a program of disease specificity. In fact, in many respects they were dead set against it; such a doctrine seemed hopelessly metaphysical. Initially, their main aim was not to find true answers to the eternal questions about the nature of fevers but to purge medicine of those very questions. The approach was outlined in *The Degree of Certainty in Medicine* (1789), the manifesto of the skeptical Pierre Cabanis, the chief antitheoretical theorist of medicine.

Cabanis, like Currie, challenged the adequacy of words, and behind them concepts, to represent the states of bodies known as diseases. In his view, from Galen on, the mountain of medical treatises testified only to the fertility (or fatuity) of the human imagination. Theorists had packed the body with explananda, attributing fevers to states of "humors," "spirits," "ferments," and "acrimony" and, more recently, to "lentor," "debility," or perhaps "Nature" itself (as an autonomous health-restoring force). Yet one could hardly bottle these entities, measure them independently, see them in a microscopic field, or manipulate them in a laboratory. What could be directly encountered—tongue, pulse, urine, expression—had only a hypothetical relation to these entities. Why then add to these houses of cards or build new ones? It would be better simply to admit that much of the medical domain, particularly the details of pathophysiological processes, was irrevocably inaccessible.[8]

Like the ancient Empiricists, these French medical philosophers were hoping that knowing *how* might not require knowing *what*. Distrust of words would be institutionalized in perception-based, hospital-centered medical education. Since words could not convey what the ears heard and the eyes saw (and, sometimes, what could be smelled or felt), medical learning must be learning how to perceive. Hence A. F. de Fourcroy's foundational command to the medical students of revolutionary Paris to read little but to see and do much.[9] Those who observed together would still need words to express common perceptions, but they should restrict themselves as far as possible to descriptive terms. Thus *intermittent fever* referred to fever that was intermittent. Theories of how it had come to be so came and went without impairing the utility of that designation.

Getting by without nouns (and theories) turned out to be tricky. Adjectives modify nouns; it was hard to teach without naming. Fevers vary. Recognition of their distinctiveness required criteria for grouping, for distinguishing the important from the incidental, for deciding when to transform concurrent attributes into a noun. Earlier fever discourse had been predominantly adjectival because most writers were imagining a single process that varied along many axes. Labels that we might see as imprecise—such as Sydenham's "bilious" or "choleraic" constitutions or "putrid" or "camp" fevers—might have therapeutic or situational utility as shorthand for sets of symptoms, even if they were not strong nouns distinguishing essentially different diseases.

The creation of diagnostic boxes was the business of nosology, a prominent focus of post-Boerhaavian eighteenth-century theorists. Particularly problematic were the so-called continued fevers. By contrast, the exanthems were reasonably distinct, as were the classic intermittents, quotidian, tertian, and quartan. Leaving aside the philosophical question of whether there was a natural, nonarbitrary way to organize the higher taxa of diseases, there remained the practical questions of how many forms of continuous fever there were and how (and why) they were to be differentiated.

To answer these questions there were three main approaches. The pioneering nosologist François Bossier de Sauvages (1706–1767), a contemporary and correspondent of Linnaeus's, privileged inclusivity. Assuming Latin to be the universal solvent of medical description, Sauvages generally trusted doctors to recognize distinct diseases: the nosologist's job was not to critique but to compile. But in successive rewritings long lists got longer, revealing disagreement about definitions and the inescapability of theory. A high road to order became a labyrinth of confusion.[10]

Sauvages's successor, William Cullen, emphasized pedagogic and practical utility. For practitioners, right name mattered less than right cure. For the continued fevers, Cullen used a single, therapeutically significant axis—nervous energy—and three ancient terms—the Hippocratic *typhus* (resurrected by Sauvages for "nervous" fevers, i.e., stuporous fevers in which exhaustion was the dominant symptom); *synocha,* for fevers that were active and inflammatory; and *synochus,* for fevers with features of both typhus and synocha. The stuporous, sinking patient needed stimulants; the hot and restless victim of synocha usually required bleeding. Cullen's scheme would be widely used in Britain, notwithstanding limitations and criticisms. It did not assume species stability; a single illness might evolve from synocha to typhus. Some saw synocha more as a theoretical placeholder than as a real disease. Others complained that the hybrid synochus, which nevertheless embraced a great many ordinary cases, could not be a real disease, and Cullen admitted as much.[11]

French medical revolutionaries generally welcomed that practical emphasis but found Cullen too theoretical. On the grounds that most therapy consisted in treating symptoms, Cabanis's colleague Philippe Pinel would label fevers with the adjective describing their most prominent symptom. *Typhoide* (prominent stupor) was one of Pinel's litter of adjectival fevers. Others were *ataxique* (fever with prominent delirium) and *adynamique* (with prominent exhaustion), with each term corresponding to an aspect of nervous fevers that might challenge the practitioner in a particular case. Since each of these dominant symptoms demanded a distinct therapeutic approach, they should, from Pinel's practice-based standpoint, signify distinct diseases. But to the extent that they might refer only to stages or aspects of a single disease, these labels were misleading, and their diagnostic utility would continue to be questioned. It is hardly surprising that some, like the romantic Currie, rejected nosology entirely. "Questions respecting nosological arrangement have a tendency to degenerate into verbal disputes," he declared; instead one must read "the volume of nature."[12] And why worry about names if buckets of cold water quenched the need for verbosity?

Pinel's terms were still around in 1829, when P. C. A. Louis presented his famous analysis of fever cases at the Charité. They help to explain his curious title, *Anatomical, Pathological, and Therapeutic Researches upon the Disease known under the Name of Gastro-enterite, Putrid, Adynamic, Ataxic, or Typhoid Fever, etc.*[13] The phrase "known under the Name" was not meant to imply that any of the terms were wrong. And the list was less a translational aid than an advertisement of the absurdity of theory-based names. For Louis, unlike Cullen, was aggressively agnostic with regard to the theory of the fever(s) he studied. In his view, one should eschew explanation and stick to precise and quantitative description. In fact, his list had theoretical implications. It comprehended and thus coopted both the ancient "putrid" fevers (*putrid* was already vanishing as disease designation and characterization of symptoms) and the new "gastro-entirite," the signature term of Louis's great rival, F. J. V. Broussais.[14]

Given Louis's contempt for names, one might ask why he settled on *typhoid.* For Louis was not claiming stupor as the disease's most important feature. The term did connect with Cullen's *typhus,* but as a nonexclusive adjective: a "typhoid state" occurred in many diseases. For the next half-century one could speak of a "typhoid typhus," a "typhoid" pneumonia or cholera, or even, presumably, a "typhoid typhoid," for the adjective became a noun without ceasing to be an adjective. Yet the term would outlast alternatives that did imply particular anatomical or pathological priority: Pierre Bretonneau's infelicitous *dothienteritis* (the referent is to a pimple on the intestine), *enteric fever* (implying the unproved primacy of intestinal lesions), or Charles Murchison's *pythogenic* (filth-caused) *fever.*[15]

It is ironic, then, that typhoid, which so loudly advertises its adjectivity, should become the first great victory of the specificity revolution. Louis himself was cautious about the extension, and from the "just the facts" standpoint that Bartlett so vehemently espoused, it will be safer not to make the leap from clinical facts to hypothetical abstractions. To an earlier star of Paris positivism, the stethoscopic pioneer R. T. H. Laennec, the inference that typhoid was a true species rather than a contingent assembly of attributes would have been an audacious conjecture.[16] Yet that is what occurred. By the early 1840s, self-conscious (and often self-righteous) positivism had evolved significantly toward a practice in which clusters of characters were indeed being regarded as fixed disease entities.

Many intellectual and institutional factors contributed to this practice. Some that we might have expected to contribute did not. One that did not was the argument from therapeutic specificity. To Sydenham, the fact of the variable effectiveness of treatments was proof that the several types of fevers had not yet been adequately distinguished from one another: a true and complete classification would be the route to a surer cure.[17] Usually, the paradigm case of that view has been cinchona bark, which worked dramatically in some (intermittent) fevers. But so many variables were in play—regimen, dose, mode, stage of disease, quality of medicament—that even the isolation of quinine, cinchona's active ingredient, in 1820, did not significantly enhance the specificity of intermittent fevers. It would continue to be used generically in fever treatment.

Nor did acquired immunity particularly impress these medical philosophers. While it might be true for some diseases, notably smallpox, that having once had it, one would not get it again, there seemed to be no evidence that this was true for fevers generally. Nor was there any explanatory framework to make acquired immunity anything more than an interesting sidelight.

Also less decisive than we might expect was clinical-anatomical correlation itself. Pathological anatomy was not new in the early nineteenth century, but Paris medicine became eponymous with its systematic practice. The great Parisian hospitals, especially the

Hôtel-Dieu, a virtual cadaver factory, offered ample opportunity to link lesions to symptoms, but most resisted the temptation to make lesions the foundation of disease specificity.[18] A move from correlation to causation would have been speculative; it was better to leave lesions as merely one more type of clinical fact. Nor was there much expectation of causal specificity. Causation had rarely been a nosological issue, because few held that a particular effect indicated a unique cause. Sauvages had repudiated cause-based classification, and Cabanis and Pinel followed suit.[19] Cullen thought many fevers were contagious but saw little value in positing specific contagions. He suspected that typhus and synochus had the same contagion; they differed clinically because of other factors.[20] Moreover, causal inquiry was often viewed as incoherent, exhausting, and unproductive. And what would be gained by sifting through the innumerable unhealthful ways that people lived? Only in the peripheral area of toxicology would causal thinking contribute to disease specificity.

Even collectively these factors did not displace the ancient practice of seeing each fever biographically, as the culmination of a body's struggle against forces antagonistic to its function and comfort. Boerhaave, Cullen, John Brown, and the American stalwart Benjamin Rush all took such a view. The exanthems were exceptions, but even the exemplary smallpox itself might be merely another variant of putrid disease.[21]

BROUSSAIS VERSUS LOUIS

In fact, the push toward specificity had much to do with politics and personalities. To admirers like Bartlett, Paris medicine was a utopia of rationality. The historian E. H. Ackerknecht would later suggest that "anarchy and chaos" more aptly characterized it.[22] Again, ironies. The focus on facts, which was to be the way beyond controversy, seemed merely to exacerbate controversy. The observation-based pedagogy of Cabanis and Fourcroy did not drive out rhetoric but loosed a flood of medical journals, reference works, monographs, and theses, which were no less polemical for being statistical.

For fever, the key controversialists were Louis (1787–1872) and F. J. V. Broussais (1772–1838). Each was, in his own way, a follower of Cabanis and Pinel. Much of nineteenth-century medicine bears the stamp of their rivalry. Louis, the younger, went to extremes to distance himself from Broussais's rule over a "physiological" medicine. Their battleground was the lower end of the small intestine.[23]

It was the iconoclast Broussais who would respond most powerfully to the antinosological frustration that Currie expresses. After a career as a surgeon in Napoleon's army, Broussais became a physician-professor at the Val-de-Grâce, Paris's military hospital. There, for nearly two decades Broussais, like Currie a radical and romantic, resisted the winds of reaction, launching vicious attacks on the metaphysical failings of his colleagues in successive editions of his *Examen de la doctrine médicale génèralment adoptée.* He targeted even the symptom-based distinctions of Pinel. Broussais considered disease names to be partial and pathetic representations of complex, seamless nature. Access points to disease were few and poor; as in the story of the blind men and the elephant, doctors were trying vainly to trap biological reality in a feeble web of words. It was scarcely surprising that they disagreed; physicians who relied on such abstractions did not even understand their own terms.[24]

As for fever, Broussais objected not only to the vacuous designations of Pinel but also to the prevailing debilitationist theories of Cullen and Brown. He denied any correlation between fever and debility. Intense heat, a rapid pulse, and manic strength were incompatible with any defensible concept of debility, while the mechanisms that were to explain the move from spasm to reaction were not only conjectural but arbitrary.[25] Practice among robust, outdoorsy soldiers, whose fevers often occurred after a binge, suggested overstimulation, not understimulation.

Behind these concerns lay a deeper objection: conventional fever explanations relied on unknowable entities, which were merely advertisements of ignorance. Each inadequate abstraction generated another to explain it: lentor led to spasm, spasm to debility. "Habeas corpus," thundered Broussais, demanding that someone show him the seat of this debility.[26] The undefinable had no place in science. Disease by disease, doctrine by doctrine, author by author, he sarcastically pummeled learned medicine, past and present.

In fact, what Broussais was objecting to was not theorizing per se but the vacuity of past concepts and categories.[27] Before becoming a destroyer, Broussais had been a synthesizer. In place of *debility*, he had explained fever in terms of "irritation" and "inflammation." He thought that most fever, and many other diseases, began with an irritated stomach, often from something one ate. Unrelieved irritation led to inflammation, which would move first to the small intestine and then lodge elsewhere, even in the lungs or the brain.[28] All fever symptoms were merely systemic manifestations of these local inflammations beginning in the stomach and the gut. The implications for therapy were simple: fevers could be quickly cured by relieving local inflammations directly, usually with leeches. By contrast, the wine and broth recommended by Cullen and Brown would simply re-irritate a sensitive stomach, as would the sudorifics or expulsive forms of physick, like cathartics or emetics. Vegetable gruels and cool and acidic drinks might supplement leeching; Currie's cold water was also beneficial.[29]

It was Broussais who labeled his approach "physiological," in contradistinction to the "ontological" medicine of disease specificity, which, in his view, characterized the views of his predecessors and opponents. In fact, the characterization is a misleading one, especially for his predecessors. With the exception of the Helmontians, almost all authorities had made a comprehensive pathophysiological account a sine qua non of medical explanation: hence humors, fibers, vascular hydraulics, acrid blood, mysterious fluctuations of nervous force. And despite his contempt for antiquity, often Broussais was retreading ancient concepts and practices.[30] Hippocratic writers had recognized many fevers as inflammatory. Beginning as local infections, they roamed the body and must be shown the door. Like Broussais, the ancients had been fascinated by sympathetic relations, especially between the pit of the stomach (the hypochondrium) and the brain.[31] Galen too had often regarded fever as the fruit of excess; his aristocratic youths shared habits with Broussais's Napoleonic soldiers.

But in the early modern period, inflammatory fevers had lost their centrality. No one denied that inflammation led to fever, but such fevers had been medically uninteresting. More a surgeon's matter, they lacked the diagnostic mysteries of the intermittents or the romance of the new nervous fevers. The return of interest reflected the rising stature of surgery. Others besides Broussais, notably Henry Clutterbuck and John Armstrong in England, developed inflammation-centered fever theories too. Some, like the toxicologist Christison, were almost persuaded, but no united front arose; Broussais's imperiousness probably prevented that.[32]

A focus on inflammation might seem especially appropriate given the climate of empiricism and the emphasis on pathological anatomy: one need not imagine inflammations, one could *see* them—in swelling, redness, and local heat. Many who rejected Broussais's gastrointestinal origins theory still focused on the local inflammatory lesions in many feverish diseases, wound infection and puerperal (or childbed) "fever" among them. According to the medical lexicographer James Copland, inflammatory fevers were more likely to have external particulate causes than to be caused by conditions that "exist within the [human] frame."[33]

By simplifying fever lore, Broussais's antiessentialist campaign, like Currie's cold water, had a leveling and universalizing effect. Long or short, mild or deadly, all fevers had a common basis and cure. The scheme could embrace pulmonary tuberculosis (chronic destructive inflammation in the lungs with continual low-grade fever) and diseases like erysipelas, in which systemic internal inflammation manifests on the surface of the skin.[34]

In one of the many ironies of Paris medicine, the archphysiologist Broussais, by focusing attention on an all-powerful intestinal lesion, would pave the way for the specificity of typhoid. His core idea—a gastrointestinal lesion triggered by something one ingested, leading to a long systemic fever, often with prominent effects on the brain and mind—is also the core of the "typhoid" concept that would solidify in the next generation. Yet typhoid would emerge as a rejection of Broussaisianism. Peculiar lesions of the Peyer's patches of the ileum, noted by Marc Antoine Petit and Etienne Serres in 1807 (and possibly by others decades earlier), would be accepted as the anatomical signature of a unique disease, typhoid, and not just another generic effect of "gastro-enterite."[35]

By the late 1820s, Broussais had managed to offend most of the Paris medical community. Often, the response was to reject the ideas along with the person—in this case to damn Broussais and all his works—rather than to engage critically.[36] Broussais, it was claimed, had violated his own precepts: he speculated. Pleading guilty, Broussais seized the moral high ground. Facts were not therapies. Treatment required judgment, and judgment required theory. Merely to describe was tantamount to medical murder.

With regard to fever, the most important anti-Broussaisian was Louis. Though Louis never held a professorship or senior hospital appointment, he would be chief developer of the so-called numerical method, which is best known as a set of strategies for empirical investigation of therapeutic efficacy. Using the hospital as a laboratory, the clinical scientist could compare treatments. Thus in 1834 Louis famously established the futility of bleeding in treating consumption. He is still recognized as a pioneer of clinical trials and of evidence-based medicine.[37]

The approach was also a way to define diseases. Given a population of patients and a variety of systematically observed circumstances, symptoms, signs, and anatomical states, one might discern clusters of concurrent elements by selecting some index feature and determining the frequency with which other features accompanied it. The clusters were, empirically, diseases: questions of the elements' relations to one another and of how they interacted to create symptoms and pathological effects were moot. Louis's more-empirical-than-thou approach is clear in the disclaimer with which he begins *Researches:* his book is not a typhus "treatise" but a compendium of facts. The insinuation is that treatises are verbose and valueless.[38]

In his book Louis reports on 138 cases of the characteristic

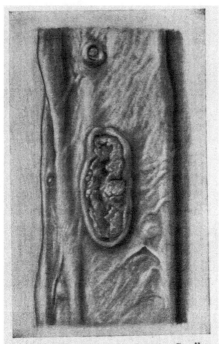

Fig. 288.—Typhoid fever. Swollen Peyer's patches with beginning separation of the slough.

The Peyer's patches of the small intestine at an early stage of the pathological process of typhoid fever.

From W. G. MacCallum, *A Text-Book of Pathology*, 2nd ed. (Philadelphia: W. B. Saunders, 1921).

continued nervous fever of Paris and 83 of other acute diseases encountered in five years at the Charité. He uses a series of index features. He begins, for example, with characteristics common to persons who die of fevers from the sixteenth through the thirtieth day. Presenting first anatomical features and then symptoms, he arrives ultimately at a list of concurring features. Keeping company with the peculiar lesions of the Peyer's patches are nose-bleeds, rose-colored spots, sudamina (skin reddened by sweating), a coated tongue, meteorism (abdominal gurgling), drowsiness, stupor, extreme debility, eschars (sloughing of skin) of the scrotum, and spasmodic movement.[39] None of these occurs invariably in this disease, as we might guess from the scrotal eschars, and each occurs in other diseases as well. But "the combination of many of them" marks the disease. Louis does not ask why they coexist, nor does he weight them. Even what would become the signatures of typhoid, the rose-colored spots and the lesions of the Peyer's patches, are given no logical priority.

Louis's caution carried over to therapeutics and etiology. He was not as oblivious to therapeutics as some Paris professors were reputed to be. Louis chronicled the therapeutic modalities used in fatal and recovered cases, charting, where possible, when in the disease they were used and to what extent. His chief finding was that early and moderate bleeding modestly improved chances of survival. But Louis was less interested in the arts of therapy than in diagnostic rigor. Without confident diagnosis of real diseases, there was no foundation for comparative study of treatment efficacy; if each case was *sui generis,* comparison was impossible. With his own patients, Louis, like his predecessors and colleagues (and Broussais), practiced conventional medicine, treating symptoms as they occurred, sometimes modestly anticipating forthcoming symptoms, basing his decisions on close monitoring of a patient's comfort, behavior, and strength. Research might require a rigorous policing of the borders of fact, but disease existed in individual persons, and practice required interpretation and speculation.

"Cause" too would come under the rule of fact. Enumeration would overthrow the blather of academics and solicitous bedside physicians, with their invocations of hypothetical pathological processes that one might hope to tame through scrupulous management of the non-naturals. For Louis, cause was not an explanatory narrative but merely the set of antecedents of the symptomatic cluster. Assembled without reference to mechanism or avoidability, his causes look to us like risk factors. Thus age and sex were "causes" of typhoid. Though there was an acclimatization effect—newcomers to Paris suffered unduly—neither the proxies of debility nor the ancient "passions of the mind" seemed significant: "The facts . . . do not lead me to place excess of labor, sorrow, and anxiety of any kind among the causes of typhoid affection." Only one in seven victims had a history of "trouble of mind" or overwork. And only one in eighteen lived in poor housing. Louis recognized that absence of evidence was not evidence of absence. Without comparative data on the distribution of such conditions in the catchment population, these findings could be suggestive only.[40]

But his dismissal of such inquiries makes clear how far Louis had already departed from longstanding conventions of causality. Undertaken as a statistical exercise and outside the contexts of individual hygiene or public preventive measures, Louis's approach to causation had no implicit political or moral valency. Moreover, from the perspective of classical medicine, with its multicausal etiologies, Louis's approach was unsound because it considered variables in isolation: that only one in seven typhoid victims had a history of overwork or troubled mind was unimportant. If one regarded debilitating causes as interchangeable and cumulative, it followed that damage to a single non-natural in a seventh of cases was intrinsically neither small nor large. That a disease only arose from many combined causes did not negate the harm done by each—by overwork or by a troubled mind.

Louis had already moved sharply away from older, individualistic notions of illness. The subject's apprehension of illness, bound up in a moral tale about self, fate, and environment, no longer mattered. His initial redefinition—hospital-defined typhoid replacing sufferer- and physician-defined nervous fever—would ease the way for a series of expert-based redefinitions over the next century. A half-century later, diagnostic confidence, built initially on the linkage of the rose-colored spots to the lesions of the Peyer's patches, would make epidemiological generalizations possible. Typhoid (and typhus) diagnoses would sanction state actions to disinfect persons and goods and to regulate and improve water, sewers, and milk. After the mid-1880s, laboratories took over. Typhoid would be defined in terms of appearances on a culture plate and then, after 1899, by a positive Widal test, a reaction showing a body to have been challenged by a unique typhoid antigen. Typhoid would become a broader entity through this transformation. The features of a disease defined alternatively in terms of a lesion of the Peyer's patches, a culture of *Bacillus typhosus,* or a positive Widal test varied so widely (including a benign form wholly unnoticed by its "victim") that some, including the eminent William Osler, would regard typhoid as a disease lacking in any clinical distinctiveness whatsoever.[41] Increasingly, diseases were decoupled from the bodies that housed them. That would have been the height of absurdity in an earlier era, and it would be an outrage still in the 1920s to those sympathizing with the asymptomatic "Typhoid Mary" Mallon, interned for more than two decades as an asymptomatic carrier of typhoid. She might be healthy, but her body was diseased.[42]

But Louis was not trying to launch any such revolution of specificity. His "typhoid" was more stable and detailed than the heuristic disease designations of predecessors, but Louis avoided suggestions that he had captured the disease's essence or that his findings implied a single necessary cause, which would later become the proof of disease identity. A species interpretation arose only after others failed to confirm his observations, finding similar fevers without lesions of the Peyer's patches and with more active delirium. For some, particularly British and Irish doctors, it became plausible to think of the Parisian typhoid as merely a regular variant of the common Cullenian typhus. The alternative, two-species view arose from the comparative investigations of the Philadelphian William Gerhard and the Londoner William Jenner and from Bartlett, who presented himself as Louis's champion. Using the exanthems as their model, they would make typhus the analogue of typhoid. But it was an asymmetrical pairing. Though the petechiae sometimes present in typhus might be distinct from the rose-colored spots sometimes present in typhoid, the absence of any distinct lesion often left *typhus* as a catchall for whatever was not typhoid.

Since neither American nor French clinicians saw much typhus, they had to rely on British and Irish reports, however much these might be colored by residual Cullenianism. Understanding typhoid's changing status requires a closer look at America and Ireland before reviewing the consummation of the species concept in England.

BARTLETT'S AMERICA

The expense, length, and difficulty of a transatlantic journey makes the early nineteenth-century pilgrimage to Paris of around a thousand American medical students remarkable indeed. Though their recorded experiences are often ambivalent—they could not understand the lectures or see the operations and were appalled by the callousness they witnessed toward patients—they became, almost uniformly, champions of a Parisian medical science, founded equally on abundance of data and parsimony of conclusions.[43] Importing that science was another matter. In France medical debate took place within a united professional community. In America, regionalism, medical sectarianism, and states' unwillingness to regulate medical training or practice worked to defeat any common context. A few with access to large hospitals, such as Austin Flint in Buffalo, New York, might emulate Parisian research, but often appeals to Paris, and later to Germany, were about what American medicine should be, not what it was.[44]

Such was the context of Elisha Bartlett's polemics. In 1827 the New Englander Bartlett (1804–1855) was in Paris, where he became a follower of Louis's. Returning to America, he practiced briefly in Lowell, Massachusetts, but spent the bulk of his career as a writer and teacher in medical schools in Vermont and Maryland and at Transylvania University, in Lexington, Kentucky (two stints). In his 1842 *History, Diagnosis, and Treatment of Typhoid and Typhus Fever* Bartlett sought to make American medicine more cosmopolitan by supplementing Louis's findings with American experience. In addition, he sought to demarcate Louis's typhoid more definitively from the lesionless typhus of the British and Irish writers. But having asserted that fact must obviate any need for argument, Bartlett found himself arguing that typhoid and typhus were not merely variants of some common fever but true species. Two years later, in his three-hundred-page *Essay on the Philosophy of Medical Science,* he created a warrant for Louisian medicine. Both works are polemical; where Louis summarizes, Bartlett declaims.

Bartlett's typhoid book is unmistakably a treatise. Like Louis, he reviews the character and frequency of each symptom and the postmortem appearances in the several organs before moving to cause (again, merely antecedents), varieties, course, diagnosis, prognosis, theory (pointedly minimal), and treatment. Defining the disease only at the end suggests that an induction has occurred: typhoid is the disease whose properties have been outlined in the previous 170 pages. That a definition was required in choosing cases for consideration is overlooked.

To defend typhoid's distinctiveness from typhus and its claim to species status, Bartlett had to do more than list and summarize: he had to interpret and evaluate. The great challenges were symptomatology and pathology. By Bartlett's writing, the list of typhoid signs and symptoms had lengthened; the disease was even more variable. "There is not one symptom, there are not two or three symptoms, which, in themselves, are characteristic . . . ," Bartlett admits. "Our diagnosis can never be founded . . . on a few positive physical signs. It must always be rational, not absolute. The evidence . . . is wholly circumstantial." That said, Bartlett happily declares that "there are few general diseases, the diagnosis of which is so well established and so certain." Being rational meant forsaking particular symptoms in favor of general features. Long-lasting febrile symptoms with prostration in young adults, diseases that did not evolve or respond to medication and in which there were no obvious local causes, were typhoid fever.[45]

Partly from Bartlett's efforts, typhoid fever would become a favorite diagnosis in America before being briefly eclipsed in the 1860s by an even more comprehensive entity, typhomalarial fever.[46] But so variable would typhoid presentations become that it is hard to imagine any common symptoms that would rule it out. It comprehended diarrhea and costiveness alike; it might involve frenzy or stupor, or not affect the mind at all. Even the rose-colored spots appeared in only two-thirds of cases (more careful examination would surely raise that rate, Bartlett insists).[47] Underwriting this confidence were the lesions of the Peyer's patches, but of course diagnosis and treatment could not wait for a post-mortem confirmation. Bartlett claimed that there were few false positives among diagnoses confirmed by autopsy (though others would point out that anyone looking for lesions in a decomposing intestine would likely find them). False negatives, in which lesions were found in cases diagnosed as something else, were generally grounds for rediagnosis and therefore had the effect of increasing typhoid's clinical variability.

Remarkably, while intestinal lesions might be the defining sign of typhoid, Bartlett still followed Louis's agnosticism with regard to their significance: constant presence did not imply a significant pathological role. Indeed, Bartlett goes further: it would be "more philosophical" to see the lesions "not as the local cause of all the other appreciable phenomena of typhoid fever" but as one of the many elements due to "some morbific agent, or influence, or process, the nature, sources and operation of which are wholly unknown to us." Bartlett notes that there is no correlation between a fever's severity and the extent of the lesion, but his larger purpose is to head off any pathophysiological discussion in which typhoid might appear to be a variant of a generic fever. Grudgingly he accepts that typhoid may be seen as a "peculiar enteritis" and that the lesions of the Peyer's patches are inflammatory. But it is a special inflammation.[48]

Instead, Bartlett would appeal in another direction, suggesting that the proper analogue of the distinctive lesion is the unique rash in measles, scarlet fever, and smallpox. Surely, a rash on an inner surface—a "pox on the ileum"—was just as likely as a rash on an outer surface.[49] In the exanthems, the characteristic rashes might not significantly affect the patient's experience or the disease's danger, yet it is by them that we know the disease.

Spelled out in his *Essay on the Philosophy of Medical Science,*

Bartlett's antipathy to pathological explanation had far-reaching consequences. The *Essay* is not just a catalog of errors in the medical past; it is an attack on three general sites of unsound inference: the founding of physiology on anatomy, of pathology on physiology, and of therapeutics (and etiology) on pathology. Bartlett insists first that seeing the interior of the body gives one no sense of how its parts work. Second, he holds that save for the trivial truth that nonoccurrence of vital processes (e.g., breathing, eating) is deadly, physiology is unrelated to pathology. A healthy body contains nothing like pus, the signature of so many pathological states. Third, he notes that the action of many therapeutic agents, like quinine or an emetic, is ultimately mysterious: we know *what* they do but not *how.*[50]

Bartlett's embracing of inexplicability was not merely an admonition to be cautious; it had substantive implications. Most important was the severing of the health-disease continuum. Diseases were simply quaint assemblages of odd appearances and states. Ideally they would be distinguished by induction, but this would be hard, for diseases may differ, Bartlett insists, even when they seem to be the same: "Different diseases may possess certain very important elements *in common,* without hindering, in any degree, their separation into perfectly distinct individual species. They may also agree, in many respects; in regard to their causes; their march; and duration; their relations to remedial measures; and in other respects, and still be susceptible of clear and positive diagnostic distinction."[51]

If Bartlett's *Essay* ostensibly puts forth a general philosophy of medicine, fever is its focus and typhoid is the model disease. In Bartlett's ontological medicine, each disease must have an essence even if one cannot say what it is.[52] Bartlett enlists the nosologists' master analogy, botanical classification. Classification can only distinguish, not explain, but to Bartlett, science is the proper ordering of facts. That some diseases—"apoplexy, pneumonia, pericarditis, phthisis, typhoid fever, measles, scarlatina, and small pox"—seemed as distinct as species in any other domain of natural history (e.g., botany) warranted the approach. (For his part, Broussais had held that flowers were really not very much like pain.)[53]

One may admire Bartlett's prescience while being astonished at the exuberance of his pruning. *Species* did not imply commonality of causes (proximate or remote), of symptoms, or of therapeutic responses. What it did imply was less clear. Bartlett's commitment to distinct diseases had little relation to the experience of illness or even to the practice of medicine. What was clear was that diseases, however demarcated, would be more important than the persons who suffered them.

Bartlett's natural-historical approach was a common and attractive approach to ordering America's fevers, but it was not the only one. Perpetuating the heritage of German romanticism, his homeopathic rivals would continue to fit the fever into the life, closely interrogating their patients to find the mix of miniscule doses that would address their complex pains and sensations. Others doubted that the truths of urbane Paris would apply to the wilds of North America, tropical and temperate. Even if there were fever species, climatic modification was more important. The most important exponent of that view was Daniel Drake (1785–1852), the medical authority on the North American interior. There typhoid had become the fashionable diagnosis, Drake would complain, diagnosed wherever fevers were more abdominal than cerebral. But many diagnosers were ignorantly aping Paris. Few had carefully read Louis; instead they were simply spouting Bartlett, "the finished, elegant, and popular commentator of the French pathologist."[54]

Bartlett, Drake, and others viewed Ireland as the key to the mysteries of continued fever. There was apparently much more fever there; it did not fully conform to fevers elsewhere, and yet it had been studied fully and well—clinically, anatomically, and epidemiologically. Additionally, Ireland had an impressive set of public institutions to cure fever and a very high cure rate. Finally, Irish fever appeared to be peculiarly but indirectly linked to famine. In August 1847 Drake traveled down the Saint Lawrence River to the quarantine station at Grosse Isle, near Quebec, to study the fevers of disembarking famine refugees. He saw both petechial typhus and typhoid, with its "rose-colored spots," but much of the deadly "fever" was neither. Patients suffered the prostration (and emaciation) characteristic of late-stage nervous fevers, but they were not otherwise fevered. They exhibited "scarcely . . . a single characteristic symptom of typhous, or any other form of fever."[55] Irish and Canadian doctors called this disease *famine fever*, even though it seemed not to fit Louis's and Bartlett's expectations of clearly demarcated disease. Well-fed Americans did not get it. Irish doctors knew all about it.

FEVER'S HOME

If Paris and Bartlett point to the future, Dublin perpetuated the classical medicine of the past. At least in its dreams, medical Dublin was the alter Paris; for fever, it was also the anti-Paris. The reign of the Dublin clinical school coincides closely with the golden age of Paris medicine. If it never evolved a comprehensive system of ward-based clinical teaching, Dublin too was a center of medical education. The city was a magnet for a mobile and impoverished population; its many hospitals housed patients with a variety of illnesses. Irish surgeons (and physicians) practiced pathological anatomy but did not fetishize it; cadavers were cheap, even an export industry. And Irish doctors squabbled too.[56]

But the organization of medical institutions in the two cities was vastly different. Dublin's medical elite were less concerned with establishing biomedical science than with cultivating the arts of practice. Medicine and even surgery were founded in the genteel liberal arts. Some published, yet the model of medical authority remained remarkably generalist, with the critical consumption of research being more valued than its production. Medicine in Dublin and Paris differed most strikingly, however, in the purposes and organization of urban hospitals. After 1794 Paris's great warehouses for the wretched had been converted into medical, often specialist institutions. Hospital physicians could study acute diseases without having to cope with masses suffering the complications of indigence.[57] Before the passage of a poor law in 1838, Irish hospitals were more likely to serve as refuges.

Despite Henry Kennedy's critique, fever hospitals had flourished; there were one hundred by the time of the Great Famine in 1845. The threshold of fever was low. Preventing contagion might mean opening hospitals to persons with more routine fevers, which never evolved into the serious typhus, though one could not know that at the outset. Hence the vast numbers of recorded fevers and an image of Ireland as "the grand mart of fever," as the Belfast surgeon Henry MacCormac put it in 1835: "In both town and country the seeds of the malady are always rife, only requiring a few of the exciting causes such as famine, misery, damp and cold, to produce a fearful increase."[58] Another result was Irish doctors' arrogance: one did not know fever until one had seen it in Ireland.[59]

As for the so-called typhoid-typhus distinction, the common view in Ireland was that while the theory might fit all the cases that fit the theory, it had little to do with fever, which according to the great William Stokes varied infinitely in "appearance and character."[60] As clinicians, Irish physicians prided themselves on being acute readers of signs and able navigators of a fever's course. They performed autopsies with modest expectations, looking more for local damage than for diagnostic confirmation. And since all fevers were essentially one, there would be no discrete diagnoses to confirm.

Experience with the interplay of poverty and disease led most Irish doctors to view French medicine as oversimplified and arbitrary. To Louis, the causes of (typhoid) fever might be mysterious. To Irish practitioners the causes of typhus were omnipresent and overwhelming. The perspective is evident in the annual reports of fever hospitals, and particularly Dublin's House of Recovery on Cork Street, completed in 1804.[61] The reporters—the hospital's physicians, John O'Brien, Francis Barker, William Stoker, and Richard Grattan—might weigh variables differently, but they shared a common framework. Fever could arise from contagion (many of them had suffered hospital-based fevers), but contagion was not the exclusive, necessary, or most important cause. Managing propagation might be a matter mainly of isolation, but because contagiousness could arise spontaneously in a debilitated population, fever's causes were jointly "physical and moral" or, at a greater remove, cultural, socioeconomic, and political.

In considering how those causes acted, Irish medics continued to appeal to the Boerhaavian/Scottish "common sensory." "Those causes especially which act on the system through the medium of the mind, more than any others either give security to the individual, or render him liable to the effects of the contagion," noted Grattan. "Hope, confidence, enthusiasm, and an ardent, enterprising spirit, will enable men to support great fatigues and privations. . . . When disappointment, apprehension and despondency seize upon the mind, disease is produced, and contagious fevers and dysentery break out."[62] Though no romantic himself, Grattan was taking the romantic fascination with fever as an expression of the conditions of one's being far beyond the pages of novels.

From the early teens to late 1830s, Irish fever writers had focused more on poverty, and particularly on episodic hunger, than on climate, race, air, or filth. Most agreed that poverty (particularly hunger) was neither a sufficient nor a necessary cause, while insisting that it was nevertheless the most important cause. Again Grattan, in 1819: "Next to contagion, the great exciting cause of fever, I consider a *distressed state* of the general population . . . the most common, and . . . most extensive source of typhus fever."[63]

The attribution of fever to hunger-induced debility became increasingly important as it became clear that fever hospitals had failed to halt epidemics. Originally, Grattan blamed Irish "manners and habits," but increasingly he and others would hold political and economic policies accountable. England would be the benchmark: why was there so little fever there? Writing in 1816, during the first of the many famine fevers, Stoker noted that Cork Street's *monthly* intake was roughly the same as the *total* intake of the London Fever Hospital during its first eight years. Cork Street had served 24,000 in the previous decade, compared with 600 served at the London hospital. Dublin's population was roughly one fifth of greater London's; Cork Street's catchment area was roughly a quarter of Dublin. Edward Percival's 1819 *Practical Observations on the Treatment, Pathology, and Prevention of Typhous Fever* was based on 6,242 cases at Dublin's Hardwicke Fever Hospital from 1813 to 1815. Louis's great claims had been based on 221 cases over five years.[64]

Irish doctors were casual about nosology. Though many preferred Cullen's catchall *synochus* or referred simply to the "common continued fever," *typhus* was widely though loosely used. Grattan protested that the term no longer implied stupor and was "so indefinite, and so universally misapplied, that its abuse has created the greatest confusion, and contributed materially to retard the improvement of medical practice."[65] (We, who use *flu* no less indiscriminately should not grouse.) Among the typhuses, a significantly cerebral fever of sudden onset, severe headaches, exhaustion, stupor, and delirium, sometimes with petechiae, lasting two to three weeks, would sometimes be designated as "true typhus" or "perfect typhus." But those same symptoms might simply register as "malignant" or be labeled "macular" fever—Robert Graves's term to highlight the reddish hue of the skin.[66]

But greater precision in species definitions was not the obvious goal. Names must not deflect attention from the more immediate issue of therapy-relevant symptoms. "A methodical division of fevers into species and varieties, has been a serious evil in medicine," wrote John Murray of Cavan. "It is a dangerous practice to treat fevers according to their names, which are and ever will be arbitrary. It is also a great source of error to make accidental symptoms the essential difference of distinct species."[67] O'Brien noted that typhus varied seasonally: it was a chest disease in the spring, a gut disease in summer. There could be "low or simple typhus," "rheumatic typhus," pneumonic or gastric typhus (both bilious and dysenteric varieties), as well as typhus *gravior, mitior,* or *lenta.* I have noted the belief in class-specific typhuses, cerebral in the rich and abdominal in the poor. Some of the latter may have been what today we term *typhoid,* a term little used in Ireland except as an adjective.[68] In keeping with ancient and Cullenian principles,

No. I.—ANNUAL ADMISSIONS INTO VARIOUS HOSPITALS.

	Kendal.	Leeds.	Man-chester.	Glas-gow.	Lon-don.	Dublin.	Cork.	Lime-rick.	Water-ford.	Kil-kenny.
1795	275, v.	...	...	18	...	...	...	...	...	...
1797	84, o.	...	371	83	...	...	...	...	...	...
1798	103, xii.	...	339	45	...	...	...	...	...	...
1799	52, xiii.	...	398	128	...	...	...	...	146	...
1800	137, ii.	...	364	104	...	...	...	...	409	...
1801	263, iv.	...	747	63	...	...	...	446	875	...
1802	92, iv.	...	1070	104	164	...	...	86	419	...
1803	60, iv.	...	601	85	176	...	254	95	188	73
1804	76, vii.	...	256	97	80	497	190	90	223	80
1805	86, vi.	66	184	99	66	1733	200	86	297	69
1806	57, ii.	75	268	75	93	2540	441	84	165	56
1807	53, o.	32	311	25	63	2389	192	110	166	81
1808	68, ix.	80	208	27	69	2544	232	99	157	96
1809	116, xix.	93	260	76	29	2180	278	105	222	116
1810	47, vi.	75	278	82	52	3162	432	164	410	135
1811	61, viii.	92	172	45	43	2689	646	196	331	153
1812	16, ix.	80	140	16	61	4271	617	146	323	156
1813	79, ii.	137	126	35	85	4497	550	127	252	183
1814	73, o.	79	226	90	59	4418	845	221	175	236
1815	206, iii.	146	379	230	80	6231	717	394	403	249
1816	133, o.	121	185	399	118	4432	1026	659	307	162
1817	291, i.	178	172	714	760	6996	5325	2474	930	1100
1818	270, o.	303	...	...	...	27226	10199	6307	2729	1924
1819	145, i.	231	...	...	...	12888	2788	1502	2656	683
1820	177, o.	...	...	...	...	...	...	...	...	...
1821	209, i.	...	...	...	...	...	...	...	...	...

This table reflects the excessive numbers of fever cases in Irish towns compared with those in towns in England and Scotland. Glasgow's profile would resemble those of Irish cities by 1840.

From Francis Boott, *Memoir of the Life and Medical Opinions of John Armstrong, M.D. to Which Is Added an Inquiry into the Facts Connected with Those Forms of Fever Attributed to Malaria or Marsh Effluvium* (London: Baldwin & Craddock, 1833), volume 2, after William Harty, Historic Sketch of the Causes, Progress, Extent, and Mortality of the Contagious Fever Epidemic in Ireland during the years 1817, 1818, and 1819 (Dublin, 1820).

fevers could evolve, with changes occurring in their site and character.

That Cullen's typhus was the quintessential disease of debility made it easy to understand how famine could lead to fever, but it also highlighted the problem of distinguishing severe hunger from fever.[69] The problem was both conceptual and practical. Following Huxham, the great Robert Graves told his students that starvation and fever *could not be* reliably distinguished. After a few days without food, hunger would evolve into "epigastric tenderness, fever, and delirium." Long fasting and "the worst forms of typhus" produced identical symptoms: "pain of the stomach, epigastric tenderness, thirst, vomiting, determination of blood to the brain, effusion to the eyes, headache, sleeplessness, and, finally furious delirium."[70]

The term *famine fever* would refer both to occasion and, more rarely, to a clinical entity. In famine-ridden Skibereen in 1847, a reporter described "a lingering . . . fever, [characterized by] great prostration, thirst, a dry, chaffy, hot feel of the skin; a weak, feeble pulse; the intellect generally clear." Unique to the poor, it showed no particular stages and did not involve delirium. Recovery depended on food, which, unlike in the anorexia typical of fevers, was much sought. The fever was accompanied by dropsy and foul odor. Elsewhere, it probably would have registered as typhoid (though it was atypical in its lack of mental or abdominal involvement and the presence of dropsy), yet the reporter distinguishes it from typhoid and typhus, diseases that struck the better-off. In Ireland it was a product of the totality of depressing causes.[71]

Given that the fever hospital's mission was to prevent epidemics, and given the belief that hunger fevers could evolve into contagious fevers, the practical problem was how to determine when the economic-ethical problem of hunger became a medical problem, a hospitalizable condition or even a public-health emergency. Especially in the early stages, it would be impossible to tell whether a fever would be a brief, local, inflammatory fever or a dangerous contagious one.[72] By the time its dangerous nature became clear, it was usually too late to stop its spread; hence it was better to open the hospital to all febrile cases.[73]

Often cases were lengthy and complicated. Dublin's doctors did not expect to confront what Christison called a "pure" fever.[74] Concurrent and supervening conditions were the rule, not the exception. It is hardly surprisingly that "putrid" symptoms lived on in Ireland well after they had disappeared in other parts of western Europe. (They would continue to be admitted in tropical fevers—yellow fever and "pernicious" malaria—even after being written out of the European febrile experience.) The blackened and foul tongue was not merely coated, and if the great ulcerating purple blotches on swollen legs were not the textbook petechiae, they might yet be perversions of them.[75] Such conditions might warrant labels such as "purple" or "gastro-purpuric" fever or simply be designated "scorbutic." In Scotland and England, by contrast, "scurvy," which did sporadically appear on the margins of the Great Famine during the late 1840s, would be a discrete disease, perhaps with feverish complications.[76]

By the end of the century such anomalous fever would be impermissible. Cases would be regarded as atypical presentations of one of the known agent-defined fevers or perhaps dismissed as problematic. Civilized exanthem-specific diagnosis required bodies to express themselves in clear and distinct spots, like the petechiae of typhus or the rose-colored abdominal spots of typhoid. Hemorrhaging blotches didn't count.

Those changes would contribute to the eclipse of Irish medical authority. Ireland's fever record would be reinterpreted, with Irish fever medicine written out of the march of medical progress. The Irish had failed to properly police the key terms, *typhus* and *petechiae*. Blessed with widespread fever (or at least feverishness), they had squandered that clinical opportunity through their inability to distinguish signal from noise.[77] Bartlett extracted from their writings the few bits that met his needs. In 1862 Charles Murchison respectfully but pervasively reframed their achievement in an imposed settlement that would wind up the affairs of classical fever theory and effect its transition into something recognizably modern.

In fact, Irish doctors produced excellent clinical descriptions of typhus and other fevers, but they were contemptuous of a medicine that would substitute categorization for careful appraisal of each case. Hence, Ireland was, arguably, the last stronghold of classical medicine, also of Norman Jewson's "bedside medicine." But the unique social (and political) character of Irish fever medicine would not survive the famine era. Political, professional, and sectarian tensions over the administration of medical charities by poor-law officials split the medical community in the mid-1840s, with the Anglican Tory Graves distancing himself from the Catholic Whig Dominic Corrigan over the relation of poverty to fever.[78] But Irish perspectives would linger in British colonial medicine in the second half of the century, as we shall see.

CONSUMMATION AND SYNTHESIS

In Paris, one collected specific lesions of the ileum among the conveniently dead; in Dublin, one faced crowds of hungry persons whose bad colds might become a public danger. Could they be reconciled? Not without reopening the explanatory agenda. Mere symptom clustering had lowered expectations. Traditional questions, such as how a disease did its work or what began (or ended) it, had been too speculative, though they kept creeping in.

The first phase of reconciliation is better seen as conquest: William Jenner's *On the Identity or Non-Identity of Typhoid and Typhus Fevers* (1850) consolidated the typhus-typhoid distinction. Louis's recognition of a suite of features that constituted typhoid had remained controversial despite William Gerhard's investigations in Philadelphia, where the two species of dangerous continued fever had briefly coexisted, and Bartlett's overview.[79] But were the diseases essentially different or accidently so? Perhaps this typhoid was no more than Cullen's typhus localized in the bowel or an evolutionary form of it.[80] Or perhaps there were many forms. Whether the two low nervous fevers were species or varieties was not merely a matter of degree or persistence of differences, however; it was tied to a fundamental philosophical division between those for whom diseases were discrete species and those who emphasized a continuum of physiological malfunction.

Jenner's approach was not novel, though his vehemence was. Often it outstrips Bartlett's. Working at the London Fever Hospital, Jenner (1815–1898) was attacking the older perspective on its home turf.[81] While there was hardly a party line, eminent predecessors at that hospital—J. F. Bateman, John Armstrong, Alexander Tweedie, Thomas Southwood Smith—had sought a universalizing framework that would unite the fevers of the Americas and the tropics with the nervous fevers of Europe (and the symptomatic fevers of the surgeons). Some had been strident inflammationists, seeing fevers as contingent responses to local injuries. To Jenner, such exercises in analogy reeked of a medical past of forcing facts into preconceived structures.

Jenner would impose on London's fevers the new rigor of enumeration and dissection, the practices of Paris two decades earlier. Over roughly a hundred pages he applies the numerical method to 66 fatal cases (23 typhoid, 43 typhus) in the London Fever Hospital during 1848–49, beginning and ending the work with a few pages of argument. While he does not explain how the cases had been diagnosed, Jenner asserts that the question of "*essentially* different diseases" rests on correlating characteristic spots (petechiae or rose-colored abdominal spots) with the presence or absence of the particular lesion of the Peyer's patches. The distinction is here an axiom, but underwriting it is the exanthem analogy. Though each varies greatly in severity and though they have many symptoms in common, smallpox and scarletina are essentially different, Jenner declares: their exanthems are distinctive. So too with the nervous fevers.

The utility of the hitherto neglected exanthems to Jenner's argument is clear. It required a new hermeneutics of spots in fevers (rash judgments, literally and figuratively). Cullen had explicitly distinguished the signature rashes of smallpox and measles from petechiae, but he had grouped the latter with generic miliary eruptions (a key feature of the very dangerous "putrid" fevers of his day) and the changing hue of the skin and considered them incidental features.[82] Jenner's petechiae, by contrast, were the pathognomic typhus spots of the modern textbook, a key basis on which later writers would equate "spotted fever" with typhus. Whether Londoners' bodies failed to exhibit the palette of spots and splotches of many sizes and shapes, colors and textures, that Irish doctors saw is not clear, but Jenner would project the idea that fevered skin was easy to read.

Like Bartlett and Louis, Jenner could not say how the other element of correlation, the signature lesions of the intestine, related to the disease. Yet those lesions were not just incidental members of a cluster of symptoms but were instead to be a defining feature, an invariable "concomitant." Like earlier proponents of specificity, Jenner presses the claim that the distinction is vital because it may have therapeutic implications. "If [there are] two diseases, then the essential treatment *may be* totally different." He does not, however, say that it will be totally different or why that would be likely. Cullenian physicians, with their rest and wine, had been operating reasonably successfully on the assumption of a single, if varied, nervous fever. Indeed, in relegating symptoms to a secondary status and admitting that the characteristic lesion was not *necessarily* implicated in the disease process, Jenner undercut any rational therapeutics. He does hint, however, at what will become the most important implication of anatomical differentiation: that uniqueness of lesion implies uniqueness of cause.[83]

While the only *consistent* distinction between typhus and typhoid is the lesions on the Peyer's patches (the spots don't always occur), Jenner lists others, qualitative or quantitative. Typhoid is a longer-lasting disease, more common in younger persons. It may involve hemorrhage; during delirium the patient will be more mobile. While many symptoms are common to both diseases, they occur at different stages.

But why do these differences matter, and what makes them more significant than similarities? Did suites of consistently observed differences signify anything more than consistent variations within a single disease? More clearly than Louis, Jenner had raised expectations in elevating some differences over others. Spots and lesions would become proxies for all the rest. Jenner does not address the question of whether other variables, more important for a symptom-based therapeutics (e.g., the presence or absence of diarrhea, headache, or delirium), might have produced tighter and more useful discriminators. In fact, a mathematics to master such issues was well in the future.

What is plain is Jenner's fascination with the walls of the ileum. Over six pages he is a virtual Virgil of the lower upper bowel.

Approaching the ileo-caecal valve from above, one meets mounting signs of chaos in the typhoidic gut. At six feet out, about three-quarters of the way down the small intestine, the Peyer's patches are thickened and pinkish; at three feet out we find the "mucous membrane . . . destroyed at several spots, by as many minute ulcers, each the size of a pin's head." The lowest patch, three inches long, has multiple ulcers, their "floors formed of submucous cellular tissue of a deep yellow colour." The membrane itself is "purplish . . . and softened" with "minute pits." Probing such pits, Jenner finds "an opaque yellow structureless mass." On its floor are the "transverse muscular fibres of the intestine; to its [jagged] edges . . . were attached shreds of tough, opaque, deep yellow, sloughy looking matter."[84] Water injected into the mesenteric arteries pours from these holes. If these lesions were not the primary pathology of typhoid, such "perforation" seemed likely to account for the hemorrhaging in about a third of fatal cases.[85]

As an application of the numerical method, a sharp marshaling of evidence, Jenner's monograph remains powerful. The Irish response was a common motif of medical diatribe—"hotshot young careerist doc rushes into print; we elders know that art is long, life short; have seen in long career . . ." Further, Jenner generalized about living disease from cadavers, which told tales only of death. And based on *so few* cases![86] Ireland, however, was already fading; there would be no turning back after Jenner's achievement.

Charles Murchison (1830–1879), Jenner's successor at the London Fever Hospital and author of the repeatedly revised and much translated *Treatise on the Continued Fevers of Great Britain* (1862), would complete the paradigm shift. Murchison converted chaos into order by reclassifying, sometimes in cavalier fashion, the continued fevers of the previous century. It is he who, most conspicuously, first presents the modern attitude toward the medical past: not only had predecessors been wrong in theory but they had been unreliable observers, failing to recognize the diseases before them. Covering not only Britain but all temperate lands where civilized medical reporting existed, Murchison's work would be authoritative for the prebacteriological generation. As late as 1896, well into the bacteriological era, a new French edition appeared. And Murchison is likely with us still. An investigation of the genealogy of the short statements about a disease's history with which medical authors often begin journal articles would probably show that many of the retrospective identifications of early fever epidemics descend from him.[87]

Unlike Jenner, Murchison seems to have been oblivious to any possibility that his razing and reconstructing of the field of fevers might be controversial. In fact, he was fortunate in his timing: most who might have contested his approach were dead or had retired. Only 32 when the first edition of his *Treatise* appeared, Murchison had already supplemented his Edinburgh MD with study in Dublin and Paris, and had practiced and taught in India. His methods too helped disguise the revolutionary character of the book. Merely to apply the numerical method required radical reinterpretation as one translated qualitative complexity into countable units, ignoring what could not be made to fit. Murchison was also an astute consumer of physiological research, which he trusted would explain much pathological variation.[88]

Murchison accepted the distinction between typhus and typhoid but reopened the question of causation, which Louis, Bartlett, and Jenner had dodged. He assumed, as we do now, that essentially different diseases will have different (remote) causes. He expected too that, in contrast to the descriptive statistics on antecedents offered by Louis or Bartlett, recognition of (necessary) causes should imply some obvious preventive strategies, an increasingly salient issue in sanitation-obsessed Britain.[89] Murchison accepted what Irish (and Scottish) physicians had long claimed: that fever was preeminently associated with poverty. But unlike the Cullenians, who usually appealed to composite debilitation, coupled perhaps with a precipitating contagion, Murchison suggested that each form of fever reflected a particular aspect of poverty. One of the nervous fevers, typhoid, arose from inadequate disposal of excrement. Another, typhus, was associated with crowdedness and poor ventilation—here Murchison drew on Pringle, the black hole of Calcutta, and the heritage of jail fevers.

But following Irish and Scottish doctors who had been front-line monitors of famine epidemics, he added a third disease, "relapsing or famine fever." Here the proxy was hunger, though it did not follow that hunger, even for an extended period, led invariably to the disease. Dublin's doctors had occasionally noted a milder famine-related fever occurring in conjunction with the rare but dangerous "ordinary typhus of this country." It was a fever lasting five to nine days, with frequent jaundice, ending without crisis, but with as many as three relapses.[90]

This new disease entity would sharply reveal the clash between old and new, between the incommensurable ways of reading a disease in the early nineteenth century. Irish doctors had been accustomed to relapses. If fever resulted directly from poverty, relapses among discharged hospital patients, returning to their normal impoverished lives, were hardly surprising. "All the circumstances, which concur with contagion in producing fever, . . . are equally productive of relapse," John King Bracken of Waterford had noted in 1818. "Deep and progressive poverty, filth of persons and dwellings, and minds depressed and cheerless, soon caused many to return to the hospital."[91] That relapses sometimes occurred among those who were not impoverished was an anomaly, but, as Sydenham had shown, epidemics varied. Francis Barker and John Cheyne, chroniclers of the great Irish fever epidemic of 1816–18, acknowledged that relapses were common to "this variety of epidemic fever." But that did not imply making *relapsing fever* a noun. The designation of relapsing fever as a distinct disease would be credited to Scottish doctors describing the epidemic of 1843.[92]

In positing a new species of continuous fever, Murchison may seem to have been a proponent of specificity. But he had not so much left behind the Cullenian agenda as superimposed specificity onto older concepts of debility-based fevers: three fevers common among the poor, three pathological aspects of poverty—what could be simpler? But after Broussais's assault on its coherence, *debility* no longer adequately explained pathological processes. Increasingly, Murchison's standard was German laboratory science, and in Germany, as we will see, the unified concept of fever remained very much alive.

In appealing to the physiological research in German laboratories, Murchison was insisting that the task of etiology involved more than identifying an entity necessary to the production of the disease; one must explain *how* that entity induced the changes that constituted the illness. Cabanis, Louis, and, later, Bartlett had lowered expectations; those that Murchison reimposed were traditional, not merely Cullenian but Galenic. If less acutely perhaps, than Claude Bernard, whose famous *Introduction to the Study of Experimental Medicine* would appear three years later, Murchison sensed that reliance on statistical generalization meant surrendering precision.[93]

Three aspects of contemporary laboratory research particularly interested Murchison. The first was heat. While nothing in his first edition suggests how central temperature taking would become, Murchison did recognize excess heat as a common and important feature of fevers, if not quite a defining one. He took the modern line, discounting the subject's sense of hotness in favor of objective measurement: patients might *feel* cold but *be* hot. Second was the application of physiological chemistry to fever, the legacy of Antoine-Laurent Lavoisier, Justus von Liebig, and particularly Rudolph Virchow: excess heat indicated metabolic excess—a fevered body was indeed burning itself up. Third was the attribution of some common symptoms, particularly mental confusion and prostration, to the buildup of toxins. A consequence of rapid metabolism would be the overtaxing of the kidneys and the resulting generation of uremic toxicity. For Murchison this was a conclusion equally of clinic and laboratory. As a clinician he noted the common misdiagnosis of kidney failure as typhus or typhoid fever; as a consumer of chemistry he was making use of Alfred Vogel's findings of high levels of uremic poisoning in fever patients, a common cause of death or complications.[94] Thus rather than representing confusing factors of diagnosis, the "typhoid" states that appeared in many diseases should suggest pathological unity.

Murchison could only introduce these themes. He foresaw a laboratory science of fever pathology but could not yet give it sustained or substantive treatment. Even if typhoid, typhus, and relapsing fever were essentially different, much could be learned from what they shared with one another and sometimes with remarkably dissimilar diseases. But recognition of their pathological unity brought with it the complementary question of what made these laboriously differentiated fevers different?

The escape from the tautology of defining fevers in terms of putative unique agents, and distinguishing those agents in terms of the presumably distinct fevers they caused, was toxicology. Here Murchison was returning to Broussais and the heritage of inflammatory fever, if through the intermediary of his Edinburgh mentor, the pioneering toxicologist Robert Christison. Broussais had deemed irritation the universal initiator of pathological processes. There were many irritants, however, and he had grudgingly opened the door to specificity by speculating that the different exanthems might reflect the action of different irritants.[95]

Most of the putative external agents—contagia, miasms, the oft-used proxy *virus*—had been irremediably hypothetical. Poisons too had been for Boerhaave and for the followers of van Helmont more a logical category or a metaphor for an inaccessible something that accounted for the specific character of a disease than an isolatable entity. But irritant poisons were material entities, ideal for experimental study. Along with physiological research, experimental toxicology was popular in Paris. It never gained the prestige of clinical-anatomical correlation, even though the most prominent of the new toxicologists was the long-serving Dean of the Paris medical faculty, the Minorcan-Catalan émigré Mateu Orfila (1783–1853).[96]

While the French toxicologists initially focused on common inorganic poisons that might occur in an urban environment (e.g., hydrogen sulfide as a product of the decomposition of organic matter), they quickly expanded to exotic organic substances— snake venom, prussic acid (cyanide), and the active ingredient of nux vomica, strychnine, extracted in 1818.[97] The variety and mys-

tery of these poisons would strengthen their analogy to contagia
or other forms of pathogenic agents.

Inadvertently, Broussais had created a structure for conceiving
pathological processes in terms of a particular agent. Translated
back into pathology via the metaphorical bridge of irritation, the
new toxicology suggested that infectious fevers too resulted from
ingestion or inhalation of a distinct toxic substance. Beginning in
the 1840s there would be much talk of a disease's *materies morbi,*
or "morbid poison"; reference to such an entity would become
a common expectation of epidemiological accountability.[98] This
toxicological initiative, more than ancient speculations of *contagium vivum* or the placeholder concepts of pathological *seminaria*
(seeds of disease) that were somehow to explain disease specificity,
was the immediate precedent for agent-based disease definition.

Murchison too would find toxicology the most accessible
framework for thinking about agents, modes of access, and the
multiple forms of havoc committed within the body. His explanations of both typhoid and typhus were based on poisons. That
hunger did not always cause relapsing fever suggested to Murchison (as it had to Cullen) that its effects were indirect: it must be
facilitating the access or operation of some poison.

But Murchison remained fascinated with how poisons
worked. In failing to explain how an initiating gastric irritation
could bring about the panoply of diseases, Broussais had revealed
the powerlessness of *his* physiological medicine. But it did not follow that the project was fundamentally futile. It is for his attempt
at toxicological explanation, misunderstood by later generations,
that Murchison is best known. He is known as one of the last "mi-
asmatists" because he assumed that the "pythogenic," or decaying-
excrement toxin, the agent of typhoid, entered the body in the air
rather than being ingested.

In fact, route of access was less important than mode of action.
Typhoid would be an exemplar of the "zymotic" theory of pathology, of which Murchison was a major exponent: fecal molecules
experiencing a kinetically unique mode of putrefactive decomposition—a unique form of shaking themselves apart—would

induce that special mode of decomposition to susceptible tissues, in the case of typhoid chiefly those in the gut.[99] Suggested by the eminent German chemist Justus von Liebig (1803–1873), refined by others, and widely shared, the zymotic model was in many ways a throwback to metaphorical explanations in pathology. Poisoning had been a key site of such metaphors. Earlier writers had referred to the sharp particles of irritant poisons, which pricked one's insides. Zymosis was equally easy to imagine.

That compound orientation is reflected in Murchison's active involvement in both London's Epidemiological and Pathological Societies. Comprehensive explanation involved *both* the external and the internal. But usually it was much less clear how the new exotic toxins operated; applied to the phenomena of living bodies, the chemists' entities retained the taint of the occult. Ultimately, one might know no more about how uremic poisoning worked than about how a miasm or contagion worked or, for that matter, how age or sex were causally implicated in typhoid fever, but systematic experimentation could reveal the determinate relations of cause and effect. Only slowly would those be accepted as explanation enough; imagining flabby fibers, tiny needles, obstructing lentors, or shaking molecules would no longer be necessary.

Although physiologists and, later, biochemists and geneticists would pursue Murchison's ambitious agenda *sans* zymotic theory, others would reject it: for them, to correlate a clinical entity with an agent would be explanation enough. Understanding that agent's doings within the body was superfluous if one could stamp it out or block its access. And the taking of such imperious actions was becoming quite plausible during an age of imperialism.

CHAPTER SEVEN

Naming the Wild

❖ ❖ ❖

The story of nineteenth-century medicine is usually told as a story of the civilizing of disease. At century's end the consummation of bacteriology would finally vindicate the commitment to ontology, the boxing and naming of diseases, the legacy of Pierre Louis. That consummation was most striking in the "long-established category of fevers," noted the eminent medical historian William Coleman.[1] In Europe, it would be the source of science-based public-health institutions, which combined surveillance, rapid response, and infrastructural reform. While typhoid or smallpox might still occur with disturbing frequency, they represented reparable errors, matters of resource allocation rather than soul-searching.

Confident diagnosis was extraordinarily comforting—to diagnosing doctor, worried victim, and concerned society. Consider the smug Alexander Collie, a physician at the Homerton Fever Hospital in East London and author of the 1887 treatise *On Fevers: Their History, Etiology, Diagnosis, Prognosis, and Treatment.* Though no typhus agent had been discovered, Collie confidently elevates loose clinical categories—typhus as *simplex, gravior,* or *gravissimus*—into specific diseases.[2] The labels will reassure. Say one is diagnosed with *typhus gravior.* True, that's more serious than *typhus simplex,* but be glad that it is not the rare *typhus gravisimus,* from which few escape. Even today a label can transform ignorance into a semblance of knowledge and control. A single esoteric acronym, FUO (fever of unknown origin), tames all unclassifiable fevers.[3]

Usually, the civilizing of fevers is seen to spread from center to periphery, from European hospitals and laboratories to the colonized ends of the world. In the metropole, expertise is concentrated, and sick bodies may be isolated and analyzed. Then that order may be exported to the waiting world. Such was the case with anthrax, Bruno Latour explains: what was true in Pasteur's laboratory would be applied first to the fields of France and then to all beyond.[4] But Paris did not always conquer. Sent to Gibraltar to master the yellow fever rare in Europe, the great Louis was himself mastered. Symptom charting and postmortems resulted in nothing conclusive or applicable.[5]

For the wild resisted. Throughout the nineteenth century colonial doctors complain repeatedly that European knowledge doesn't work in their particular exotic places.[6] A few years before Collie, Sir Joseph Fayrer, the London-based overseer of Indian medical affairs, had sneered at the confident characterizations of typhoid, exemplar of European medical order: "I came out [to India] imbued . . . with a belief in the truth of the views of European pathologists, but Indian experience has compelled me to recognise that those views . . . are too exclusive, and quite inadequate."[7] Fayrer was representing a trope of tropical practice, touting one's irreducible local expertise even while asserting the uniqueness (and often the unimaginably horrific and deadly character) of tropical fevers. Ultimately medical authority was at issue. If the fevers of Jamaica or Bengal failed to conform to those of cold and damp northwestern Europe, perhaps what was being taught in Leiden, Edinburgh, or Paris was not a foundation for universal medical understanding.[8]

Eventually that protest would become the foundational motif of a "tropical" medicine. An alternative framework would emerge, one dictated by the ecology of vector species and the differential experience of race far more than by the regularities of hospital wards.[9] Tropical medicine would be more than a geographically designated specialty, however; it would be a practical and ideological component of imperialism.

Fevers were a significant problem for colonial administrators. There were garrisons to protect and indigenes or slaves who must get a harvest in. A malarial coolie with a spleen swollen to the size of two heads could still work, explained the pioneering malariologist and tropical-medicine institution builder Patrick Manson (1844–1922) in 1898, but one must beat such persons carefully lest the spleen rupture, the coolie die, and one be charged with manslaughter. (Nor, he adds, should a person with a swollen spleen be allowed to play cricket.) Fever also impeded settlement. Manson advised against settling or even camping where the natives had enlarged spleens.[10]

Knowing (and naming) the unique fever of each wild place was a form of flag planting; it signified mastery over that place. Such named fevers included "Levant fever," "Bukowina fever," and "Smyrna fever," as well as, from India, "jungle fever, Terai fever, Bengal fever, Deccan fever, Peshawur fever, Mysore fever, Moultan fever, Nagpore fever, Scinde fever, Arracan fever." There were also "Punkah fever" and, from America, "Chickahominy fever."[11] (The practice lives on in American military campaigns against new viral fevers like Ebola Sudan and Ebola Zaire, notes the evocator of "hot zones," Richard Preston.)[12]

But the expectation of specificity developed only slowly. Well into the nineteenth century a common view was that tropical fevers were accidents of climate and diet, exacerbations of a universal febrility. That view became increasingly untenable as newcomers pushed beyond ports and hill stations into the "interior," the hostile hearts of darkness.[13] The horrific fevers they met, with multiple hosts and complicated insect- or arthropod-facilitated life cycles, were not only fascinating scientific problems but called for an even harsher version of the white man's burden, a for-your-own-good approach to prophylaxis that reinforced the grim realism of imperial rule.

For wild fevers were not only a practical obstacle to colonial rule; they were also a moral justification for it. Malaria had "profoundly modified the world's history by tending to render the whole of the tropics comparatively unsuitable for the full development of civilization," declared Ronald Ross, winner of a Nobel Prize for his discovery of malaria's mosquito vector. But he was sure that that could be changed.[14] The paradigmatic achievements in tropical medicine—recognition and control of the means of transmitting yellow fever and malaria—were means of empire building as well. American children long learned to thank Walter Reed for the Panama Canal; a trope of American exceptionalism is that he mastered yellow fever when Spain and France had not. What better justification for ruling Cuba, long blamed as the chief exporter of that disease.[15]

A focus on rule usually led to a preoccupation with race. Recognition of differential racial susceptibility to local fevers was longstanding. The new tropical medicine incorporated it. Natives or long-term colonial residents might be distinguished by their relative immunity, or they might be represented as a residuum of survivors, diminished by local fevers and perhaps even unusually susceptible to renewed attacks.[16] *Race* might include socioeconomic factors and cultural practices. Fayrer blamed Indian fevers on conditions and vices, "early marriages, and sexual excesses and abuse," the effects of poor nutrition in an overpopulated land, as well as bad air.[17]

Concepts of race (and species) did help demystify exotic places. Heat and humidity do challenge feeble white bodies, Manson explains. A "seasoning" fever was often seen as a necessary physiological rite of passage to a region, but usually one's "machinery" could adapt. But these climatic discomforts, even when they included febrility, were not properly a "disease" and did not normally kill. Instead, it was the beings that threatened: you died because little native bastards wanted to kill you. Agent-centered conceptions of fever flourished in a world full of venom, of tiny blood-sucking fiends and hole-boring worms, which would get us if we did not get them first. Manson himself still saw the swampy environment as the immediate source of the malarial plasmodium; he believed it entered a body through inhalation or ingestion and returned to nature in the dying bodies of sated mosquitoes. Ross would correct him: the mosquito's bite both deposited and withdrew the plasmodium.

Recognition that the true reservoirs of infectivity were the bodies, sometimes symptom-free, of the inhabitants of malarious regions, and especially of the children in those regions, sharpened the racial divide: if what killed "us" was normal for "them," it was their ally.[18] Again, the tropes persist. In many places, notes Sonia Shah, malaria is not a significant problem for those who live with it; it is a crisis only for resource-exploiting newcomers.[19]

I focus first on the medical-advice literature for tropical travelers. I move then to essential tensions having to do with place and race in explaining tropical fevers and on to the key criterion in distinguishing civilization from chaos: whether fevers will be accidental, transitory "epidemics" or "endemic"—inherent to place. The chapter ends with the emergence of agent-based causality and the full-scale militarization of fever response in the First World War.

SAME OR DIFFERENT, SAFE OR DANGEROUS?

Demographic reconstructions by Philip Curtin and John McNeill remind us how extraordinarily deadly the tropics were for European settlers and especially for soldiers and sailors. The *annual* mortality rate of British troops in Sierra Leone during the years 1819–36 was 483 per 1,000. The annual mortality from fevers was 1.4 per 1,000 in home postings, compared with 101.9 per 1,000 in Jamaica, a difference of 7,300 percent. In the West Indies fevers accounted for 71 percent of soldiers' deaths, compared with 10 percent in Britain.[20] The Black Death would be Fayrer's reference point for the ordinary fevers of India, where deaths from fever (three times as many as deaths from *all other causes combined*) vastly overshadowed cholera deaths.[21]

That deadliness was not initially evident. Sporadic reports of deadly illnesses from sea rovers in the Indian Ocean, slavers in

West Africa, or plunderers in the Caribbean were more likely to be interpreted as tall tales or as isolated events than as general laws of tropical fever. Longtime settlement by observant practitioners might have allowed the distinguishing of extraordinary visitations from regular features of place, but there were few such practitioners, nor could commerce and conquest stop while the doctors compiled their notes. In the Western Hemisphere, the most formidable disease of place, yellow fever, would only acquire plague-like status in the 1790s, three centuries after European conquest. Thus, only retrospectively would it become an inherent aspect of Caribbean wildness.[22]

Theory might substitute while the empiricists organized data. An advice market in tropical medicine existed long before there was a specialty with that name. For English readers a key work was James Lind's *Essay on Diseases Incidental to Europeans in Hot Climates* (1768). Fever was the great menace. Lind confidently declared that 95 percent of Europeans who succumbed in such hot places died of "fevers and fluxes."[23] Lind's book would be succeeded by James Johnson's *The Influence of Tropical Climates on European Constitutions . . .* (1827), which in turn would be followed by James Ranald Martin's 1856 work with the same main title. The genre survived into the bacteriological era: Manson's *Tropical Diseases* is subtitled *A Manual of the Diseases of Warm Climates.* Usually, the authors of these works claimed experience of the tropics, but a single person's experience was partial and often brief. Military and naval surgeons pioneered in sharing knowledge, but comprehensive views would long be hampered by skepticism about claims from remote observers.[24]

For readers the chief issue was simple: "will I die if I go"? Books on staying healthy in the tropics were aimed at those who could choose. Many, McNeill believes, chose to stay home.[25] But authors applied their medical knowledge to lubricate fate, chance, or providence in order that prudent persons might decide that they could (relatively) safely follow the dictates of greed, adventure, or reasons of state. Often authors did so by making tropical fevers familiar. According to Henry Warren of Barbados, writing in 1740, yellow fever was just a milder plague, having "many Appearances, peculiar to the Plague itself; the Plague-boil and the Rapidity and Ferocity of the Symptoms only excepted."[26] Or it might be the lost Hippocratic *kaûsos,* a "*febris ardens biliosa.*"[27] Many sought to unite tropical and European fevers on a continuum of variation on the grounds that of course bodies would react differently to different climates.

To mitigate deadliness, authors suggested ways of coping, preventing, or merely managing apprehensions. True, "Strangers and New-comers" were especially at risk, Warren admitted, but from avoidable acts: the local fever came on "most commonly, 1st, After hard Drinking and sitting up late o'Nights, and then exposing the Body to the damp, chilly, Night Air; 2dly, Upon any too violent Exercises of the Body, by Labour, Walking, Running, Dancing, and the like, and then cooling in the Air, too suddenly, without sufficiently defending the open Pores." Or it might be deadlier in someone else's demographic. The "poor common Sailors . . . notoriously suffer most"; they "have perhaps less Conduct in their Way of living than any Sett of People in the World, drinking ever hard of the vilest and cheapest strong Liquors . . . then going off upon the Water with Breasts open, and their Bodies poorly covered." And, too, the docks where those sailors lingered would be prime sites for concentrating the island's contagious impurities.[28] Along the Bengal coast, rotting intertidal mud bred a deadly "fen fever." While it struck Europeans unused to tropical heat and weakened by shipboard food and crowdedness, ordinary sailors were particularly prone because of their long stints in wet clothes. Other causal factors were sleeping ashore, especially on the ground; excessive eating and drinking; and grief, fear, and fright. (The phases of the moon were significant too.) The fever could be contagious, however; officers might get it from sailors.[29]

Like Collie, these authors managed fear by breaking the danger down into units, most of which were really safe. Surely acclimatization would lessen the danger. And the knowledgeable settler could accentuate the positive. Settlers arriving in Arkansas and Missouri in the first half of the nineteenth century regarded themselves as sophisticated assessors of the healthfulness of places. Fever lore, rooted in classical medicine, was important in their assessment, as Conevery Bolton Valenčius has shown. Manson advised his readers to read the bodies of "natives" as well as landscapes and to be guided by their spleens as well as by their advice.[30]

FINDING THE GLOBAL FEVER EQUATION

Below I review four theorists who illustrate two of the many axes along which attempts were made to relate tropical fevers to European fevers. One such axis, with John Macculloch at one end and R. D. Lyons at the other, is the possibility of prophylaxis. Is there a deadly "malaria" inherent to place or are tropical fevers merely the result of reparable social conditions? The second, with Francis Boott's climatic hypothesis at one end and Daniel Drake's living-agent hypothesis at the other, concerns cause. Do tropical fevers reflect composite geographic force fields or incidental contact with hostile life forms? All four authors challenged the authority of Paris, with its hospital medicine. However detailed the observations, conclusions drawn from one's own cases—like those Jenner or Louis offered—were valid only for one time and place. Instead these authors sought not tidy boxes but laws of febrile variation.[31] Not only malaria but fevers known as *relapsing* and *remittent,* including what today we know as yellow fever, would be central to their inquiry.

Macculloch

"The Thames is not the Congo," but "the disease is the same, the poison the same."[32] To John Macculloch (1773–1835), writing in 1827, Europe too was tropical. On one's country estate or on a leisurely Continental sojourn one met the fever of West Africa or Bengal. Macculloch would apply the term *malaria* to the poison of that fever. He was following G. M. Lancisi (1654–1720), who had attributed both intermittents and continued fevers to marsh effluvia. But as an Italian commonplace, the term had highlighted a period of local danger more than a specific disease or its cause.

Though physician to a royal duke, Macculloch was an outsider in medicine and in geology, his main fields of activity. He was, however, chiefly responsible for popularizing the term *malaria*. His 1827 *Malaria* focused on a widely diffused atmospheric disease agent, the malarial poison. His succeeding two-volume *Essay on the Remittent and Intermittent Diseases* (1828) treated the multiple illnesses it caused, not only the familiar intermittent fevers but also neuralgia (including migraines), a nervous tic, dysentery, and cholera.[33] Not only was malaria "probably the chief . . . source of the most painful diseases to which mankind is subject", it was also the source of a vast array of subacute conditions, known "in the ordinary language of society [as] ill health" (and, later, as "masked malaria"). Macculloch considered it a travesty that these latter conditions were often dismissed as inherent weakness or malingering or ascribed to race, character, nerves, invalidism, or "any other convenient and fashionable cause."[34] While he was careful to reserve *malaria* for a poison rather than for the fever it caused, others were not. Arguably, *malaria* was the first etiology-based disease name, though it is difficult to say when it clearly became that. That the term would persist even after it became clear that swampy air was only a proxy for its cause suggests the primacy of the spatial (and colonial) over the clinical aspects of such diseases.[35]

While admitting the greater deadliness of tropical malaria, Macculloch wanted English readers to know that they were *not* safe. Not only did the poison scourge the tropics, it also caused 90 percent of fever deaths in temperate latitudes, he asserted. In the latter it simply killed more slowly: "To live a living death, to be cut off from more than half of even that life, to be placed in the midst of wealth and enjoyment, yet not to enjoy, such is the fate of man in the lands of Europe where Malaria holds its chief seat."[36]

Macculloch wrote to awaken readers to the fact that not just marshes reeking of putrefaction but most water and much soil in both woods and meadows harbored the poison. A long chapter treats "soils and situations less conspicuously productive of malaria or yet unsuspected of it." There were gradients of danger:

running water was safer than still; drained soils were generally better, but almost any place could be malarial. Too many courted danger in foolish pursuit of nature's beauty. The fad for water gardens had empoisoned Versailles, London's St. James's Park, and many a country estate. An evening "saunter among wet groves while the moon rises, listening to the nightingale" would expose one to the "true night air, the Malaria, the fever." Travel was even more hazardous. Italy's blue skies were the "canopy [of] a pest house"; there, to the knowing traveler, "the sweetest breeze of summer is attended by an unavoidable sense of fear." His message was the ubiquity of danger.[37]

More compiler-theorist than field investigator, Macculloch mined accounts of outbreaks for details to which he, as a geologist, was peculiarly sensitive: soil type, drainage patterns, the relation of valley axes to prevailing winds, the breadth of the intertidal zone, flooding patterns, changing riverscapes.[38] The payoff for the worried reader was a fifty-two-page guidebook to the dangers of malarial Europe (as for the Americas and the tropics, everyone knew their deadliness). Facing a common problem of travel writing, Macculloch struggles for superlatives to express the greater deadliness of each succeeding place. Italy and Greece occupy a third of the text, and as for Russia, there was (in Europe) no place "more pestiferous." It was as bad as Africa and Asia.[39] Fen dwellers knew they lived in a malarial place, and precautions were taken in many parts of Italy, yet most of Europe was in denial. The Dutch and particularly the French blamed the fever on foreigners and would not admit its seriousness.[40] Worst were English tourists, who were sure they were safe; in Rome, writes Macculloch, "no one but an Englishman sneers at Malaria."[41]

"Avoidance is prevention," Macculloch declares. But how was one to avoid a pervasive poison? Macculloch advised "change of habits; change of air, change of climate, change of every thing. . . . Thus is the disease sometimes extirpated." But acclimatization was untrustworthy; repeated exposure might simply make one more susceptible and the disease harder to cure.[42] As a natural theologian and Malthusian, he saw malaria as a providential check to overpopulation. Humanity would prosper only when a land had been well settled, drained, and dried, but some places, like parts of India and parts of the southern United States, might never be habitable.[43] Macculloch's paranoia is hard to miss, but so too is his prescience. He not only recognized many aspects of the habitat and behavior of plasmodium-bearing Anopheline mosquitoes but appreciated malaria's impact on populations.[44] Influential in the nineteenth century, his work would be respectfully cited even into the twentieth.[45]

Certainly the term *malaria* was astonishingly successful. After 1830 malaria became not only the reigning fever of the swamp-ridden tropics but, as the signature disease of disorder, the yardstick of uncivilization. This was ironic. For Galen, the intermittent fevers had been exemplars of order. But though plentiful in France, intermittents had been underrepresented in Paris medicine. They had resisted the anatomist's gaze, and even Broussais had puzzled to understand how gastrointestinal inflammation might register as periodicity. Already Cullen's followers were finding their paradigm in the new *typhus*.[46] The migration of power and civilization to northern and western capitals after 1800 and greater emphasis on urban over rural contributed to malaria's becoming the fever of nature, while typhoid would be the fever of civilization. That there might be typhoid in the tropics and malaria in Europe didn't matter. Typhoid, well understood (if clinically variable), would signify reparable error; malaria, the deadliness of exotic places. Likewise, it mattered little that there was a specific (quinine) for malaria and none for typhoid; if anything, the anomaly of a prodigious mass of a disease for which there was a known cure simply accentuated malaria's reputation as the antithesis of civilization. Nor did it matter that the civilized typhoid was associated with willful filth, while malaria was associated with the unspoiled wild. Evidently, when it came to rot, nature was filthier than any city. And tropical malaria was no respecter of the neat Italian categories of intermittents, declared Manson. The disease varied infinitely, and often quinine was ineffective.[47]

Boott

Macculloch did not concern himself with how a universal poison could produce different effects. However, the American expatriate Francis Boott (1792–1863), author of a painstaking review of fevers of the Atlantic Seaboard from South Carolina to Maine from roughly 1780 to 1820 and of recent fevers in Europe, did.[48] Boott agreed that malaria was a "very fertile source of disease," probably existing everywhere that was not sea or icecap, but he sought the geographic determinants of a universal fever continuum, chiefly a north-south axis that could unite tropical and temperate febrility while explaining regional differences.[49]

Boott's review appeared in an unusual venue, as a vindication appended to a two-volume memoir of the maverick inflammationist John Armstrong (1784–1829).[50] Armstrong, like James Currie, was a would-be medical reformer. He was popular in America as the "new Sydenham," and his views meshed with those of the great Benjamin Rush.[51] Rejecting Cullen's spasm-reaction model, he emphasized a "congestive" and inflammatory pathology in which debilitating causes dangerously concentrated blood in the brain. Prodigious bleeding might be required; in one case, "one hundred and eight ounces . . . before the violence . . . was subdued" (he reported that the fever ran a "mild course afterwards").[52]

Boott's great meta-analysis was a synthesis of military and naval sources and local practitioners' reports on prevailing and episodic fevers. He found that in the cooler North typhus predominated; further south, intermittents. But serving as a conceptual (and sometimes geographic) bridge was *remittent* fever. The term referred to fevers that intensified on a daily cycle without fully disappearing. (In a quotidian intermittent, by contrast, 16-hour bouts of fever alternated with 8-hour afebrile periods.)

Modern writers usually assume that "remittents" were compound malarial infections, yet remittency is a feature of typhoid and of many continuous fevers. For continuum theorists like Boott, the term *remittent* was useful precisely because it obscured the boundary between continuous and intermittent fevers. Always there had been intermittents that defied the expected schedule. As malaria became the signature fever of the wild, its variety and instability would often seem more prominent than its earlier regularity. His colleagues, Macculloch had complained, wished only to discuss diseases in "their more perfect forms," even though "ill-defined" cases were far more common.[53] A "well-marked tertian" might be distinct from a remittent or a continued fever, but there was gradation in between, Fayrer would insist: "In England] there may be little difficulty in diagnosing . . . in India there is often great difficulty."[54] Or victims might suffer from both. By midcentury, remittent fever would be complemented by an American nosological entity, "typhomalarial" fever. Manson, while objecting to the easy multiplication of species, accepted that typhoid might reawaken a latent malarial infection: patients might really have both diseases.[55]

In the Americas and India, a common variant of remittent fever was epidemic "bilious remittent fever," *bilious* referring to aspects of vomit and excreta. By the mid-nineteenth century some Atlantic bilious remittents were being recognized as a specific yellow fever. But to James Lind a half-century earlier, the late-stage symptoms of high jaundice and blackish vomit had simply been "accidental appearances in the common fever of the West Indies," which resembled the bilious remittent of India.[56] Even after the spate of Caribbean and Atlantic epidemics in the 1790s, the status of this fever remained ambiguous. Of the Philadelphia epidemic of 1794, Benjamin Rush had written that there had been "but one reigning disease . . . a bilious remitting or intermitting, and sometimes a yellow fever."[57] Another had written: "It may be called a simple remittent; a bilious remittent; a malignant bilious, or a yellow fever. They are only different grades of the same disease." A North Carolina practitioner quoted by Boott suggested the deeper hermeneutic problem: "In 1799, when we had a true yellow fever in Edenton, I saw the genuine black vomit . . . I do not believe I have ever seen *exactly* the same thing since." "*A* vomiting of black matter [was] no unfrequent occurrence," but it was not "*the* black vomit."[58] That this yellow fever occurred as epidemics in port towns, while other malarial fevers were endemic and rural, was no reason to separate them: it was presumed that a disease's presentation would vary according to environmental conditions.

Many of Boott's sources shared his view of continuum and annual variation. Fevers were "mongrel" diseases requiring "mongrel" cures. They occurred mainly in late summer and fall. The reigning character of each year's autumnal fever dictated the character of winter and spring fevers, à la Sydenham's constitutions. Heat was the primary generator, moisture the modifier; wetness promoted intermittency, while dryness promoted remittency. Both affected the generation of the poison from rotting filth and the susceptibility of recipient bodies.[59] Yellow fever's unusual ferocity might stem from its urbanity, noted one of Boott's sources; somehow "confined atmosphere, the heated walls and pavement, the dirt of the streets, alleys, common-sewers, docks, and manufactories" led to a powerful disease. Yet this yellow fever was only "a higher degree of the common autumnal bilious fever of our country," noted another; "the disease producing principle, . . . I consider the same."[60]

Lyons

Long associated with the intermittents, remittent fever carried one lane of traffic over the bridge linking tropical (malarial) to European (nervous) fevers. After midcentury, Murchison's relapsing fever, a periodic nervous fever often accompanied by jaundice, carried the other lane. Macculloch had represented Europe as tropical; a half-century later the Anglo-Indian surgeon R. T. Lyons (1834–1903), author of *A Treatise on Relapsing or Famine Fever* (1872), represented tropical fevers as European.[61] Besides the more common intermittents, ancient authors had recognized fevers with longer periods—*quintans, septans,* and *nonans,* for fevers recurring every fifth, seventh, and ninth day, respectively. Relapsing fever had an afebrile period of roughly a week. Since regular relapse was the flip side of periodic paroxysm, it could be seen simply as a long-period intermittent. (Yellow fever, which had a major midcourse mitigation before a deadly relapse, could also be included in the pattern of intermittency.) Unrecognized in 1840, relapsing fever would be a major Asian fever by 1900. Presumably it had been present earlier but had not been distinguished from tropical intermittents.[62]

For Macculloch, *malaria* had explained everything. For Lyons it explained nothing. He viewed the term as a fatalistic fetish of tropical exceptionalism and an excuse for colonial irresponsibility. For the young Lyons, a jail surgeon in the Indian Medical Service and an admirer of Charles Murchison, what governments did in Europe to prevent relapsing fever they could and should do in India. There were ample records of famine-associated fevers in famine-ridden India. But these diseases varied in character, and Lyons's "famine fever," like Macculloch's malaria and, increasingly, like typhoid, would be more an etiological than a clinical entity. "It may be intermittent, remittent, or continued," he observed.[63] British and Indian doctors had given it many different names, while missing its link to hunger. Jaundice, which for Murchison was merely an incidental symptom, would come closest to a pathognomic for Lyons and plausibly linked the disease to yellow fever.[64] In fact, however, his collection of outbreaks, most of them campaign- or jail-related, included not only fevers without jaundice or relapse but also diseases without fever, as well as occasional exanthems.[65] It comprehended most of what was being called *malaria*. But the inclusivity is understandable: there was no obvious boundary between "recurrent" and "intermittent," and Manson too would emphasize the same social triggers to account for the conversion of latent to active malaria.[66]

Lyons's antipathy to so-called malaria was both ethical and epistemic. On grounds that "the human race is essentially the same," he objected to the racist implications of uniquely tropical fevers, especially when such concepts were arbitrary and untestable. "The malaria theory" required ascribing fevers "to something unknown." It was word pretending to be thing, a scientized version of fate, providence, or atmospheric constitution.[67] That both humid Bengal and the dry Punjab could produce the same harmful air stretched credulity. Lyons allied himself to medical progress. Just as the clinically defined typhoid of Paris had superseded the loosely defined Cullenian *typhus,* so *malaria,* placeholder of ignorance, must yield to Murchison's relapsing fever.[68]

In equating (and effectively replacing) the Irish "famine fever" with the apolitical "relapsing fever," Murchison had blunted the question of culpability. Convinced that nosology must guide prevention, Lyons resharpened it: a contagious disease generated by hunger (chiefly among institutionalized persons) was plainly a state responsibility. The disease responded quickly to improved military and penal conditions. It had been chased from Punjab jails once the "malarious etiology" had been given up and "homely plans" for better provisions implemented.[69] Extra-institutional sources of the disease, like debt peonage (usury with 75% interest rates), were much less tractable.[70]

Like Macculloch and like other epidemiologists of his day, Lyons was beginning to define disease in terms of cause and using evidence of prevention (and cure) to identify cause. Though he insisted that relapsing fever was not merely hunger or prolonged malnutrition, the fact that more and better food cured and prevented it was a key foundation of its integrity as a disease concept. We are apt to see in his broad "relapsing fever" symptoms of scurvy, beriberi, or generic malnutrition, listed either as features of the disease or as common sequelae. Thus Lyons notes limb and joint pain; hemorrhaging (nose, gums, bowels, etc.); gums "swollen and spongy"; numbness with "tottering gait"; edema in limbs and a general dropsy; ophthalmia; dysentery; and cardiac problems leading to syncope.[71] There being no complementary category of deficiency diseases, a critic's suggestion that many of Lyons's cases were better understood as the Sinhalese *beriberi* effectively rendered them as mysterious as ascribing them to "malaria."[72] Lyons simply argued that prolonged hunger had transformed a mild disease in Europe (2% mortality) into a sometimes deadly one in India (53% mortality in one outbreak).[73] And if *relapsing fever* comprehended too much variability, *malaria* did too.

Lyons's reinterpretation of malaria and his critique of imperial administration were not welcome. Though his was the first systematic work to recognize a large and underappreciated Asian disease, most later writers preferred the more modern epidemiology of his Bombay (Mumbai) colleague Vandyke Carter, which was narrower but rigorous.[74] Manson ignores Lyons's book, while to Charles Creighton, pioneer of social explanations of diseases, it was an eye-opener.[75]

Drake

In the interior of North America, in the densely wooded bottomlands of the Mississippi basin and the Great Lakes, Daniel Drake too was concerned about the causes of fevers. His project was Boott's shifted west a thousand miles, but his compilation reflected as well his own long experience of the region. Trained in Philadelphia in the pre-Paris generation of American medicine, Drake spent most of his career in Cincinnati, where he was a leader in medical education and publishing.[76] A primary goal of his great compendium, *A Systematic Treatise Historical, Etiological, and Practical of the Principal Diseases of the Interior Valley* (1850), was to map the boundary between the intermittents of the South and the continuous fever of the North in order to understand the mysterious "bilious remittent." He too confronted the problem of whether there was one fever or many. The gap between tertian and remittent was as great as that between measles and scarletina, he held, but the former pair were united by periodicity and one often evolved into the other.[77] In the South, *autumnal fever*, sometimes known as "malaria" and correlated with biomass, water, and heat, was "the *great* cause of mortality, or infirmity of constitution." But as one traveled north that fever gradually ceded to "typhus" (or, more likely, to typhoid, also known as *autumnal fever*). Except along the southern shores of the Great Lakes, that southern fever petered out in the northern Midwest, and there was none above the 47th parallel.[78]

But unlike Boott, rather than seeking a general law of variation to account for these changes, Drake, reasoning from a mix of analysis and analogy, posited unique animate agents. In principle, endemic fevers might be due to meteorological factors, to

some "malarial" product of rotting vegetation, or to a minute living agent. The products of decomposition were indeed toxic, but
none caused autumnal fever. The regularity of atmospheric processes and vegetative decay were inconsistent with fever's variability, thus Drake tentatively ruled them out. As for living agents,
most were not harmful, but those that were harmful caused a distinct form of harm, just as diseases were distinct.

Drake's analogies were the dangerous fauna of the American
South—rattlesnakes, wasps and bees, the sandflies of the Gulf,
and gnats. For his was not a generic germ or animalcular theory
of disease causation but a tropical one: "We must not forget the
fact that nearly all the animals and plants which secrete a poisonous fluid, grow in the southern regions," he noted. "We may,
analogically, suppose that the microscopic beings in those regions
are more pernicious than those of higher latitudes."

The hot, wet rot of the South nurtured these tiny beings.
Drake was not departing from the linkage between climate and
fever but merely using habitat variability to give it a mechanism.
However ad hoc, the model fit the spatial and temporal variability
of autumnal fevers. A digitized poison accommodated the sporadic appearance of disease, for even in their well-defined habitat
the tiny predators were not present everywhere. It also accounted
for the varying character of fevers. Just as different species produced different chemical products, so too one type might cause
intermittents, another remittents. Immunity might be a process
akin to acclimation to insect bites. Drake doubted that the malign
agents would ever be isolated or eradicated, but their effects could
be mitigated. After the dangerous first turning of the soil, cultiva-
tion would bring health.[79]

While the views of these authors may seem incompatible, all were
ingredients in the incipient tropical medicine outlined in Joseph
Fayrer's 1882 Croonian Lectures, titled *The Climate and Fevers of
India.* The well-connected Fayrer (1824–1907) had come to rule
Indian medical affairs after a stellar career in the Indian Medical
Service from 1850 to 1872.[80] In his lectures, Fayrer highlighted

the recognition by Armstrong, Boott, and Macculloch of a malarial poison variable in its effects; considered the role of hunger
in Indian fevers (though he too rejected Lyons's claim of pervasive relapsing fever); and addressed Drake's problem of defining a
boundary between malaria and typhoid. India's fevers resembled
those of the hot American South, and for Fayrer too climate centrism was hardly antithetical to agent-based explanations.[81]

Fayrer was also fascinated by the search for the malarial agent.
The Philadelphian J. K. Mitchell had posited a cryptogam in 1859.
Two decades later Edwin Klebs and Corrado Tommasi-Crudelli
had found a bacillus pervading the bodies of malaria victims in
the Italian Campagna and now Alphonse Laveran had found fila-
mentous entities in malarial blood. Fayrer rightly sensed being on
"the threshold of the discovery of unknown and almost unsuspected disease-causes." But none of these hypothesized entities
had produced malaria in an animal subject, he noted.[82] And he
doubted that the discovery of an agent would have clinical utility,
for he still questioned the twin expectations that clinical regularity implied common cause and that a common agent implied
a common course. In Fayrer's view, Louis's greatest achievement
had not passed the test of Indian epidemiology. Coexisting with
the European typhoid was an Indian enteric fever, having typhoid
symptoms but lacking the distinct intestinal lesions and rose-
colored spots and having no association with filth.[83]

Manson, writing two decades later, on the other side of the
etiological revolution and after recognition of the mosquito vector, would admit all this variability but would put his confidence
in the microscope. Surely the ability to discover plasmodia in the
blood of fever victims would clear up the confusion. But he too
would be overly sanguine, for the plasmodium proved elusive. It
might hide in remote parts of the body, and finding it was a matter of skill, perseverance, and luck.[84] Even when the new science
failed to provide certainty, it might still bring an authority to imperial rule, one based on the process of science rather than on its
particular achievements.

EPIDEMIOLOGY TO THE RESCUE?

These authors, Macculloch, Boott, Lyons, and Drake, are recognizable as epidemiologists. They sought patterns in order to tame the wildness of disease, intellectually if not practically. If it was not yet a discipline, epidemiology and its companion, vital statistics, were becoming increasingly important progressive practices in the mid-nineteenth century. Their ascent reflected rising public expectations and the reciprocity of state growth with the growth of medical authority.

The simplest patterns were those of epidemics themselves, the temporary visit of a particular fever to a well-bounded space for a discrete period of time. Just as epidemics signified the radical loss of normality, their absence correspondingly signified normality, a state of health and civilized order. Samuel Kline Cohn notes that in the Italian city-states it was important that the dates both of the beginning of a plague epidemic and of the "liberation" of the city from it be clearly marked; they were public events.[85] Whereas plague history is a history of its epidemics, prior to the nineteenth century fever writers had been more interested in individual diseases or in the regional maladies of season. The new epidemiology of the nineteenth century arose in conjunction with new visitations, of cholera and yellow fever.

Unlike modern epidemiology, in which mathematics and computing power make it possible to explore the overall effect on health of multiple interacting variables, much nineteenth-century epidemiology was tied to the central assumptions of the new doctrine of specificity. No longer could fevers be amorphous, evolving, and individual. If patterns were to be discerned, practitioners everywhere would have to use a common diagnostic vocabulary. Then, by tracking activities and movements and connecting the dots, it would be possible to show the origins of each person's disease and thus to make these diseases fully explicable. But to find the pattern, one had to believe that it was there to be found among cases of a real disease. Yet "secure case identification [was] rarely an easy accomplishment," notes Coleman.[86]

A well-understood epidemic, well responded to, was a banner achievement of civilization. But often diseases defied order. They struck without warning, failed to conform to type, or jumped quarantines. In the nineteenth century, one front line in the war against wild disease was the Atlantic Seaboard, where tropical yellow fever, America's plague, occasionally ranged north into the domain of civilization. There was no reliable cure. A third of those infected might die, and gruesomely. At the end of the century, even the confident Manson was advising flight.[87]

To Boott's sources, including Benjamin Rush, yellow fever had been a deadly variant of seasonal fever, but by midcentury it was more commonly the specific epidemic disease that Carlos Finlay and Walter Reed would later explicate. Exacerbating confusion, however, was that fever's failure to conform to either of the main alternative etiological frameworks, communicability from infected persons or direct generation from a critical state of the environment.[88] Only later would the mosquito model unite the two. The two epidemics discussed below, both mixtures of farce and tragedy, highlight the importance and the difficulty of imposing epidemiological order. There is irony: in the first, the order may be illusory; in the second, important insights arise only after the abandonment of expectations of order.

William Tully's studies of yellow fever in the port towns of Knowles Landing (1796) and Middletown, Connecticut (1820), are masterpieces of village epidemiology—epidemiology on so small a scale that the facts could clearly speak.[89] Tully (1785–1859) belonged to the pre-Paris generation of home-educated American physicians. At issue in his retrospective reconstructions was his authority in the local medical community. It is as a medical scientist, elucidator of the long controversy over the contagiousness of yellow fever, that he will distinguish himself from fellow practitioners.

The Knowles' Landing epidemic seems to exemplify contagious transmission. A crew member on a local brig returning from the West Indies dies en route, perhaps of yellow fever. Within days of the brig's docking, the fever attacks two ship cleaners who have handled a sail on which the sick sailor died, a crew member's wife, and a woman who washed sailors' clothing. All die, as do two family members of one of the victims, one of whom had mended sailors' clothing. A few others in that family and persons in contact with the ship become ill, but no one else. Tully's sources, witnesses whose records and memories he has consulted, have traced each case. One can almost see the disease move, Tully suggests.

The Middletown epidemic was more complicated. There were multiple ships, but most cases followed the arrival of the *Sea-Island* from Cuba with sickness aboard (earlier there had been deaths from fever). In mid-June 1820 yellow fever struck the Saybrook, Connecticut, customs officer and his son, who had boarded the ship. In Middletown, where it docked, "there was the freest communication" with the shore. The next cases (17 June) were a mate of another ship from the Caribbean who had visited the *Sea-Island* and a woman employed in cleaning sailors' clothes who had been visited by the captain and an invalid crewman. Next (19 June) was John Wild, associated with a cotton mill several miles upstream. Wild had visited another ship from the Caribbean and perhaps the *Sea-Island* too (Tully says that there were "strong circumstantial reasons, . . . [though] no absolute proof"). Three young women at Wild's factory then took the disease. Of the ninth case (22 June), Tully had to confess that "Mrs. Child was not known to have had any connexion with the *Sea-Island*," but she lived near its dock, and there had been, after all, "free communication." Nor could he explain the twelfth case, Simmons (27 June), but he lived close both to Child and to the second case, Captain Vail, whose yellow fever probably had a separate origin. Also nearby was the thirteenth case, Cotton, who had visited Child during her illness. The fourteenth and fifteenth cases (4 and 6 July) were keepers of lodging houses where *Sea-Island* crew had boarded.[90]

Here marshaled facts were to transcend argument: the conclusion that yellow fever was as specific and contagious as any febrile disease would be "incontrovertible, and irresistible." Representing himself as responsible for "management" of the epidemic, Tully reviews the courses and treatments of individual cases with even greater thoroughness. His epidemiological insights had led to a therapeutic plan that he claims stopped many cases of the fever in the early stages.[91]

But Tully had found the pattern he wanted to find. Concentrating on contacts, he had ignored any alternative explanations involving urban filth or meteorological aberration, both staples of the explanatory repertoire. He had defined the epidemic temporally, based on the arrival of the ships, and assumed that yellow fever was a distinct disease, reliably diagnosed. Yet, notes Coleman, even a quarter-century later there was no single diagnostic indicator, and establishing the duration of yellow fever epidemics remained problematic.[92] Tully was bootstrapping; by seeking connections between cases of what *might* be a specific disease, he hoped to establish its specificity. The relations between the cases indeed suggest contagion, but as Tully occasionally admits, in places like Middletown there were too many contacts.

Thus a hostile reviewer in the *North American Review* wondered whether there had even been an epidemic. Tully had implied that he, along with his associate, Miner, had been in charge. But many practitioners had treated cases, diagnosing them variously, not always linking them to one another or to the ships.[93] Tully's imagining of an epidemic that he alone could see and conquer was self-aggrandizement, according to the reviewer. Anticipating such doubters, Tully had warned readers that commercial interests would distort his findings, mislabeling yellow fever as the "common remittent" or some such to deprive it of the "terror it ought to produce."[94] In fact, the reviewer accepts the probability of importation while doubting contagion and specificity. He suspects a "sick ship": putrid matter in the hold has generated a local "miasma."[95] The reviewer's main objection is that Tully's case tracing was actually fact selection, the form rather than the substance of authority.

Tully seeks to order Middletown's fevers by converting them from normal background into epidemic crisis and then reassures his readers that a knowledgeable expert, himself, has matters

well in hand. Similar issues arose in the much larger yellow fever epidemic at Norfolk and Portsmouth, Virginia, in 1855. Was
the episode utterly aberrant or comprehensible and therefore
somehow normal? There, however, doctors' narratives of the epidemic would be marginal. In these towns, overwhelmed by the
epidemic, scientific accountability would matter much less than
moral accountability. Against a foe so incomprehensible and unmanageable, defeat was inevitable. Sacrifice might be celebrated,
but not prudent management or careful analysis. But just as Tully's confident ordering of Middletown's fevers had been contested,
Norfolk's representation of heroism in the face of chaos would
have unforeseen implications. It would help to sharpen a division
between the civilized North and the wild tropical South.[96]

From early July to October, yellow fever killed around 3,000
people in Norfolk and Portsmouth. By mid-August, flight (mostly
of whites) had lowered the pre-epidemic population of 26,000
in the two cities to around 14,000. Most who stayed caught the
fever, though mortality rates were much lower among the 6,000
or so blacks—5–8 percent compared with 35–42 percent.[97] The
etiological options were familiar: yellow fever might be locally
generated, an accident of importation, or some combination of
the two. Doctors disagreed, but respectfully. Most viewed the origin, or at least the spark, of the epidemic as the arrival of the *Ben
Franklin* from St. Thomas with yellow fever aboard (though the
captain denied this). Without adequate quarantine, the ship had
been allowed to discharge ballast, and locals had boarded the ship
to repair the engines. Though Portsmouth's leading physician believed that yellow fever was locally generated, there was plenty of
middle ground: an imported entity, meteorological conditions,
filth, and poor drainage might all be complicit, and the season
was right for fever.[98] But however much towns or region might be
susceptible, the epidemic itself was aberrant: there had been no
yellow fever since 1826.[99]

There was in fact considerable recognition of local transmission. Yellow fever had moved from the docks where the ship was
being repaired into Portsmouth and across the river into Norfolk.

Norfolk investigators recognized a northeasterly movement of the fever of 40–60 yards per day. But once cordons had failed to stop it, interest lapsed in accounting for its spread. The majority view, that the wind carried whatever malign entity the *Ben Franklin* had introduced, had multiple advantages: it exonerated the area and, while acknowledging communicability, suggested that surrounding towns had nothing to fear from refugees.[100]

But that knowledge had translated into neither prevention nor cure. Instead, the emphasis was on simply coping. To have stayed and tried to help was warrant enough for saintly commemoration. Victims were the "fallen"; order came not from epidemiology but as medals, testimonials, and obelisks. Such concerns dominate the several retrospective works on the epidemic—the reports of the relief committees in each town, which had coordinated what response there was; William Forrest's *The Great Pestilence in Virginia;* and, a partial exception, Rev. George Dod Armstrong's *The Summer of the Pestilence.* For unlike the others, Armstrong had allowed the epidemic little tutelary status; he had simply witnessed. Though the epidemic also occasioned medical and religious treatments, both are secondary to the battlefield genre of remembrance: Forrest sandwiches a brief discussion of fever theory between commemorative poetry and plans for Norfolk's yellow fever memorial.[101]

These works addressed a national audience. The press had made the Norfolk-Portsmouth epidemic a media event, and beginning in mid-August local relief committees were overwhelmed with unsolicited contributions from almost every state—money, goods and services (coffins were wanted)—and by medical volunteers with varied motives.[102] Accountability became important. Donors wanted to know that they had helped to restore order. At stake was (white) civil order itself. However horrible things had been, it was important to know that Norfolk had not become Caribbean.

But as in the contemporary Crimean War, to which the epidemic was compared, the concomitant of heroic sacrifice was incompetence and chaos.[103] The disaster was societal and institutional as well as natural. The most patent failure of account-ability was in accounting itself. Both stately Norfolk and artisanal Portsmouth received far more money than they could effectively spend; still, despite sky-high prices due to the suspension of trade, Norfolk wanted cash, not provisions. And while Norfolk's relief committee was meticulous in listing income, it could not account for outgo. The deaths of community leaders confused things, but a postepidemic reporter held that the "correct amount of receipts" was rightly a low priority amid the "poverty and suffering." By contrast, Portsmouth's bookkeeping is an auditor's delight, de-tailed down to cigar expenditures. The city returned a large sur-plus to donors.[104]

There were also hints of anarchy, crime, and opportunism. Not only did food prices spike but so did the cost of necessary labor—nursing, transporting patients, digging graves.[105] Some worried that volunteer nurses had come to "plunder."[106] Fearful people hid indoors, shops closed, churches canceled services, and newspapers ceased publication. Doctors, clergy, and undertakers still roamed, but there were no funerals, and bodies were lost.[107] Equally upsetting was flight. Authors tread carefully, but within the predominant military motif flight was desertion: nurses were needed to replace "run away" family members. Armstrong ini-tially treats flight as "unmanly" and "unchristian" and notes that clergy and doctors stayed. (It was not, apparently, unwomanly: aside from a few Sisters of Charity, there are conspicuously few heroines in these accounts; male nursing is emphasized.) But after the deaths of his wife, his daughter, and other household mem-bers, Armstrong relents. Had all stayed, there would have been more deaths, and more stricken than could be cared for. "Panic" had been providential—"God's means for scattering them that they might be saved." Finally, Armstrong leaves too; there is little a convalescent can do. And in Baltimore and Richmond it seemed obvious that the towns should be evacuated.[108]

Nor do authors admit the epidemic's real and potential racial impact. For persons of color did not leave. On Portsmouth's main street, Armstrong met "but one white person." He explains this as a matter of poverty but also as a rational choice: the presumed differential immunity meant that blacks had, "comparatively, very little to fear."[109] Reassuring readers that the social order was well, other authors include token male slaves among the epidemic's heroes. John Jones, a hearse driver, could heft a coffin singlehandedly and seems to be always on duty. "Uncle" Bob Butt, the slave leader of the grave-digging crew, copes with up to eighty deaths per day.[110] The image of common sacrifice is misleading. The epidemic afforded rare opportunities to participate in a labor market.[111] No one mentions slaves left masterless or recalls that in 1802 yellow fever had ended white rule in Haiti.[112] (There was, however, no corresponding delicacy about Irish slum dwellers, who embodied "intemperance, poverty, and filth" and were the "proper food for any such disease.")[113]

Nor could the peculiar horrors of yellow fever itself be wholly suppressed. It violated genteel ways of dying. Notably, Forrest's sole example of wild delirium is "Bill, the well-known cake-boy (colored)." Bill escapes from the hospital, but his noisy delirium is "disturbing to some of the citizens." Recaptured, he is tied to his hospital bed but ends up on the floor, "where the writer noticed him in the agonies of death."[114] Surely no white person would have died in so undignified a way is the insinuation (though of course, as honest Armstrong admits, they did).[115]

What the Virginians saw as valiant failure in the face of an inconceivable disaster was to others simply a record of ineptitude. Recognizing the obligation of accountability to its own citizenry, Philadelphia's relief committee published selections from its frustrated correspondence with Norfolk. Would someone at least acknowledge receipt of the many checks they had sent? And say what was needed? (Hearing nothing, the Philadelphians sent ice cream, which they thought would be good for convalescents.) Surely the epidemic could be stopped by sanitary measures. Philadelphia, "seat and fountain head of Medical Science," would explain exactly how and supply the means. Conspicuously appended to the Philadelphia relief committee's report were Philadelphia's own ample regulations for responding to yellow fever.[116] (In Norfolk,

Yellow Fever Medal, presented to the Portsmouth Naval Hospital by the town council of Portsmouth, Virginia, 1856.

Library of Congress.

by contrast, the most conspicuous sanitary action would be the vigilante burning of a tenement, less to avoid contagion than to prevent the landlord from renting to the wretched Irish.)[117]

That Armstrong's and Forrest's books and the Norfolk report were published in Philadelphia suggests the importance of Philadelphia's verdict. But Norfolk's reputation did not recover, notes David Goldfield.[118] Norfolkers might claim that it wouldn't happen again, but yellow fever was a part of southern life. The Philadelphians were well aware that it might reach them—there were rumors of its new northward march—but refused to see it as normal to their region.[119]

The Virginians themselves helped create the division. Citing concerns about nonacclimation, they began politely rejecting Philadelphia medical volunteers in early September. While telling the northerners that they had doctors enough, they were soliciting Charleston for more.[120] Norfolk and Portsmouth might not be the Caribbean, but nor were they part of the realm of epidemic order.

THE REALM OF THE ENDEMIC

After 1855 the border between the controlled North and the wild, fevered South lay somewhere between Norfolk and Baltimore. But such lines were being drawn all over the world. Most dramatic was the splitting of malarial Italy. Macculloch had represented Italy as tropical. The progressives who took over after unification in 1867 would seek to rescue it, making malaria control central in their efforts to transform Italy into a viable European state and in the process making Rome into a world center of malaria research. But they succeeded only in sharpening a divide between the modern north and the malarial south, where the barriers to improvement were cultural as well as biotic. In some regions malaria, it was noted, was "not a disease but a state inherent in the lot of the peasant."[121]

As a complement to *epidemic* the term *endemic* captures situations in which fever was not an interruption but rather normality itself. If *epidemic* suggests a pattern, *endemic* suggests its absence, a monotonous perpetual random morbidity. Any discreteness is in regard to place or its proxy, race, rather than time. Hence, as in parts of southern Italy, endemic fever might be unappreciable to those living where it occurred. Randomness need not imply evenness of incidence. In the Caribbean, notes Alan Bewell, yellow fever *epidemics were endemic.*[122] Nor does it imply low morbidity and mortality: endemic places were not necessarily healthier than those prone to epidemics. Epidemic and endemic reflect different modes of apprehending disease, however. The advance of an epidemic often produced terror, the fear that normal life, perhaps life itself, would soon disappear. If endemic diseases were noticed at all, the response was not terror but fatalism.

If yellow fever was the great invader, malaria was the archetypal fever of place. Today, malaria is regarded as a major cause of mortality, especially among children. Late nineteenth-century and early twentieth-century commentators were more apt to see it as a debilitating and dehumanizing disease. Like Macculloch, some had a keen sense that much of its morbidity and mortality would be indirect, operating through complex cascades of cause and effect. Among the key impediments to responding to malaria was malaria itself, for the exhaustion it produced eroded the ability to act.[123] James's bourgeois Londoners expected fever to be transitory; his powder expedited their return to normal "business." But the malarial parasites hid deep within the body and were regularly reinforced by reinfection. Their periodic bursting from red blood cells left a sticky mess that clogged blood vessels and impeded circulation—the old speculations of Hoffmann and Bellini given new life. The result, notes Shah, is a "mildly hobbled human."[124] And, because malarial invasion often began in early childhood, it was easy to infer congenital, developmental, and even racial effects.

The author of the Hippocratic text *Airs, Waters, and Places* was certainly aware of lands of the spleen-swollen, but regional human diversity had signified mainly as natural history, perhaps useful to the itinerant healer. Likewise, aguish places and their peoples would later be recognized, but ague rarely stood out above the many other aspects of regional color. The tinge of menace was new in the nineteenth century, due presumably to the greater mobility and heightened expectations of bourgeois travelers.

The effects of chronic malaria would come to be recognized not only in swollen spleens and livers but also in subtle characteristics, ranging from the behavioral to the physiognomic—sallow complexion and lank hair. Drake reflected that "standing before the medical classes of Lexington, Louisville, or Cincinnati," he had "seen very few with plump and rosy cheeks. . . . Their physiology is not sound, although they may regard themselves as in health." Even those who had never had clinical malaria were "not vigorous." The regional poison had stunted their growth; they would never reach "perfect manhood."[125]

Increasingly the traditional adjective *aguish* would give way to a technical term, *malarial cachexia*. Medicalization did not replace the place-race equation but gave it a new authoritative foundation. Additionally it opened the door to explaining culture and character in terms of malarial pathology—what W. H. S. Jones, reinforced by Macculloch, would do for ancient Greece. In the view of such writers, malaria begat crime and apathy, murder and suicide.[126] Commentators would come to explain malarial cultures as a complex product of traditions and learned behaviors, diminished economic performance (and hence reduced economic options), prejudice from outsiders, and institutional failures but also of the physical effects of infection, like a low-grade delirium in which despair and night fear were prominent. "Depression is the most persistent feature of malaria," W. K. Anderson would declare in 1927.[127]

One apt analogy is the junkie: cachectics were strung out by the malarial monkey on the back, shiftless yet dangerous, wrenched by regular withdrawal-like paroxysms that were sometimes kept at bay by opium itself. Another is vampirism. Bram Stoker's *Dracula* appeared in 1897, as details of mosquito-borne malaria were being disclosed, but it reflects familiar scenarios. The Count's victims become anemic, pallid, and apathetic and waste away, while becoming themselves reservoirs of infection. Dracula himself comes from the malarial Balkans and, like the Anopheles mosquito, drinks blood by night.[128]

CAUSATION AND PREVENTION

While the enormity of the Norfolk-Portsmouth epidemic undermined the coherence of the public response, it did not wholly extinguish epidemiological inquiry.[129] The most intriguing inferences had to do with cause. In their search for proto germ theorists, medical historians have found recurrent suggestions of animate agents of disease from isolated southerners—Drake, John Crawford in Baltimore, Josiah Nott in Mobile, J. K. Mitchell in Cincinnati.[130] Later, Carlos Finley in Havana would suggest a mosquito vector for yellow fever. Like Drake, they combined induction and analogy.

In Norfolk, the focus was the "plague-fly" hypothesis. Armstrong noted that local African Americans associated infestation by a type of small fly with the end of yellow fever epidemics. They thought it consumed pathogenic putridity. But might it be the

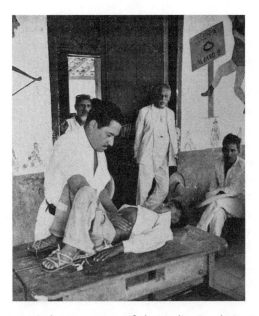

Early in the twentieth century, use of the qualitative designation *malarial cachexia* was giving way to quantitative studies of splenomegaly in children as a measure of the presence of chronic malaria.

From Paul Russell, *Malaria: Basic Principles Briefly Stated* (Oxford: Blackwell, 1952).

cause of yellow fever?[131] Ultimately Armstrong dismissed the idea, yet it fit the facts. Norfolkers had noted gradual movement of the fever that could not be explained by contagion, and they appreciated the danger of night exposure and the effect of frost in ending the epidemic.[132]

There was an even longer catalog of circumstances of malaria incidence, much of it collated by Macculloch. Malaria needed moisture, but not much: hoofprint puddles, dew, and open water barrels were evidently enough. The poison might enter via the skin, since sleeping beneath gauze helped to prevent malaria. Some suggested that Rome's sudden vulnerability to malaria might have been a result of a fashion change, from thick togas to flimsier garments, which left the rich relatively more susceptible. But Macculloch foreswore speculation, convinced that the fever poison would never be isolated even though the laws of its activity might be inferred.[133]

The romance of discovery is a key theme of the revolution in causality that occurred in late nineteenth-century medicine. Serendipitous observations in backwater places reveal disease vectors. In Norfolk, the plague-fly speculations reflect neither profound erudition nor refined method, but only attention to nature.[134] In China a generation later, Patrick Manson unexpectedly finds that filariasis is transmitted by mosquitoes and extends the analogy to malaria. Stuck in Algeria, Laveran keeps to his microscope and by chance observes the tiny malarial parasite bursting from red blood cells.[135]

But discoveries of agents belonged to a sea change in expectations of causal explanation as well. In 1850, textbooks still carried quasi-Galenic discussions of cause. They distinguished proximate cause, that is, the immediate concomitants of symptoms, from multiple remote causes.[136] The latter included both factors affecting susceptibility (predisposing causes) and those that precipitated actual disease (exciting causes). Rarely were remote causes necessary causes. A "contagion" might be included as a component of an explanation, though usually to account for the inception of a case more than to explain its characteristics. In Norfolk, while there was interest in minute agents, the most popular explanation was a composite comprising an introduced agent plus meteorological and geophysical factors, sanitary conditions, race, ethnicity, and individual predisposition.

The developing etiological revolution involved scrapping explanatory elements and the causal questions they addressed. It also involved decouplings. Increasingly the causes of distinct fevers would imply nothing about the causes of the pathological state of febrility. Except to a handful of German physiologists, the dynamics of fever, hitherto so central, would not matter; the constant correlation of agent with disease was "cause" enough. If one half of that correlation was the notion of specific febrile disease, the heritage of Paris, the other was the notion of a single morbid agent, the analogy from toxicology. As for how the malarial toxin worked, one might as well ask how the smallpox virus (increasingly the relevant model) made pustules, noted Drake.[137] Macculloch appealed to Newton:

> Fevers abound in certain climates, places, and seasons, where vegetable decomposition proceeds in a rapid or peculiar manner, and they are proved to be produced by exposure to the atmosphere of those places, . . . The causes productive of Malaria being demonstrated, such fevers are proportioned in number and severity to the power of these causes, increasing as they increase, diminishing when they diminish, . . . Malaria therefore, however unknown it may be, possesses all the philosophical properties of a cause.[138]

Even Boott hypothesized a special toxin. "Of the existence of some morbid agent [that acts as] the remote cause of disease to those exposed . . . , there can be no question."[139] The form—contagion, miasm, or gas; molecule, fungus, or animalcule—was less important than the fact of these discrete fever-causing agents in wild places.

Boott, Macculloch, Lyons, and Drake had no logical basis for subordinating some remote causes to others, but each was highlighting a single necessary agent. The non-naturals, hitherto considered sufficient to explain fevers, would subsequently have a bit part only, as modifiers of intoxication. "Dissipated habits" and "anxious . . . minds" of London's poor generated "latent weaknesses about the internal organs," Boott explained. But when the malarial toxin was added, the result was "visceral inflammation," presenting as remittent or continued fever.[140] Tropical parasites too were single agents, but there were closer links to toxicology.[141] Long before parasitism became paradigmatic of tropical medicine, the rattlesnake- and insect-ridden South had moved Drake to think of animate agents, while Fayrer was a world authority on snake venom.

The new etiology may be summarized in two axioms. One,

conventional enough, is, "No effect without (remote) cause." The toxicological analogy suggests quantum poisons, active in infinitesimally small quantities (malaria, with greater range than any contagion, must affect at much greater dilution, Macculloch insisted). They were undetectable in vitro, though one could appeal to the analogy of smell, for human bodies could sometimes discriminate entities that could not necessarily be isolated.[142] Their existence could be inferred only from their effects, effects that appeals to mismanaged non-naturals or mysterious states of atmosphere could no longer credibly explain.

The second axiom, "No cause without effect," was more controversial. No one could "hide from the universal atmosphere, or refuse to breathe the wide air," Macculloch declared; if there was poison in it, there must be poisoning. While the malaria fevers only some, it must harm all.[143] For Macculloch, what would be called *malarial cachexia* and all forms of "masked malaria" were not merely observations but deductions. Such a perspective was especially powerful before the emergence of immunology, which would explain how some were unaffected by conditions that killed others.

Earlier, disease specificity had been considered, by Sydenham, for example, as an important precondition of any effective therapeutics. In the late nineteenth century, recognition of distinct routes of transmission would become the foundation of prevention. And yet, except in Lyons's case, these speculations about agents were products more of deduction than of therapeutic or prophylactic optimism. Having concluded that the fevers of the American South had living agents, Drake saw little "practical importance" in that conclusion: "Did we know the particular meteoric condition, the gas, or the organized species . . . we should not probably be able to defend ourselves against it."[144] In colonial contexts, new knowledge did not rapidly displace old responses like flight, a recourse that clearly separated colonizers from indigenes. Manson told recoverees from West African blackwater fever to return to Europe and not come back.[145] And for Africans?

The messages "Don't go" and "Leave" had never been the same as "Don't exist there in the first place."

The pessimism was not absolute. Fevers of place slowly succumbed to drainage and deforestation, often after initial intensification.[146] An agent's invisibility did not preclude effective disinfection. Like the sanitarians who would follow, Boott and Macculloch had worried about the "taint" of drains and the doorstep dung heaps of English cottages.[147] But in most colonial settings the utopianism so conspicuous in European public health was absent. The tentativeness of postepidemic discussions in Norfolk suggests weariness with etiological debates. Whatever its causes, yellow fever would overpower. Understanding it would make no difference.

Then as now, one moral of causal inquiry was humility. Macculloch wrote in part to oppose British global militarism. There was no safe way to campaign overseas; the febrile debacles in the Caribbean, West Africa, and on the Dutch island of Walcheren, where thousands sickened in 1809 as they mobilized to march on Napoleon, proved that. Given what was known of malaria, whoever ordered, advised, or even permitted travel to a malarial district—even the doctors who failed to stop Lord Byron from making his fatal trip to malarial Greece—bore responsibility for the consequences. Malaria taught a simple lesson: one should improve the domestic environment and forsake "foreign colonization."[148]

But Macculloch also noted that *a single cause,* such as an agent or toxin, constituted the most powerful medical mandate for change. "If, of an effect, or of many effects, there is but one cause, we have attained a mastery by knowing that, which becomes materially reduced in value should there be more than one: while should we even suspect additional causes that we cannot prove, there is excited a want of confidence in our philosophical principles, which materially interferes with the results that we might otherwise have derived from them."[149]

SMALL-GAME HUNTING

By 1885 Macculloch's malarial toxin had become a microbe, Laveran's plasmodium. The triumphant age of bacteriology involved
new instruments (better microscopes and microtomes), new
methods (culturing, inoculation), and rigorous protocols to in-
tegrate them, but it was equally metaphysical, sociological, and
ideological. The new model doctor was a microbe hunter, not an
avuncular adviser. One stained rather than stuffed one's trophies,
but however tiny, they were difficult and deadly game. The enterprise, like its age, was militaristic as well as imperialistic. The
"enemy" typhus was likened to "a hostile nation-state, a colonial
territory, a devious villain, or—to invoke another biological set
of myths—a parasitic and alien, marauding race," notes Paul
Weindling, and the "microbe hunter" was imagined as a dragon-
slaying white knight.[150] And certainly in its early days the new microbiology was hardly the truth factory of popular imagination.
Among accepted agents were many microbes that had, at least for
their proposers, satisfied protocols, later to be displaced by some
other agent that did a better job. Usually left off the lists are the
Klebs–Tommasi-Crudelli bacillus of malaria, the multiple bacteria declared responsible for viral yellow fever and influenza, and
the presumed "germ" of beriberi.[151] Yet microbiology and tropical medicine quickly became paradigmatic biomedical sciences
as methodological conventions and well-marked career paths replaced isolated discovery.[152]

Quite quickly the new etiological revolution bypassed the
premises on which it was ostensibly based. To Bartlett, for example, a clinical species, inferred from a clustering of symptoms, had
implied a common cause. Macculloch, however, began with the
cause and inverted the reasoning. A single agent, he insisted, did
not imply a particular clinical species. In the etiological revolution
a disease would be whatever clinical effects its agent produced. In
his magisterial textbook, the great William Osler would struggle,
as had Fayrer, with such things as typhoid fevers that produced no
lesions of the Peyer's patches. One might allow immense clinical

variation precisely because it no longer mattered. Similarly, epidemiology often became agent tracking rather than case tracing. With the rise of serology-based immunology, agents too would become superfluous; diseases could be defined by a fingerprint left in the victim's immune system. The distinction between typhoid and the new paratyphoid would not be a bedside distinction, but rather a microbiological and then serological one: a variety of *E. coli* caused the latter.[153] For Vandyke Carter, Lyons's successor as the authority on Indian relapsing fever, that disease continued to have as wide a range of symptoms and courses, but common to all was the distinct spirillum.[154]

After 1900, following recognition of their microbes and complex life cycles by Ronald Ross, Walter Reed, G. B. Grassi, and others, malaria and yellow fever would quickly come to exemplify civilization's power. Modern science-based militaries would raze their swampy habitats. Epitomizing the totalitarian possibilities of disease control was Mussolini's transformation of the deadly (and hence largely uninhabited) Pontine Marshes near Rome, in the 1920s. The "military analogy" of a grand offensive was central, notes Snowden. So too was race: malaria-free regions would be home to an "Italian superrace."[155]

By the First World War the ordering of fever had expanded to involve the natural history of lice and ticks as well as mosquitoes; it also involved agglutination reactions, mass campaigns of vaccination and quinine administration, thoroughgoing disinfection, and monitoring and control of human movement.

The border between civilization and wildness was not exclusively a malarial one. In early twentieth-century Europe the great concern was invasion by wild Asian infections, led by typhus. Delicate, fastidious Aedes mosquitoes had been easy prey to global public health crusaders, their hardy Anopheline cousins somewhat less so. But the gregarious hard-shelled lice (linked to relapsing fever in 1907 and to typhus in 1909) were another matter, riding along in dirty clothing, awaiting the opportunity to excrete fever-causing microbes in their gambols over human heads, armpits, and groins.

"Typhus" was the site of greatest transformation. Cullen had made his generic "typhus" paradigmatic of continued fevers, but its spinoff typhoid would become the flagship fever of the nineteenth century. I have noted its exemplary features: a distinct lesion and clear modes of fecal-oral transmission—hands, water, milk, flies. In 1880 its microbe would be discovered (by Carl Eberth). By 1900 there was a specific serological test, the Widal test, and a vaccine. All the while, notwithstanding Collie's confident ordering, what we know as louse-borne typhus was increasingly coming to be identified with margins—spatial, racial, and moral. By midcentury, it had become the mobile toxin of the starving, wandering Irish. As typhus waned in postfamine Ireland, typhus revulsion would concentrate on Europe's Slavic corners. Mysterious and massive Russia would become its great homeland. There, across some nebulous border, the "typhus-ridden" lay "hovering, ghostlike . . . awaiting the moment when the guards would be dropped," notes Kim Pelis.[156] Hence to western Europeans, Russia would be as Norfolk was to the Philadelphians, needy and dangerous, neither fully accountable nor directly fixable. In Germany the eastern threats would be racial, Semitic—in certain quarters typhus would become known as *Judenfieber*.[157]

In 1848 the radical young German doctor Rudolph Virchow had issued a manifesto calling for a socioeconomic and political form of public health. He attributed fever in Silesia to the starvation of the peasantry under illiberal Austrian rule.[158] In Ireland too, and later in India, typhus (and relapsing, or "famine," fever) had been products of deprivation. As recognition of the mosquito vector had transformed the politics of malaria, recognition of the interplay of spirochetes and lice would transform the political significance of these diseases.

Hence it is ironic that Murchison's relapsing fever would be the first of the discrete fevers to have its microbe observed. Even more ironically, the discovery of its spirochete (Carter's spirillum) would be made in November 1867, in Virchow's Berlin laboratory, by his assistant, Otto Obermaier. Proof of its causal role would require reproducing the disease in an experimental animal, Virchow insisted. Though no friend to the doctrine of disease specificity, he was merely enforcing rules for experimental demonstration outlined a generation earlier by his colleague Jacob Henle, rules that would become famous as "Koch's postulates."[159] While Obermaier was unable to reproduce the disease either in animals or in himself, the organism would come to be accepted as the cause of a disease not even recognized a half-century earlier.[160]

In principle, in relapsing fever as in malaria, a microscope and a blood smear should be sufficient to confirm a diagnosis. In other fevers the more arcane culture plate or agglutination tube might be needed, but in most of these diseases there would arise a great gap between the possibility of sure diagnoses and routine laboratory confirmation of those diagnoses. Yet even in the absence of routine confirmation, the power to pronounce fed the fetish for quantification. For governments and, later, international health organizations, statistics supported the illusion of control and rationality. And the fact that cases could be counted deflected attention from who was counting them and how. The result was a mirage of precision: who counted the 1,570,604 relapsing fever cases in European Russia in a single year?[161] Clinically, relapsing fever mimicked pneumonia or malaria. How far its status as a fad diagnosis impacted such figures is hard to say; laboratory diagnosis too was often tricky to perform. Even today, notes Shah, in much of the world no one bothers to confirm malaria; it would be too costly and troublesome to do so. "Presumptive diagnosis" suffices.[162]

As governments paid greater attention to the secret agents of fevers, they often paid less to the matrix of febrility, the diffused social constituents of health. Virchow's predominantly physiological understanding of disease had meshed easily with the recognition of sociopolitical causation. In a February 1868 lecture on "Hungertyphus," a fund-raiser for people suffering from a hunger fever in East Prussia, he had reiterated the themes of 1848. Bad government was still to blame. But he had read Murchison: malnutrition and conditions that might favor the spread of fever were now distinct from fever itself.[163] Later writers would further foreground the agent. They would declare, often gratuitously, that famine was incidental to relapsing fever (or typhus), admitting all the while the very common association of the two. Human movement and congregation would become the greater concern, the reasons for that movement less so. Except perhaps in India, where "famine fever" (and famines) hung on well into the twentieth century, deprivation would largely be passed over in etiological discussions, regarded as a feeble proxy of more direct causal processes. But still today, relapsing fever (louse-borne) occurs mainly in central Africa and remains a concomitant of deprivation.[164]

CONQUEST?

In many respects the struggle between order and wildness was consummated in the First World War. Certainly in terms of fevers, the decade after 1914 brought an eruption of wildness to the homeland of order. Typhus and relapsing fever flourished. In a war of trenches there was a new "trench" fever, a clinically variable condition, but one given its own microbe after the war. As it shared its means of transmission—the louse—with typhus and relapsing fever, its diagnosis brought no new prophylactic leverage.

The response to all these diseases was both military and militaristic. This was most evident in the antityphus "campaigns" in eastern Europe, in Serbia and then Poland, where epidemics were not just matters of fact but threats. Invasions themselves, they warranted counterinvasions. The microbe hunters would be countervectors, agents of civilization, notes Pelis.[165]

Dwarfed by the great states around it, "valiant little" Serbia had nonetheless distracted Austro-Hungarian armies from the major fronts of the Great War.[166] Hence, when it was overwhelmed by typhus in 1916, eager teams of health workers from Britain, France, and sympathetic but not yet belligerent America rushed in. Then in 1919 American antityphus brigades entered the new Polish state. On their thirty-two trains were "40,000 sheets, blankets, and pillow slips; 40,000 towels; 100 tons of soap; 17 motorized bath plants; 300 portable bathing plants; 4000 Serbian barrels; 50 tons of washing soda; 1 million suits of cotton underclothing; 160

5-ton Packard trucks; 324 Ford ambulances; 160 Ford touring cars; 3 mobile machine shops." The massive "cleaning-up campaign in the epidemic territory, and in all the counties of Poland" was undertaken by 294 "mobile field sanitary columns." According to Polish statistics, the expedition "dealt with" 311,374 people, disinfected 72,731 habitations, and bathed 19,400 in the "four bathing motorized columns of Colonel Gilchrist's American Unit."[167]

It is ironic that a war ostensibly against lice should involve so much contempt for institutions and cultures, nations and persons. Serbia and Poland were as abject as any colonizable place, the American reporters suggested. One held that defeating typhus would require "uplift[ing] the whole race of Serbian people and . . . chang[ing] their entire habits and methods of living." For Poland "it would be necessary to elevate the entire moral tone of the people and their mode of living." But they asserted a geopolitical mandate: "If the world does not fight typhus in Poland, it may soon have to be fought . . . in other [countries]." Writers worried not only about persons and their louse-infested belongings but about "the industrial and social unrest which accompanies [typhus]." That disorder, they feared, would cross the Atlantic; it would be better and cheaper to contain it in Poland.[168]

But the typhus these modern doctors encountered defied both military discipline and their textbook expectations. In Serbia the Allies faced not a single well-defined disease but the kinds of compound cases that Pringle would have labeled "putrid" or "scorbutic," with emaciation, fetor, and hemorrhagic maculae. In their reports, they seek to reduce all this complexity to distinct concurrent diseases—"scurvy with typhus," perhaps, or coinfection of typhus with "typhoid, dysentery, diphtheria, relapsing fever, or the exanthemata."[169] Some, defining typhus as petechial, rejected the label for spot-free cases. There "should be no difficulty in distinguishing typhoid from typhus when the eruption is well marked and typical," complains one author, but many Serbian bodies failed to conform. Or atypical typhus might be influenza (whose ostensible "bacilli" they found and which they claimed sometimes had spots) or pneumonia or Pappataci fever, one of the few regional fevers of the great age of imperialism that remain in the modern medical canon. Some puzzled over phenomena observed in the past, such as cardiac failure during an uncomplicated convalescence, very much like what Drake had observed in Irish famine recoverees.[170]

All the while these authors remain smug about being on the modern side of a profound divide in medical history. They occasionally cite Murchison and other "older authors," misrepresenting and chastising even as they rediscover the variability that had been plain to every pre-Louisian fever doctor. In therapeutics too they were rediscovering empirical approaches common a century earlier; for example, that wine plus nutriment plus quinine had an "almost magical" effect on typhus.[171]

They objectified too. *Delirium* had invited subjectivity; the clinical *hyperemotionalism* did not. "Ferrety eyes" were now effects of pigment. Even had there been a shared language, there was no need for fever narration: patients' answers to questions about fever onset "were not considered of much value." Instead, American doctors would read the rash, often finding "a probable duration which did not correspond with the patient's story." And as for subjective sensations of hot or cold, "the temperatures were all taken by rectum by the nurses and are, therefore, reliable."[172]

One need not undervalue the considerable achievements of these doctors and the public-health campaigns they led to be struck by the sharpness of the divide between chaotic reality and the pretense of order. Colonialism has gone, but the prospect of control by classification remains a seductive one. From high on the bridge of the global-health and NGO flagship, the world's malaria problem is repeatedly solved and ever soluble, provided the cash flows freely. The demand for simple final solutions has led to the evolution of more resistant plasmodia.[173] We still invest authority in numbers, oblivious to the conditions of their creation. In December 2011 malaria seemed finally to be yielding to improved antimalarials and bed nets: there were 655,000 world deaths from malaria in 2010, boasted the World Health Organization, down from 800,000 six years earlier. Only three months later, the *Lancet* protested, claiming that there had been 1.24 million malaria deaths in that year and noting the significant problem of fraudulent antimalarials on the market.[174] Yet half a million missed deaths did not visibly shake confidence in the statistical order. The romance of microbe hunting persists among the new virus hunters, for whom the world is split between wild zones, which are always "hot," and the cooler world, which must be protected from them.

But the epidemiologist's imperial gaze was not the only site of action. In the same period there were other revolutions in fever—in its definition (temperature), in fever physiology (homeostatic regulation), in therapeutics (febrifuges), and in care (fever nursing). Their effects were profound but ambivalent. Together they would help to produce a split between private febrility, the realm of safe self-medication, and public febrility, in which one is a datum for the statisticians.

CHAPTER EIGHT

Numbers and Nurses

❖ ❖ ❖

Presenting a paper entitled "Proper Nursing Absolutely Necessary for the Successful Treatment of Typhoid Fever" to his county medical society at the end of the nineteenth century, the rural Illinois physician J. B. Coleman confessed that in his early career he had had "trouble in treating typhoid." Its length, complications, stupor, and delirium "all were dreadful . . . I did not always have success in their treatment." "Nearly in despair," Coleman was called to a case in which a "professional" nurse had been engaged. He took up the practice. "Since then the worst of my troubles in treating typhoid have disappeared."

This (female) nurse controlled fever in ways familiar to us, lowering the patient's temperature with cold baths or phenacetin, or a little whiskey in water. The nurse understood medicines in terms of their "action," watched unceasingly, could give hypodermics as needed, and could "use intelligently the ice bag—water-coils—the catheter—or administer rectal feeding." In typhoid, a rapid response to signs of incipient collapse or perforation could make all the difference. The nurse knew those signs but also knew how to secure adequate rest by preventing the annoying "typhoid cough." She would serve public health by ensuring proper disinfection and limiting the sickroom access of well-meaning but dangerous friends. The skilled nurse, Coleman declared, was "a godsend to the physician, as well as to the patient."

Happy medics like Coleman present themselves as delegating authority in the name of physician efficiency. In great hospitals, such delegation might seem a great boon in increasing productivity, but Coleman's context is private-duty fever nursing. A new technology, the telephone, allowed Coleman to retain control. For one of the skilled fever nurse's most valuable skills would be knowing when to phone (and also when not to phone): "In the first typhoid case I ever attended with a professional nurse in charge, I was called by the telephone at 2 o'clock (morning) to find the patient just beginning to rally from collapse. The nurse had recognized the condition and given stimulants, and at the same time (very properly) called the Doctor. I am very certain that this case would have died had he not been carefully watched by a trained nurse."[1]

Coleman was writing during a period of increasing specialization in responses to fever. At one end of an increasingly wide spectrum were career researchers, not just microbiologists and natural historians in search of germs and their vectors but also chemists seeking quinine analogues and physiologists unraveling the feedback mechanisms that regulated body temperature and incidentally gave rise to other febrile symptoms. Many of those involved did not expect any immediate therapeutic or prophylactic payoff; simply to learn more about medically relevant aspects of living things was worthy.

Physicians, by contrast, hitherto eager to present themselves as curers on the front line, began to adopt a more remote and magisterial role. For this was the age of therapeutic nihilism. Studies of outcomes like those of P. C. A. Louis had made clear that most fever treatments brought little benefit, since most fevers were self-limiting.[2] In those that were not, success depended, as Coleman recognized, on nursing skill, which was hard to quantify.

A key agendum in the phone conversations between nurses (or, later, parents) and physicians was the patient's temperature. Long neglected, temperature would become the definition of fever. Often, and surprisingly, its lowering alone brought relief. Conveniently, at roughly the same time the new fine-chemicals industry found in coal tar (and later willow bark) chemicals that lowered temperature with relatively few side effects.

But if the seductive simplicity of the thermometer and oral febrifuge allowed separation between head and hands, it also transformed (and disguised) the labor process of fever care. What was a "godsend" to the physician might be exhausting labor. The labor required in typhoid nursing was often extraordinary: round-the-clock care, perhaps for weeks. For not all fevers were unproblematic, nor was it at all clear which ones would be difficult to manage or dangerous. Those who tend fevers of family or friends know that with care comes responsibility and corresponding doubt. The nurse or parent may be in the ambiguous position of ostensibly operating under a physician's direction, while actually making the difficult decisions. The telephone remains a touchstone in a labor process in which worry is one of the most exhausting aspects—"The temp *hasn't* fallen . . . *how long should we wait before we call?*"

HEAT RETURNS

Even during the golden age of specificity in the late nineteenth century, with the discovery of exotic new fevers and their fearsome modes of spreading, the opposite agenda of pathophysiology was also being pursued.[3] Fever was becoming a general physiological process, perhaps also an accident of evolution, that manifested as heat. Galen, of course, had defined fever as heat. Thermometers suitable for taking skin temperatures had long been available; Santorio had used them thus in the early seventeenth century. Periodically fever writers had published temperature data, but usually temperatures were facts without signification.[4] Certainly in some stages of fever the body was hot. Yet the Germans called some fevers *Kalte* (cold), noted Daniel Sennert in the early seventeenth century. To his successor, the empiricist Hermann Boerhaave, a rapid pulse was the defining feature. Pulse or even respiration seemed better indicators of the turmoil within; heat, usually of the skin, seemed more an accident, a feature only of the surface.[5] And certainly, feverish heat bore no relation to ambient temperature. As Hippocrates had observed, fever often followed exposure to cold. Even Carl Wunderlich (1817–1877), the clinical scientist most responsible for the rule of the thermometer, initially viewed temperature as incidental. His great research project had to do with temperature in disease generally, not with fever specifically.

Longstanding interest in the hotness one felt during fever had not implied research into quantifiable, objective temperature. From antiquity fever writers had emphasized the profundity of change during fever, unwarmable chill giving way to unquenchable warmth. Some early thermometric studies had confirmed those sensations. Thus James Currie had reported cold-stage temperatures (axillary and oral) as low as 92°F, "and on the extremities many degrees lower." Gradual recognition of the discrepancy between feeling and fact would ultimately become paradigmatic of the superiority of instrument truths in clinical medicine, however: one should acknowledge the subjective, to be sure, but trust the thermometer. Even Currie on occasion admitted that parts of the body were hot even while the person felt cold.[6] But in his delirium-fixated romantic era, the transcendent insights of the fevered brain and strange sensations of the fevered body were far more interesting than that body's failure as a thermometer.

Subjective did not yet imply erroneous, as it would for late-century writers such as Patrick Manson. And even devotees of thermometric reality might allow care to be based on the patient's (illusionary) sense of comfort. The fact that temperature was actually elevated was no reason for Edward Register (1907) and George Paul (1917), authors of fever-nursing manuals, to deny patients blankets and hot-water bottles to ease their malarial chills.[7]

The coming centrality of heat, its objectification in thermometry, and the foundation of both in systematic experiment were achievements of that magnificent knowledge-generating engine, the nineteenth-century German university system. Most of the authorities Wunderlich would cite were, like him, German physicians or academics. Yet that achievement had roots in Paris and even in England. Wunderlich particularly admired the work of James Currie, which he described as "practical" and free of theory.[8] Indeed, Currie had launched a research program that would evolve into the study of homeostasis a century later. He had begun experiments on the relation of internal states to exposure—to cold, wet, and heat—during his student days in Edinburgh. A key context would be survival after shipwreck; he was particularly interested in why officers succumbed more readily than ordinary sailors. Industrial-revolution (and tropical-medicine) concerns would also loom large: Currie made much of auto-experiments by George Fordyce and Sir Charles Blagden, who spent time in rooms with temperatures greater than 200°F (or even 260°F) and passed back and forth between them and the cold outside. Also interesting was the custom of Glasgow glassworkers, who habitually left "the consuming heat of their furnaces, to plunge into the Clyde; a practice which they found in no respect injurious," and the Russian custom of alternating hot baths with rolls in the snow. In his first published case (1790), that of a Liverpool infirmary nurse, he was already concerned with the relation of the water temperature (44°F) to temperature reduction and in turn to slowing the pulse, which was still the measure of fever. He was interested also in the chemical qualities of water (here from the River Mersey). Its stimulating salinity helped temper its coldness, he believed.[9]

Most important were Currie's experiments on Richard Edwards, "a healthy man, twenty-eight years of age, with black hair, and a ruddy complexion." Currie had Edwards strip, acclimatize to the raw Liverpool air (44°F), and then plunge into 170 gallons of Mersey water. Once his "convulsive sobbings" had ceased, Edwards's body temperature was found to be 87°F, rising after twelve minutes to 93.5°F. In successive experiments Currie varied protocols, leaving Edwards immersed for as long as forty-five minutes; following immersion with exposure both to the cold wind and to a warm bath; and comparing Edwards's reactions with those of the "feebler" R. Sutton and on occasion with his own. Currie was interested in the subject's feelings as much as he was in the thermometer's readings. Preexisting states of mind might affect perceptions of temperature: the mad were famously insensitive to cold.[10]

These experiments showed the power of cold water to lower temperature, but Currie was no less interested in the power of heat to raise it. Though normally perspiration would kick in to cool the body, a hot bath or a hot room might cause a mini-fever, that is, a slight rise in temperature and an increase in pulse, pains, and debility, which would dissipate as the body's "heat" returned to "its natural standard." After lengthy exposure, however, "the inordinate action of the heart and arteries" would persist even after the temperature had dropped; if heat seemed to be the cause of the symptoms, pulse still defined the disease. But certainly, if bath heat caused fever symptoms, might not fever's heat cause them too? To Currie it seemed implausible that an elevation of a few degrees, "however generated," would "not have the most important effects on the system." It would "irritate the spasm of the extreme vessels," augmenting the action of the heart and arteries and leading to a "morbid association" that would continue even after "the spasm has relaxed, and the heat itself subsided."[11]

THE EMPIRE OF THE THERMOMETER

Carl Wunderlich's celebration of Currie's work was part of the historical foundation for the next medical revolution, a repudiation of the hospital-based medicine of Louis for the laboratory-centered physiological (and, secondarily, clinical) medicine that would be centered in Germany. In Paris, François Magendie's physiological research and Mateu Orfila's toxicological research had held a subordinate status to clinical and anatomical approaches. Broussais's iconoclasm on behalf of physiology had hardly helped.[12]

In the very years that Elisha Bartlett, in the name of positivism, was proclaiming the independence of pathology from physiology, Carl Wunderlich and a cadre of young German researchers were proclaiming in the name of positivism exactly the opposite: the full union of physiology with pathology. Most familiar are Jacob Henle (of Henle-Koch postulates), Virchow, and the physi-

ologist turned physicist Herman Helmholtz. Many were students
or acolytes of Johannes Muller (1801–1858), of Berlin; many, like
Wunderlich, had personal knowledge of the Paris medicine they
would overthrow. The titles of their new journals betray their mission—not just Wunderlich's *Archiv fur physiologische Heilkunde*
but also the *Zeitschrift fur rationelle Medizin* and the *Beitrage zur
experimentellen Pathologie und Physiologie.* "The members of this
school," notes Knud Faber, "resisted every endeavor to establish
definite clinical pictures, or to describe individual diseases, and
all mention of specific characters was opposed and dubbed ontology, or characterized as a relic of an unscientific and antiquated
age." There were no diseases, but only "conditions." As Virchow
would put it, "Disease is nothing but life under altered conditions." Within this perspective disease names were but "a practical make-shift borrowed from the popular, unscientific view of
things," notes Faber. What mattered was the body's functioning,
and that was discovered by experimentation on the determinants
("conditions") governing organ or cell performance. The rest of
medicine, including whatever was supposed to be happening at
the bedside, was mostly superfluous.[13]

Within this group, Wunderlich was unusual in concentrating his research in the hospital rather than in the laboratory. In
1868 there emerged from his Leipzig clinic a magisterial treatise
titled *On the Temperature in Diseases.* In it are the numbers that
define normality and hence deviance. From Wunderlich and his
assistants came the misleading precision of 98.6°F (a conversion
from 37°C) and the routine of temperature taking, that veritable
handshake of the clinical encounter. Wunderlich, like Bartlett and
Broussais, was a historian and a philosopher of medicine, a critic
and an advocate. But he was not a fever theorist in the old sense;
he was adamant about what the data *did not* show, impatient with
speculation. Predecessors had appealed to the authority of facts.
Wunderlich might respond, "Yes, but how many?" The book represented millions of observations on tens of thousands of cases,
each patient's temperature having been taken at least twice daily.[14]

A key concern was standardization, not only of the instrument but of its users. Wunderlich imagined a worldwide research program. His own protocol was formidable. He identified the armpit, "the *well-closed axilla,*" as the best site for temperature taking (a choice for which he would be much criticized). It was "convenient," did not "fatigue" the patient, and was, unlike the rectum, unobjectionable on grounds of "decency." (Oral temperatures were compromised by mouth breathing, since cool air might lower the reading. Thermometer breakage from biting was also anticipated as a problem.)[15] The long stick thermometers that Wunderlich used stabilized only slowly; twenty minutes might be required. He also recommended keeping the instrument in place for "some minutes" after the temperature had apparently leveled off. This lengthy and exacting procedure might have to be repeated several times a day. Ideally, in a hospital, despite the "immense amount of labour [required of] the medical man," this task should be performed by a doctor. Private-practice physicians should do some of the temperature taking, at least for "very acute cases," Wunderlich thought, but they would need to employ "trustworthy and well-trained attendants . . . or [even] intelligent relatives" who could be relied on once they knew the importance of these measurements. "If they err at all, [it would be] by disturbing the patient too frequently to take his temperature."[16]

In hospital practice, Wunderlich would have some temperature taking coincide with rounds. Prior to the doctor's entry into the ward all thermometers should be in place. "He should go round quickly, and ascertain that the instruments are in good position, correcting them when necessary." ("Very seldom" did conscious patients let thermometers slip out of place, Wunderlich noted; they had a "deep interest in the temperature being properly taken.") After the thermometers had been in place for twenty minutes, and while the doctor was conducting examinations, an assistant charted the temperatures, but the thermometers remained in place for the chief's final check.[17] Ultimately, of course, it would be the thermometer-bearing nurse, as iconic as the stethoscope-bearing doctor, who would do this work. Later it would descend the professional ladder further to the nurse's aide or a technician.

What did these data show? Wunderlich knew that readers would have to be convinced that temperature change was evidence of "some disturbance in the [body's] economy." Echoing Currie, he held that a rise of roughly 3°F would bring "feelings of heat, and lassitude; . . . usually with thirst and headache, as well as with increased frequency, and rapidity of the pulse." But most important was what that indicated. For changing temperature provided a glimpse of the mysterious interior, "a *scene of continual changes.*" Wunderlich imagined temperature as a pendulum. In a healthy person the oscillations were small, but the "sudden and powerful swing" in disease must indicate a "similar perturbation in the domestic economy." And temperature was, he claimed, "a *more accurate* and a *more delicate* measure . . . than other symptoms." All other indicators were "summed up, as it were, by thermometry, which presents to our judgment a *phenomenon dependent upon the whole of the vital processes of the entire body.*" Wunderlich would not quite equate fever with elevated temperature, much less define it as such, but it was certainly an indicator of fever itself and usually an index of its severity: "The height of the temperature often decides both the degree and the danger of the attack."[18]

Abnormality, of course, required normality, and Wunderlich sought to discover the small range of temperature variation consistent with health. He had searched beyond Leipzig for this "normal." Sub-Saharan Africans were said to be cooler than Europeans, Icelanders hotter. He found no evidence that variation was associated with sex, age, body type, or condition of life. Even among persons in the same environment there was substantial variability. Though there were still too few data, Wunderlich tentatively pegged the range of normality at between 36.5°C (97.7°F) and 37.25°C (99.05°F).[19]

Most remarkably, the thermometer had disclosed specificity. In tracking temperature over time, Wunderlich had come to recognize distinct "laws regulating the course of certain diseases." It was thus possible to "simplify, confirm, and certify the diagnosis." Many of the new fevers had distinct signatures, but typhoid got the most attention (distinguishing typhus from typhoid would be foundational in confirming the power of clinical thermometry). Beyond diagnosis, the thermometer would allow recognition of complications, of the effects of therapies, of passage from stage to stage. It had prognosticatory value: the thermometer "gives an absolutely fatal prognosis with great distinctness" and would also provide "certain proof of the reality of death."[20]

It was important too that the thermometer, unlike the stethoscope, produced objective, permanent, and quantitative records. "An advantage of almost priceless value, [were] results which can be *measured,* signs that can be *expressed in numbers,* . . . materials for diagnosis which are incontestable and indubitable, . . . independent of the opinion or the amount of practice or the sagacity of the observer."[21]

Wunderlich recognized that with the chart of changing temperature, the thermometer had ushered in a new form of medical perception. "The entire course of the disease, with all its fluctuations, complications, tendencies, and changes can be seen at a single glance," he boasted. Such instrument-generated images and inscriptions, whose skilled interpretation would demarcate medical specialties, are now iconic of modern medicine. Being able to see the disease graphically represented obviated the need for the troublesome integration of multiple facts.

> No memory . . . affords so [expressive] a likeness of the course of the disease as such a chart. The comparison of many such charts . . . exhibits the uniformity of the general course of diseases, lets the laws of disease promulgate themselves, so to speak, and exhibits all the variations and irregularities of the malady, and the working of therapeutic agents in so striking a manner, that no unprejudiced mind is able to resist such a method of demonstration.[22]

Arguably the greatest impact of Wunderlich's achievement was in the legitimation of illness. The thermometer would become, note Philip Mackowiak and Gretchen Worden, a "clinical divining rod for ferreting out *true disease* among the many aches and minor perturbations of other-wise healthy existence." Early on,

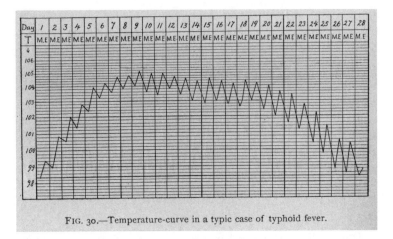

FIG. 30.—Temperature-curve in a typic case of typhoid fever.

Temperature charts like this one for a "typical" case of typhoid would contribute to differential diagnosis by graphically representing the stages of different diseases.

From Edward Register, *Practical Fever-Nursing* (Philadelphia: W. B. Saunders, 1907).

Wunderlich declared that "temperature can neither be feigned nor falsified."[23]

But of course it could be: In John Hughes's popular film *Ferris Bueller's Day Off,* made and set in the 1980s, high-school student Bueller does just that, effectively feigning fever to win a day off from school. Temperature, he knows, is the gold standard of malingering. For Wunderlich's observation suggests that the status fever was coming to have in an industrial society as a demarcator of duty. He assumed that people would want to fake fever. Whether one had to go to school, work, or the front line would depend on a thermometer reading. Doctors had become enforcers of mandates of state and society. The transformation was profound. For George Cheyne's patients a century earlier, *fever* or *feverishness* would have referred to a state of feeling akin to a modern mood disorder. One expected a physician to assess its type and probable course and then suggest the best response, not sanction its existence. There had been neither a need nor a widely accepted means for that. Thus the thermometer was part of a profound change in the objectification of disease, with the corresponding transfer of power from sufferer to physician and perhaps then to state.

In fact, fever charts, like postmortem dissections, did not always disclose simple truths. The chart might transgress as well as confirm the standard categories. And however helpful in the post hoc rationalization of a case, the completed portrait of a fever, drawn in rising and falling temperature, might come too late to help the bewildered practitioner, who, in the midst of that case, would face a familiar decision tree: What's the worst it can be? How soon must we act?

Wunderlich's legacy evolved in two directions. One was toward simplicity. Once the chart displayed the diagnosis, one had only to give the appropriate order, whose implementation devolved to the nurse. The other was toward complexity. Wunderlich had shown that temperature varied, but he had not shown why, or what that had to do with the other phenomena of fever.

THE FEVERED WORKPLACE AND THE KINDNESS OF STRANGERS

The coming of fever nursing raises the question of who had been caring for the fevered during all those centuries of theorizing. Fulsome in praise for their favored remedies, authors of fever treatises rarely say who was administering those medicines to the fevered. Sydenham was a partial exception, as was Pringle. Otherwise references to actual care are often critical, and generically so: family or friends fiddle with good physick and endanger or kill the fever patient. A turning point is evident in the reflections of the great Dublin clinician Robert Graves. Ever one to highlight the ineffable arts of the bedside, Graves told his students that proper management of a fever case required not only multiple daily visits but also means for ensuring that orders were followed exactly. In Dublin in the 1840s, that work was being done by "zealous, intelligent students, . . . young medical friends, and . . . well-educated apothecaries." "We are never at a loss for an assistant," Graves declares, but he closes with the cryptic comment, "I do not know how they manage this matter elsewhere."[24]

For nonhospitalized fevered persons, there were "watchers." In early nineteenth-century America the designation was formal, even quasi-professional. Watching was not nursing; watchers supplemented the care family members give. Their chief job was vigilance at night, when most fevers worsened. There are hints of other tasks too: keeping the delirious patient in bed, administering modest medications or sustenants, providing companionship or spiritual comfort, and summoning appropriate others should a crisis occur. Watching, usually done in pairs, was both a paid occupation and a charitable undertaking. But contemporaries made clear that there should be watchers in the case of any serious fever.[25] In Portland, Maine, Cyril Pearl was well watched by pairs of lodge brothers. So long as his disease was judged serious, at least one of each pair had had experience tending the sick, he noted.[26] At the Ohio Normal School, watchers were recruited from a prayer group of the victim's fellow students.[27] In Elizabeth Hill's melodramatic autobiography, the watchers (in early nineteenth-century New York City) were hired directly by the attending physician after Hill exhausted herself caring for her fevered husband.[28]

Generally, representations of watching are positive. An exception was a widely reprinted 1864 snippet written from a patient's standpoint. The anonymous author begins: "Buzz, buzz,—whisper, whisper,—came from an adjoining room, until I thought I should be distracted. 'Can't live long,' said one watcher. 'What will become of his poor wife and children,' said the other." As doctor and patient both realize, what he needs most is absolute rest, which the watchers' "intolerable whispering" prevents. Good sleep is a common concern. Writers note that the hyperacute senses of fevered persons make sleep especially difficult. Here, the watchers' mere breathing was a problem; much more so was their "rattling of newspapers, . . . squeaking of boots, . . . eating of the midnight lunch, and . . . snoring." All this left the victim watching the watchers. Ultimately, the author and his wife reach a compromise. She will stay with him but not try to stay awake.[29]

The golden age of fever nursing was also that of bacteriology, which had displaced pathological anatomy and even physiology as the exemplar of scientific medicine. Yet its success as knowledge did not carry over to medical practice. I have noted that this was the age of "therapeutic nihilism." It had become clear that effective specifics were few, and generics of limited value. The best strategy was simply to help the patient survive until the fever broke. Temperature, the main form of actionable intelligence, fostered the illusion that all important actions could be taken at a distance—that beguiling separation of brainwork from handwork that Coleman imagined. In fact, far from simplifying fever, the thermometer had simply altered and relocated medical work. Fever nursing was the work of both brain and hand, continual care of a person in pain, weak (though occasionally manically strong), delicate, irrational, and distressed. But the vast expansion of the nursing domain did not remarkably enhance the nurse's professional status. One can imagine significant labor troubles at the fever work site had there been less false consciousness about the professional duties of nurses.

The training of fever nurses was curiously ambivalent. Register and Paul, authors of the main American fever-nursing texts, presented what were effectively two separate courses in their books. One was elementary bacteriology—natural histories of pathogens and of the courses of different fevers, along with differential diagnosis. The other was generic fever nursing—sickroom management, temperature and pulse taking, and matters of nutrition, bathing, elimination, and so forth. Occasionally, the bacteriological knowledge was important for nursing practice, as fevers varied in degree and mode of communicability. Nurses had preventive responsibilities; they were to be frontline Anopheles and Culex (Aedes) detectors, responsible for keeping their patients from being bitten.[30] (They were not expected to become involved in swamp draining, however.) Three diseases, either new or previously poorly distinguished, presented unique nursing challenges:

diphtheria and rheumatic fever during their acute stages and scarlet fever during convalescence, when it might damage the heart.

But for practical nursing much of this science was superfluous.
It reflected nursing's subordinate status to the dominant profession of medicine. Most fever nursing was generic. Fevers heated
and exhausted, nurses cooled and restored. As diagnosis evolved
from fever chart to culture plate or test tube, clinical distinctiveness became less important. Training at the University of Pennsylvania in 1910, Edith Griscom heard her lecturer say that gastrointestinal influenza "resembles typhoid" and likewise that aspects
of tuberculosis were "sometimes the same as in typhoid fever."[31]

Not only would typhoid be the fever nurse's greatest challenge
but it was the warrant for fever nursing as a distinct specialty. Register and Paul gave it far more coverage than any other disease.[32]
It had to be fought with proven tactics and a long-term strategy.
The skilled fever nurse rejected the passivity of mere "watching"
for active command. Watchers or family members were amateurs
who might assist with heavy lifting (under direct supervision) but
could rarely be left in charge.

Many of the actual tactics and the ethos of command came
from Florence Nightingale, and before her John Pringle, and reflect the legacy of the military hospital. The fever nurse, with hospital training and techniques, would bring military order into the
home of the (necessarily relatively wealthy) patient: *shall* would
replace *should*. The logic was simple: care really mattered, and the
nurse was the expert carer. While the nurse acted on behalf of a
physician, the distance between the two hampered any challenges
to authority; the division of labor allowed each of these professionals to invoke the independent authority of the other.

The nurse's domain comprised both the sickroom and the surrounding household. Indeed, it included anything that might affect the patient. On arrival, the nurse would commandeer two
adjoining rooms, one (with two beds and a southern exposure) for
the patient, the other as a staging area. Excess furnishings would
be removed. A sash window would be adapted to allow ventilation while avoiding a draft, and all available means would be used

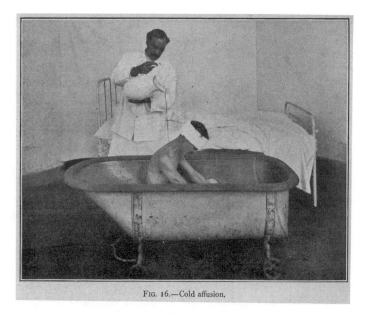

FIG. 16.—Cold affusion.

The application of cold water, in a variety of ways, remained a key means of fever control long after James Currie.

From Edward Register, *Practical Fever Nursing* (Philadelphia: W. B. Saunders, 1907).

to eliminate exogenous noise. For very contagious fevers, disinfectant curtains might be hung and disinfecting pails placed to receive excreta or soiled linens.[33]

No less important were the social arrangements. Unless stopped, family and friends would kill with kindness. At least at first, visitors, even family members, would be forbidden. They generated feelings, which elevated the patient's temperature and accelerated his or her pulse. Effectively, the patient, mind and body, was to be turned off, kept supine and largely motionless, perhaps for several weeks. As Register explained,

> The mind of a fever patient should rest. It is just as important as the quiet and rest of the body. It is unwise to let fever patients in their convalescent stage, while they have fever, even look at pictures. This will make them think, and thinking is an exertion

that should be avoided. As is well known, to allow a patient to
read is objectionable, to listen to reading is equally as bad, and
merely to listen to conversation is harmful.[34]

Certainly, the nurse must *not* amuse the patient. Not just laughing but merely sitting up might cause the patient to literally "bust
a gut"; though rare, intestinal perforation was often fatal.

Effective sickroom dictatorship required that the nurse be
a professional stranger. Anyone subject to intrafamily politics
would be unable to enforce the necessary regimen. Nevertheless,
tact was essential. No matter how competent, a nurse who was
disliked by the patient or the patient's family must resign from the
case, as success depended on trust.

Learning to command was a crucial part of a nurse's training. Interesting in this regard are the pamphlets in the Bedside
Stories series (1909–12) prepared for students in the Chautauqua
School of Nursing.[35] This correspondence school could not, of
course, teach actual nursing skills, but the pamphlets emphasize
principles, theory, background knowledge, ethics, and ethos. The
"stories," ostensibly graduates' accounts of their successful struggles, are plainly promotional. But they reflect a common anxiety, namely, whether one could muster the wisdom and will to
command the sickroom. *Nursing Typhoid Fever under a Texas Sun*
(1912) highlights poverty, insanitation, and utter hopelessness.
The nurse, hired by an outsider, must first extract both patient
and family from a filthy dugout. *My First Case* (1910) deals with
double pneumonia. The author accepts her role as both nurse and
watcher, "remaining near him that I might hear him breathe and
know if he was restless." She must also change poultices throughout the night. Ten days pass before she feels she can safely sleep
on the sickroom couch; until then she has napped in the after-
noons while others oversaw the patient's care. The theme of *My
First Case of Typhoid* (1910) is the nurse's exhaustion. The author
struggles with an embarrassed teenager's incontinence and the unwillingness of her family (of "modest" circumstances) to invest in
a rubber sheet. She learns to humor her patient during delirium,

to deflect her anxious mind, and also to resist the uncommon manic power of this delirious child, using "strength (not force) to hold her in bed." She has learned, she explains, the need for "unremitting watchfulness," for delirious patients can be cunning. Slowly, over the six weeks of the case, the author trains the mother to take on some of the watching so that she can have time for a brief rest or walk.

The most poignant of the pamphlets is *Nursing Typhoid* (1910). Disembarking at a rural station, the author treks six miles further. "I was to be alone in the heart of the country with a case of fever, which three leading physicians . . . had given up as hopeless." She faces a hostile grandmother who will not have a nurse in the house and a 27-year-old male patient who has lost two brothers to typhoid and neither expects nor wants to live. Finding him filthy and bedsore, she bathes him. But she interjects too her fury at the neglectful family: "It certainly is trying to see a human being dying from filth and neglect when there are able-bodied people whose pleasure it should be (if not pleasure it is their duty) to at least keep the helpless person clean." The doctor's return two days later comforts her: she learns that the patient's melancholia and belligerence are only the "mild insanity" of fever. But the patient then worsens, twitching, picking at the bedclothes, "jabbering day and night incessantly . . . . not quiet five minutes for four entire days." Following the doctor's hypodermic injection of an unnamed medication, the patient goes into convulsions, exhibiting cyanosis and narcosis. Finding no pulse, the nurse, on her own, injects glycerin, digitalis, and strychnine. After twelve hours in a coma the patient rallies, but it will be three months before he can sit up for more than a few minutes.

The account is remarkable in its candor. Professional detachment prevents the author from expressing her outrage at the family's behavior. Expecting greater and more authoritative guidance from the doctor, she is told that doctors do not always know what will be the effects of the medicines they order. In remote settings, virtually all important decisions reside with the on-site nurse.

The stories exemplify professional behavior. Each author commits herself to extreme acts to save her patient. Rarely are these single dramatic interventions, but instead continuous small acts of solicitude. But while they take pride in their patients' recovery and plainly seek approbation from physician and family (and through it, future assignments), these authors are also promulgating an ethos in which there is really no adequate response to fever. Could more be done? Yes. Might it make a difference? Yes. There can be more quiet, closer watching, more scrupulous moving and bathing of the patient, more assiduous disinfection, greater comforting, and, in retrospect, wiser judgment about how to juggle all these. Paul recites the case of a nurse instructed to keep her patient's temperature down with cool baths. On returning (within 24 hours), the doctor discovers that the nurse has given the patient eighteen baths. Probably, no more than six would have been necessary. Paul's point, however, is not this nurse's extraordinarily conscientious care nor her inordinate labor but rather her lapse of judgment: each bath not only cools but exhausts the patient.[36]

Increasingly, the currency of good nursing was comfort. The delicacy of the typhoid-stricken body, fragile long into convalescence, made comfort a key to recovery, since lying still was the surest way to avoid perforation. The easing of pain, along with calm reassurance, would transform dangerous restlessness into necessary rest. Hence, much more than in earlier fever literature, the fever-nursing literature was concerned with pain. Whether or not comfort was important in its own right, it was a key factor in disease management.

BEATING THE HEAT: NURSE OR PILL?

Unexpectedly, relief from pain, better sleep, and renewed appetite often turned out to be concomitants of lowered temperature. Coinciding with the golden age of fever nursing was the seductive prospect of the one good drug that would eliminate the need for that skill and vigilance. By the late 1880s the first designer febrifuges, drugs to make fevers fly, were being created through the magical techniques of coal-tar chemistry. Best known were variations on quinine, beginning with the rivals antifebrin (acetanalide) and antipyrin (phenozone), introduced in 1886. Phenacetin (acetophenitidin, equivalent to today's paracetamol and acetaminophen) followed in 1887 and would long be a junior member of the group.[37] Others were the salicylates, extracts of willow bark, including aspirin in 1899.

In an 1891 prizewinning review of these drugs, the University of Pennsylvania clinician Hobart Amory Hare glimpsed the world to come of lucrative molecules and controversial drug trials. He noted that upon its introduction, antipyrin had been "at once seized upon and used by every practitioner with feverish haste." But how were utility and safety to be evaluated? The new febrifuges had been in use for only a few years, but already the corpus of clinical studies was unmanageable. Thermometric studies were easy to do. Any hospital doctor could turn researcher. The spirit of experimentation had made room for pharmacology, but much of the research was primitive. To review the whole antipyrin literature "would be absurd," Hare declared. It was the author's duty "to gather up the good . . . rather than to bring before the reader . . . the entire crop of communications." That meant reliance on well-reputed German researchers but also on Americans. Remarkably, this "country which is at once the youngest and most pushing for money, and not science" had supplied many of the "reliable and accurate studies."[38]

Hare evaluated three factors: mechanism, dose and speed of action, and side effects. A febrifuge might work in three ways: by inhibiting heat production, augmenting heat dispersal, or both. It was important to determine which process was occurring. A drug that augmented heat dispersal without lowering heat production might mask tissue consumption. Measuring and conceptualizing a drug's efficacy was difficult; there was no standard. Antifebrin's partisans touted the lower dose necessary, but in America antipyrin was more popular. It was also hard to know whether the drugs shortened fevers or made them milder. The unexpected discovery that these drugs were analgesics as well complicated assessment, as pain was less readily measured than temperature. The most visible side effect (from antifebrin and antipyrin) was cyanosis, but were there others? Results on any of these dimensions might vary by fever type, stage, dosing, or population.[39]

Though Hare acknowledged the pressure on doctors to prescribe medications, he was generally negative: people regularly mistook relief for cure, but all the while, "the disease plows its way onward to recovery or death." He worried that antipyrin would overtax the stomach and the kidneys. In low fevers like typhoid, such drugs could reduce the temperature too much, causing collapse. Though the newer phenacetin seemed to work well without having the side effects of the others, Hare was critical of the lot. "The fever will sometimes resist all doses of antipyrin that we can give, or, at any rate, all that it is safe to give." But "no fever can resist the cold bath."[40]

Habituated as we are to a medicine cabinet stocked with cheap combination analgesic-febrifuges, Hare's denunciation of them may puzzle. He was not alone. Paul and Register, the fever-nursing text authors, agreed, as did many authors of popular fever tracts in the first half of the twentieth century.[41] But the issues were not what they would become. The feverish diseases that concerned Hare were more dangerous than those generally experienced today, and the new febrifuges too were powerful medicines requiring close supervision. On the other hand, nurses, to whom fever management had been consigned, had transformed Currie's initial idea of external application of cold into a fine art. Patients who could not be moved might be wrapped in wet sheets ("cold packs") or sponged. Or by placing a rubber sheet on the patient's bed, the nurse might even make it into a temporary tub: one could use bolsters and pillows to raise the rubber sheet on the sides and at the ends of the bed, then pour in water, draining it at a corner when finished. And for really rapid cooling there was the ice-water enema: "Carefully administered they are rather grateful than otherwise to the patient."[42] Such techniques constituted a well-recognized arsenal of temperature-control technologies to be professionally selected and administered.

SELF-FURNACING?

While nurses (and coal-tar chemists) were finding ways to conquer fever by conquering temperature, others were probing the
mysteries of overheating. Wunderlich knew that he did not know
what temperature meant. It was, at it had been for Boerhaave, a
net effect, possibly the product of multiple antagonistic factors.
Elevated temperature indicated a changed balance between heat
production and heat dispersion, but not the state of either variable. In a single case of fever the same temperature might reflect
different causes of imbalance at different times. Accordingly, temperature data really had almost no "immediate theoretic application," Wunderlich admitted. Yet framing his investigations was
the new science of thermodynamics. Perhaps the body was like a
heat engine. Wunderlich cited J. R. Mayer, a pioneer articulator
of the principle of the conservation of energy. Wunderlich too assumed that stored chemical energy could be released as heat or as
work (motion), but all must be accounted for.[43]

Systematic comparison of three states might shed light on
the body's interior workings: the heat of acute fever; that of the
healthy body able to maintain a constant temperature even when
exposed to temperature extremes; and that of a body whose resistance has been overcome and which has fallen into febricula,
the ephemeral fever so central in ancient writings. Excessive work
would heat the body if one expended energy more rapidly than
it could be dispersed through sweat or other means. In the tropics, even mental work could raise the body's temperature, Wunderlich reported. But while there were clearly limits to the body's
ability to regulate its temperature, humans differed. In a given
case it might be impossible to distinguish the effects of exogenous
disease from those of environment and activity. In Wunderlich's
practice, the highest temperatures ever recorded in a wide variety
of diseases were recorded during an exceptionally hot summer.
"The impossibility of keeping the sick ward sufficiently cool" was
partly responsible, he thought. Ultimately, it was "*impossible to*

draw a hard and fast line to indicate . . . the exact limits of health and disease."[44]

Irregular temperature might not itself constitute disease, but regularity of temperature, especially when challenged by a changing environment, was usually a sign of health, though that health was gendered. "A healthy man may have scanty or luxurious fare, he may fast or feast, drink water or stimulating liquors, he may remain quiet, or exert himself vigorously, both bodily and mentally, . . . yet have almost the same temperature." Women, by contrast, experienced "apparently causeless elevations of temperature . . . with the rapidity of the recoil of a spring." Besides the "heats and flushes" of menopause (which some practitioners failed to distinguish from the chills and heats of malaria), there was menstruation itself, as well as lying-in, or "milk," fever. Children too became overheated, from teething and growth spurts, from fatigue, "psychical depression," and similar situations. An excess of nervous excitability, which affected "many *women,* and sometimes also men, of delicate, somewhat feminine build and constitution," was still sometimes to blame. Separate was the diurnal fluctuating temperature of "chronic starvation," much of which would occur in the subnormal range and be accompanied by malaria-like perceptions: "sensations of cold, shiverings, and 'chattering' of the teeth."[45]

If "very slight influences" occurring in "many conditions in which no accurate diagnosis is possible" produced "very considerable deviations from the normal heat," was that fever? These "cases of slight illness . . . , general irritability, persistent feelings of lassitude, slight disturbances of all the . . . functions, impaired digestion, imperfect respiration . . . &c." recalled the "vapors" of George Cheyne. However trivial, to Wunderlich such phenomena marked an important border between pathology and physiology. The distinction between health and disease was real, he believed, but "neither in theory nor in any given case can we discriminate a point where health ends and sickness begins."[46]

Such symptoms correlated with temperature change. Were they its products? Wunderlich struggled with conflicting evidence. Sometimes elevated temperature was not accompanied by other symptoms, but usually it came with "a complex group of other disturbances of the general health, of functional anomalies, and . . . impaired nutrition." There were effects on pulse, urine, digestion, musculature, respiration, and perceptions, along with a sense of "heat, thirst, and loss of appetite, loss of power, malaise" and problems of "disturbed sleep" and "mental operations." The number of blood corpuscles fell, and the body lost weight. All these together constituted "fever."[47]

And yet there was no clear correlation between the severity of symptoms and the magnitude of the temperature rise. Wunderlich decried oversimplifiers (like Currie) who attributed fever symptoms to high temperature. One might conclude, on the contrary, "that the temperature is either no guide at all, or a very deceptive one," but that too was an overstatement. It was safer to say that whatever its ultimate relation to disease, changes in temperature were both prognostic and diagnostic. "Whatever the sickness . . . careful observation of the temperature . . . offers a far more reliable . . . basis for judging the progress."[48]

Though dubious about the utility of animal models, Wunderlich reviewed both experiments and hypotheses on the relation of heat regulation to febrility (and to collapse, which he saw as an underrecognized analogue). But here too the results were conflicting and, despite being experimentally grounded, redolent of the speculation of an earlier era. Ludwig Traube's attribution of rigor to "tetanic contraction of the smaller vessels" arising from "agents which . . . act with varying intensity on the vaso-motor nervous system" recalled Cullen's spasms. Traube might vivisect, yet, Wunderlich warned, much of his model remained conjectural. With facts so powerful, it was better to eschew explanation. "We cannot indicate the special chemical processes which influence the production of heat; . . . cannot say why the temperature is such and such in one form of disease, or why it differs from that of another disease. We ask in vain why in many severe disturbances in the economy, accompanied with very extensive tissue changes, the temperature, as a rule, remains normal."[49]

Wunderlich walked a fine line: the thermometer was the key to fever, yet fever was much more than temperature. Often the distinctions were too fine. As John Harley Warner notes, "Physicians began to think more in terms of temperature than of fever."[50]

"THERMIC" FEVER?

The concept of human heat engines fit well into the context of the political economy of imperialism: one may note the Marxian overtones. Work, especially in hot places, could destroy the body's thermoregulatory capacity. Was it being restored? Or was hard work in hot places pathogenic and even uncompensatable? As Wunderlich realized, the thermometer had raised the question of whether there was any difference between a hot body and a "real" fever, a matter that had both practical and moral implications. It was the very first question Patrick Manson addressed in his 1898 *Tropical Diseases.* Heat and sun, he reassured readers, did cause "sun erythema, sun headache, and symptomatic fever," along with "heat exhaustion" and "vague, ill-defined conditions of debility." Yet "none of these . . . can with justice be regarded as *disease.*" Usually a real disease implied germs.[51]

Among those to whom Manson was responding was one of the first American doctors to choose Germany over France for postgraduate training, the Philadelphian Horatio C. Wood Jr. (1841–1920), winner of the 1871 Boylston Prize for his monograph on thermic fever, what today we call heatstroke.[52] The condition was well recognized; Wood would study it to understand fever more generally. Some of his evidence came from his own Civil War experience, and some from his Philadelphia practice, but much more came from Anglo-Indian military surgeons, who reported that heatstroke often occurred on humid nights as soldiers lay in their tents, seeking sleep. Predisposers were debility or fatigue, crowding and lack of ventilation, but also "want of acclimatization." Certain of the mysterious seasoning fevers of the tropics were probably heatstroke, Wood suggested. The symptoms were common fever symptoms: insensibility, delirium, effects on eyes and conjunctivae, hot dry skin, rapid pulse, even petechiae and ecchymoses, and also, perhaps, a distinct odor. The condition was serious: "intense fever, with great nervous disturbance, . . . unconsciousness, paralysis, convulsions, etc." Sometimes, especially when insensibility and convulsions were prominent, it was recognized as sunstroke, or *coup de soleil,* but often it was diagnosed merely as common continued fever, "ardent fever," or even "phrenitis."[53]

As an admirer and consumer of German research, Wood took up pathophysiology too. "We are able . . . to induce sunstroke in animals, and, by varying its conditions, study its nature much more thoroughly than can be done at the human bedside," he wrote.[54] One focus of Wood's research was physical changes in muscle and nerve fibers, particularly the loss of responsiveness in motor nerves when they were exposed to very high temperatures—125.6°F in mammals. Cooling restored some potency, but repeated heating and cooling deteriorated the nerve. Lesions too, such as the severing of the cervical spine, caused overheating. Wood concluded that fever resulted from paralysis of the vasomotor nerves. In heatstroke, that paralysis induced even greater heat production through rapid oxidation, resulting in an "overwhelming of the cerebrum; [and] intense fever."[55]

For a brief period before physiology gave way to bacteriology, Wood's "thermic fever" garnered some interest.[56] But did it have any relevance to normal infectious fevers? Wunderlich's anomalies—heat without symptoms—remained. Nor did Wood's research suggest how an unidentified toxin of a yet unidentified microbe might have the same effect on the nervous system as external heat or the physical lesions the experimenters could so readily produce.

Others were taking up these questions. There were two main models to explain overheating. One, the "metabolic," stressed overproduction. The other, the "neurotic," stressed "under-regulation." In the first, the nervous system was itself the victim, coping unsuccessfully with overproduction of heat, which might have many causes and was effectively an independent variable. One might note the urea-laden urine in many fevers, which showed the body's furnaces burning up its very substance; heat, aches, and dizziness were all presumed to come from this overstimulated blast furnace within.[57] In the neurotic model a toxin or other subversive element was seen to be causing havoc with regulatory mechanisms. A faulty feedback allowed temperature to climb, then stabilized it at some higher level. Symptoms might stem directly from the raised temperature or from whatever had disrupted regulation. Support for this view came from the rare phenomenon of hyperpyrexia, rapid and often brief elevations of temperature without tissue metabolism, and also from the power of noninflammatory brain lesions to raise temperature.[58]

While both metabolic and neurotic processes might be involved, there was no ready way to untangle their effects. But as the Scottish practitioner T. J. Maclagan (1838–1903) observed in *Fever: A Clinical Study* (1888), the fact that no one could identify a "thermic centre," a site of regulation, did not diminish the attractiveness of temperature as an "index to the amount of febrile disturbance." One could find no "better or more convenient clinical test of the existence of fever, or one . . . so serviceable at the bedside."[59]

If less boldly than Bartlett, Maclagan, a pioneer in the use of salycilates in rheumatism and in the emerging rheumatic fever, was calling attention to the fact that *fever* had not been adequately defined.[60] The precision of temperature had exposed the imprecision of fever, a complex entity that might be recognized by sufferer and practitioner alike, even if they could not say what exactly it was. Remarkably, then, toward the end of the nineteenth century, fever, that long-lived enigma, seemed increasingly to be sinking into superfluity, its several symptoms left clinging to the raft of temperature. Maclagan would suffer the term to stay on, but only with its new thermometric identity:

> Fever is not a distinct entity, but a collection of various phenomena the co-existence of which in the system is conveniently characterised by the term 'fever.' Of these . . . the most constant

and characteristic is increased body heat. So prominent is this feature . . . that a rise of temperature is generally referred to as fever: a patient whose temperature is raised is said to be feverish, and the terms 'fever' and 'increased body heat' have come to be regarded as synonymous.[61]

In moving temperature from proxy to synonym, Maclagan had gone beyond Wunderlich to redefine fever. And yet this was not quite a return to Galen's *calor praeter naturam*. Galen had understood, at least to his own satisfaction, how the unnatural heat arose. Wunderlich had been delighted merely to find patterns in his accumulated data, but Maclagan, two decades later, would speculate. Excess heat, he thought, was the work of the germs that caused fevers, the tiny parasites that had crawled within. By thinking ecologically about what effects their rapidly expanding operations would have on the closed system of the body, Maclagan deduced the principal phenomena of fever: "All organisms have a definite action on their environment. . . . The environment of these organisms is the blood and tissues of the human body." To multiply, these tiny entities required vast quantities of nitrogen and water. Their sole source of nitrogen was the body's tissues. These "waste during fever," noted Maclagan; "the fever patient . . . emaciates . . . almost visibly before us." The body's thirst was the germs' demand for water, a demand "so great . . . that each draught has but a . . . temporary effect. . . . Very soon another is called for, and another, and another; and the demand continues so long as the fever lasts." Little of this water was eliminated, for it was being taken up by the alien host. The model had both metabolic and neurotic elements. Heat arose as the body's metabolism accelerated to meet both the body's own needs and those of its invader. Toxins affecting regulation were accidental by-products of microbial metabolism. Maclagan recognized both specific toxic effects, which were the unique markers of diseases, and the generic uremia as kidneys became overtaxed.[62]

Although Maclagan is not overtly allusive, his images, like

those considered in the previous chapter, do fit their time, one of militarism and of global anxiety, especially over nitrogen resources:

- Fever's symptoms are the effects of "attack" by "some foreign agency."
- Invading fifth columnists have subverted the body's "economy."
- "Contagium particles step in" to "convert into their own protoplasm so much as they require" of the body's nitrogen.
- They have "wants and demands identical with those of [our] tissues."
- "So close is the struggle between the two for the possession of the nitrogen . . . that the contagium seizes upon it at the very moment at which it is about to be incorporated with the tissues."[63]

It was the brain, malnourished and flooded with waste products, that suffered most. Even if the attackers were repulsed, invasion might undermine the brain's stability, and "the patient may, for weeks, say and do things which, but for the preceding febrile attack, would be regarded as evidence of insanity."[64]

FEBRICULA REDUX

Few now experience the many weeks of battle with typhoid, the odd intimacy that must arise from total dependence on a live-in nurse whose domain includes the mind as well as the body. Two profound changes accompanied that transformation.

First was a rejection of Wunderlich's, Maclagan's, and Hare's preoccupation with seeing fever as exhaustion. So what if one's fever was burning up one's body from the inside out? It wouldn't last long and might be a good way to shed extra pounds. Hence there was no need to worry about whether these coal-tar derivatives did mask a hyperactive metabolism; that they stopped pain overrode any such concern. The new drugs relieved not only feverish headaches and neuralgia but sometimes also migraines, rheumatic pain, and gout. Their analgesic power rivaled morphine's: 5 grains of antipyrin were equal to one-twentieth of a grain of morphine.[65] What one anthologist would label the "aspirin age" of American history was well under way by 1920.[66]

Second was the trivialization of fever. Besides typhoid, the fevers nurses were taught to handle in the early decades of the twentieth century were coming to be dominated by the exanthems, which had become the bourgeois fevers of American life and increasingly the fevers of childhood. The most serious of these were cerebrospinal meningitis and smallpox; less so were measles, German measles, chicken pox, mumps, and whooping cough.[67] Nurses would likely encounter such diseases, and each had its challenges, but most did not involve the long life-and-death struggle typical of typhoid.

The availability of febrifuges brought greater interest in the mild fevers they cured, fevers that might once have been called ephemera and were still sometimes referred to as febricula. These did not readily fit into the nosologies. In principle they were assumed to be microbial; in practice they were attributed to overdoing it or taking a chill. Maclagan, following Murchison, suspected them to be mild cases of a more serious disease.[68] The new serological techniques, which might not distinguish between recent and remote exposure, might suggest the same. By the 1940s, however, a view we still hold was becoming ascendant. Brief, easily medicated fevers would be ascribed to what the ancients called *virus,* placeholder for a malign agent, and what the moderns knew as a distinct filterable virus. But because the identity of the particular virus that had infected us was rarely important, most fever was again effectively generic. Some lapse in resistance might also be a contributing factor.

By 1930 *fever* had been dumbed down. To a few isolated pathophysiologists, *febrility* might still refer to complex reciprocal interactions; to everyone else it was simply a temperature too high. Fever's terror receded as typhoid retreated, while typhus was relegated to unthinkable and uncivilized places. Recovery came quickly with rest and fluids. Prevention became similarly simple: cover the mouth, wash hands, disinfect spaces and things, avoid kissing strange babies, and foster one's resistance. By the 1930s popular fever literature was mainly for mothers, who could presumably do anything that nurses had done, as long as they were not facing typhoid. The availability of an easy cure obviated any need for introspection and speculation about what was going on within. Talk of lentors lapsed, as did talk of spasming capillaries or internal putridity, while delirium was dismissed as a residuum of elevated temperature rather than a door to transcendence. Relatively quickly, an immensely long and large legacy of learned concern about the relations of mind, body, and environment was forgotten. In the new master analogy the body was prime mover and heat engine; it released energy stored in food, converting it to heat and work. Who ran the operation and was responsible for the evident slip-up termed *fever* was a recondite question for experimental physiologists.

PART IV

Fever, Modern and Post-Modern

❖ ❖ ❖

For roughly a century now, fever has been something utterly different than at any other time in human history. When I took up this project, I mused briefly about applying for a grant to the National Institute of Fever. Of course, there is no NIF among the more than twenty institutes and centers of the National Institutes of Health, nor is there need for one. Certainly, in various ways, fevers come under the gaze of researchers in many parts of the NIH, and other parts of the public medical research establishment take up fever-causing diseases, notably the Centers for Disease Control, the familiar CDC. But had there been something like the NIH in Paris in 1800, in Dublin in 1840, or in Berlin in 1880, fever would have been the banner disease.

Perhaps the most vivid testimony of fever's centrality is in textbooks. Of the seventy lectures in the two volumes of *Clinical Lectures on the Practice of Medicine* (1848), by Robert Graves, probably the greatest of the Dublin clinical teachers, fourteen lectures in the beginning of volume 1 deal with the symptomatology, treatment, and epidemiology of a generic *fever*. Another six are about associated conditions (e.g., inflammation and delirium), while another eight deal with particular fevers—yellow fever, scarletina, intermittent fever, influenza, and pneumonia. If we include lectures dealing with symptoms in fevers, such as headache, then we might conclude that half of Graves's medical curriculum is concerned with fever.[1] By contrast, there is no lecture on cancer, none on diabetes or smallpox, and there is only one each on the famous scourge diseases of the century, phthisis and cholera. Graves's volumes are an extreme case, as fever was peculiarly conspicuous in Ireland, but texts from other lands too would testify to the centrality of fever in medical practice.

In general, we know what happened. Some fevers were prevented—sometimes by precautions that, like hand washing, were simple in principle but based on profound changes in sensibility; sometimes by the capital-intensive technologies of urban sanitation or the eradication of mosquito habitats; sometimes through the sterilization and disinfection campaigns of vigilant states. Others were cured, perhaps by the many derivatives and successors of quinine, or by antibiotics or antivirals. And finally there were the readily available drugs that did not cure but relieved symptoms and restored us to normality.

Hence what was once grave danger is now pesky interruption, at least in some parts of the world. I fear cancer or dementia but not fever. Occasionally a fever ruins my weekend, but generally actual or potential fever has almost no impact on my life.

That cannot be said of the large portion of the human population living in malarial regions. For malaria, more than for any other fever, there has been hope of a science-based public-health revolution that would free whole regions from pervasive debility and childhood death. Many regions have become malaria free. Roughly a century and a half ago in the part of the world where I live, a malaria-stricken English doctor closed his practice in Lafayette, in northern Indiana, and moved north beyond the malarial line to southern Minnesota.[2] His name was William Mayo. His life is a poignant reminder to a historian of medicine of the profound impact of disease geography on regional development: Lafayette is no Rochester. That northern Indiana—or Chicago or Cleveland—could once have been rife with a "tropical" disease is, to my students, among the most unbelievable things I tell them.

They are surprised too that the CDC itself, probably the most powerful symbol of the power of science to conquer infection, began as a malaria-control institution. Nowadays, almost all U.S. cases of malaria represent infections acquired elsewhere. But there are parts of the world where it remains a familiar part of life, and others where its incidence has vacillated remarkably as antimalarial campaigns have come and gone. Certainly we cannot count on some autonomous progress to fight that fever.[3]

The main theme of the final chapter is that profound split in the experience of fever. The remission of fever as a prominent part of lives in the developed and urban world has left odd residues of images, fears, and curious practices that are increasingly detached from real disease. One of these is the unloading of most of the anxiety about fever from physicians and even from nurses onto the shoulders of parents of small children. They must make the decisions that for Graves required the most refined clinical experience (which, he sometimes hints, only he possessed). And having dragged the toddler into the emergency room in the small hours, parents often receive the message, usually indirectly, that they have wasted the formidable resources of great hospitals and deep learning on a trivial cold.

Another is the fever-as-adaptation theme, which in some cases effectively translates as "fever is your friend." Many ancient physicians recognized fevers as restorative processes but they would probably have balked at this formulation; however necessary, the condition was often deadly. But to a society with ample technologies for controlling bodies, the implications of adaptationism seemed obvious by the late nineteenth century: if high fever was the route to resolving diseases, then perhaps it should be cultivated when it did not naturally arise. That spirit would lead in the preantibiotic age to devices and procedures that may seem to occupy that nebulous border between rigorous workout and mild torture. But however much modern readers shudder at the idea of "shock" treatments, of which fever induction was one form, such approaches were merely versions of what has become familiar in modern medicine, doing drastic harm to produce necessary help.

The same era saw the enormous expansion of surgery. Making holes in bodies would be joined by poisoning them with chemotherapeutic agents or zapping cells with radiation. Raising the body's temperature by a few degrees for a few hours seems mild in comparison.

And, unlike surgery or chemotherapy, these elaborate technologies could be made to seem natural! For both the acceptance of the devices and the adaptationist trope itself relied on what has become an almost untouchable component of modern ideology, the goodness of nature. Fever evolved naturally, and nature was good. That belief underlies, I suspect, the two most profound modern transformations of the fever concept. The first, evident to any surfer on the Internet, is the deployment of fever terms in public expressions of sexuality. We can use fever language to talk about sex because both are species of "hot." But more importantly, we trust that both are natural and therefore good. Second is the image of planetary fever, whether it be the literal rising of the global temperature or the new menace of pandemic fevers that come from out-of-balance human activity or simply from overpopulation. Frequently the mechanism is one of disturbed environments. Viruses that have coevolved with nonhumans suddenly devastate human populations that have pushed into their primitive habitat. But the moral message is an ancient element of fever lore. I have noted Galen's default fever victim, the aristocratic youth who has done something stupid, usually overindulging in food and drink. Fever is his natural and just desert (or dessert). Perhaps we too, as species, have failed to moderate our urges. Pandemic fever will be both punishment and correction.

CHAPTER NINE

Machines, Mothers, Sex, and Zombies

❖ ❖ ❖

For much of the twentieth century the scourge fevers still scourged. Often it was clear what caused them and how their spread might be stopped. Concentrated by region and social situation, their presence divided human experience. For some they remained regular features of life and death. Among those who lived in the increasingly postfebrile developed world they elicited sympathy but also horror, revulsion, bewilderment, and sometimes apathy.

I have noted the resurgence of malaria, seemingly controlled in much of the world by the 1980s. The largely forgotten typhoid, now just another diarrheal disease, has long been absent from the developed world. Though preventable by sanitation and vaccination and treatable by antibiotics, it remains a major killer in parts of Asia. There, what in the nineteenth century was a disease children did not get is now primarily a disease of children. While estimates vary, typhoid still causes upwards of 200,000 deaths each year.[1] Concern about typhus has virtually vanished, yet it still devastated into the mid-twentieth century, especially in Russia. Between 1918 and 1922, the turbulent years of revolution and civil war, there were 2–3 million deaths among 25–30 million typhus cases.[2] In the Second World War too, and infamously in the death camps, it was rife. I have noted too, the unsettling resonances between typhus control and the Holocaust: the common focus on eastern European Jews, on herding persons, on taking control of hair, clothes, and possessions in the name of a greater health, even the hardware of the delousing showers.[3] That typhus, permitted to flourish in the camps or in the awful conditions of the Warsaw Ghetto, did some of the deadly work of genocide has, ironically, been taken by those who deny the Holocaust as exoneration:—nature (typhus), not human action, did the killing, they say.[4] But the episode reminds us how far modern fever is a product of policy *not* nature. Other fevers too are relatively invisible. Smallpox has been eradicated; the childhood exanthems are most familiar as occasions for routine inoculations against them. Influenza certainly rocked the world in 1918–19. We are warned regularly that it will do so again. But generally, "flu" translates as "under the weather; up in a day or two."

Hitherto, all these would have been understood as fevers, and given the long-lived belief that fevers could mutate, all would have been taken seriously. Yet the events reviewed in chapters 6–8 had effected a split. The isolation of the pathogens of major fevers and discovery of their secret means of transport had vindicated an agenda that began in Paris and lives on in the occasional announcements by global health authorities of new exotic fevers. But middle-class persons in the developed world do not expect to come down with the fevers that interest the World Health Organization. The heritage of Wunderlich, of thermometers and cheap pills, is more immediate. Living in a region where fevers are mild makes contemplating places where they not mild conspicuously exotic. "Off to the global south?" "Got the shots?" "Which antimalarial?" These too are matters of fever, but distant from ordinary practices of casual self-medication. Our fevers divide us.

FEVER REDEEMED

Fever's trivialization, evident in the mass-circulation magazines of good advice that await us at supermarket checkouts and in doctors' waiting rooms, has been its most striking feature in the last century. Though hardly uniquely American, that literature is well

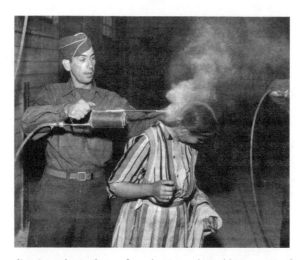

Louse eradication, shown here after the Second World War, was often done with DDT. The iconography of typhus control is ambivalent. It can readily register power and victimization, but it was a major tactic in mastering a deadly disease, and it would be a mistake to read coercion into all such images.

http://med-dept.com/unit_histories/28_fld_hosp.php.

developed in a nation in which personal freedom has long been expressed as medical self-fashioning. While there have been noteworthy variations over time, a handful of tropes dominate.

Most remarkable is that fever is often viewed positively as an adaptive response. A survivor of the Warsaw Ghetto or of any place where malaria remained rife would have been puzzled, even offended, by a 1954 assertion in the *Country Gentleman* that fever was "truly a friend, not a foe."[5] That view began to appear in the mid-1920s. "Fever is to-day regarded by physicians in a totally different light from what it was a few years ago," the *Literary Digest* announced in 1927. "Then it was thought to be a wholly bad thing . . . to be reduced at any cost. The increased heat production is now looked on as a reaction on the part of the living cells to the noxious stimulus of the micro-organisms or their soluble poisons."[6]

All this reflected a more biological agenda for understanding the workings of bodies. That the phenomenon unhelpfully known as warm-bloodedness (temperature equilibrium) had seldom figured in discussions of fever probably reflects both the lack of comprehensive data on the temperatures of human and other bodies in the pre-Wunderlich era and lack of interest in the question. But by the second decade of the twentieth century physiology was flourishing. It was the beneficiary of new patrons: scientific agriculture (concerned with the nutrition equally of farm animals and of those who ate them) and scientific management (concerned with the productivity both of the workplace and of the worker). The new orientation is reflected in the titles of two classic works accessible to lay as well as specialist readers: L. J. Henderson's *The Fitness of the Environment* (1913) and W. B. Cannon's *The Wisdom of the Body* (1932). Henderson reintegrated two aspects of the fever problem: how the body regulated combustion and how it adjusted to ambient states under varying conditions of skin exposure and activity. Cannon articulated the concept of homeostasis. H. C. Wood Jr.'s term *thermic fever* had gone out of fashion, but the problems he had considered related to the body's (limited) ability to compensate for a changing environment had become more important.[7]

The public learned of such research in the context of the celebration of evolution and adaptation.[8] If in the romantic era fever had been the apotheosis of individuality, it was now a reminder of the profound ways in which humans were connected to all beings. Even plant bodies were "smart" temperature regulators, a delighted popularizer reported: "In the case of parasites attacking trees there is a rise of temperature . . . botanists actually speak of 'fever' in plants."[9]

Explaining to its readers how the body's politburo managed its temperature, the *Literary Digest* in 1929 made much of the mysterious "glands," underappreciated organs that were fast becoming the primary focus of medical entertainment. Evidently, a number of secretions interacted in complicated ways, collectively directing the body's changing state. The *Digest* author credited the prolific

Yale pharmacologist Henry G. Barbour with having tracked the body's master heat controller to its lair in a primitive part of the brain, below the "thinking portion." There, Barbour "was able to fool the nervous watchman into an increase of heat that was not really necessary." Barbour had effectively made fever an experimental system (sometimes inducing it with cocaine, as well as by other means), but others denied that the master controller had been discovered, and Barbour himself was cautious. The concept of the hypothalamus as heat controller would only stabilize three decades later.[10]

Even while its location remained unknown, that controller would come to be known as the body's "thermostat." The metaphor of the household thermostat would be seized upon by writer after writer in the next decades. "It may help to think of the human body as a heat engine. . . . This engine works with marvelous efficiency because of a 'thermostat,' called the hypothalamus, in the brain," wrote one in 1957.[11] "Your body's heating system is not unlike the furnace and pipes in your home," declared another in 1968.[12] Differing views of the fever-as-adaptation thesis led to differing construals of that metaphor, however. Was fever a mistake or a sophisticated response? Theodore Irwin, writing in 1968, saw fever as "a sign that your heat-regulating system has gone out of whack."[13] More common was the view that fever represented a thermostat simply turned up; by whom was not clear.[14] Elevated temperature in babies might be immaturity; as one writer put it, "A baby's 'body thermostat' has not yet become well adjusted."[15] Some implied that the thermostat had a safety switch, that it foresaw dangers and would let temperature rise, but only so much. "High fever causes brain damage. . . . Because of the hypothalamus, which acts as the body's thermostat, fevers rarely exceed that level."[16]

Rivaling domestic thermostat metaphors in frequency, and wholly integrated with them, were military metaphors, holdovers from the nineteenth century made newly vivid by the revolution in immunology. These too emphasized stimulus and feedback. Not only did a body have an inertial quantum of stored health but it had a command and control center to deploy its smart weapons. Here too was adaptationist language founded on a long heritage of natural theology, for usually it was Nature who was fighting on our behalf and who should, if justice prevailed, win. Of course there was no explicit reference to the "natural" and "preter-natural" of Aristotle and Galen; the images were equally anthropocentric but far more jejune. Fever was, depending on the author, itself one of Nature's tools, or a consequence of the use of those tools. The language shifted with changing military idioms and mood:

- "When germs invade the body, Nature marshals her forces for a defensive counterattack. Fever is often one of the results."[17]
- "The invader may be the common cold, or an army of troublesome measles germs. Nature may be battling some nasty little bug which has set up an infection in the throat or in a cut finger, or it may be waging war against some hidden source of infection. . . . Whatever it is, it has no business being there and nature's protest is shown by fever."[18]
- "When a child's body is invaded by a disease germ, warfare is declared . . . between the germ and the child. . . . Hostilities begin when the germ lands somewhere in a child's body and finds conditions pleasant. . . . As soon as the germs have gained a foothold their very presence stimulates the child's body to mobilize its defenses. . . . If the child is strong and sturdy he will put up a good fight."[19]

Notably, in 1950 the enemy had become "an alien invader."[20] The Cold War introduced the idea of fever as an "early-warning" system, a sort of radar that the medical expert could interpret.[21] By the late 1960s that confidence had gone. In Vietnam "many soldiers have been struck down by an FUO (fever of unknown origin) which produces a high temperature for a few days, then leaves the victim exhausted," worried one writer, noting that in such fevers "the cause is not readily or clearly discerned."[22] Evidently, soldiers' bodies were as baffled as the nation.

Just as it was important to imagine a tiny thermostat somewhere inside one, so too it was important to visualize the body's weapons and tactics, however much the picture stretched and oversimplified what was really happening. The discovery of the immune response known as phagocytosis, the consumption of foreign bodies by so-called macrophages, was a publicist's godsend. Again, the *Literary Digest* in 1929: "The body's way of fighting such germs is to digest them." This was the "duty" of the white corpuscles.[23] More graphically, a later writer described how "these cells ooze out from the blood vessels at the spot where the germs are multiplying and literally eat up the germs."[24] (Whether phagocytosis inspired the 1980s video game Pac-Man is not clear, but the game is regularly used to illustrate the concept.)[25]

There were manifold ideas about what heat had to do with any of this. Writers in the 1930s emphasized heat's stimulating effect in getting the phagocytes "liquored up," as it were, thus to screw their courage to the sticking point. "Under the influence of heat, the vital functions of the body cells are quickened; and as success in the battle against germs depends on the vital functions of the body cells, their fighting capacity is increased by fever." Later writers held that high temperature prevented microbe reproduction or that it simply fried germs ("bacteria are simply cooked to death").[26]

By contrast, tales of *pyrogens* were disturbingly ambiguous. Evidently, phagocytes emitted chemicals that caused the hypothalamic thermostat to raise temperature. But inasmuch as these pyrogenic substances might be the residuum of destroyed phagocytes, the process seemed morally ambiguous. Adaptation perhaps, but the process sounded more like dumb luck than design. No wise controller would rely on corpse residue to do important physiological work.[27] Or perhaps the rise in temperature was incidental. Perhaps fever, like the smoke of battle, simply signaled that a great fight had occurred.

In the 1920s there was still need to explain why, if fever was your friend, it sometimes killed you. To any spectator of modern warfare the reason was plain: collateral damage. Evidently these tiny berserkers were not overly fine at distinguishing enemy from friend. At 104°F, "not only is the digestion of the invading germs accomplished, but the destructive forces, overstimulated by the heat, turn on these cells themselves and on their neighbors. . . . useful, healthy cells," noted the *Literary Digest*.[28] Still, within limits, the higher fever was the better one, John Lentz asserted in 1957: "Chances are that the child with the higher temperature will overcome the infection faster than the one with 'just a touch of fever.'" It was "putting up a better fight to get well." A fever above 105°F might need to be lowered, but only modestly, and chiefly so the patient could rest.[29]

Accordingly, even in this great age of powerful febrifuges, many writers counseled against medicating fever: "When we realize that fever is the best thing that can happen under these circumstances, we will know enough not to meddle with it by indiscriminately knocking it down," noted Bernard Fantus in 1935. He worried too that by conquering aches and pains, the use of these "fever narcotics" would fool us with the illusion of recovery: "A serious and even fatal condition [may] become established when, without their use, the ill feeling would have compelled rest in bed, the really important remedy to secure prompt recovery."[30] Another held that while aspirin was itself unproblematic, it encouraged drugging: "When children see their parents taking aspirin for whatever ails them, they are tempted to do likewise."[31] Curiously, the arrival of sulfa drugs in the 1930s and of antibiotics a decade later had little impact on the language of fever. Fever was still one's friend, the wise body doing unpleasant but necessary work. But already the bacterial fevers, for which these drugs would be useful, were giving way to a virocentric fever culture.

These metaphors and sensibilities persist—in "fighting" disease and in imagining our bodies as full of thermostat-like self-correcting devices. Notwithstanding Susan Sontag's famous instruction that we should not read meaning into disease, we still do so.[32] Phagocytosis may be too overtly videogamish, but the unavoidable sacrifice of healthy cells remains a prominent theme

of cancer therapeutics, even as techniques become ever more precise. Curing disease is not for the squeamish.

BAKED GOODS

If fever was friend, why not cultivate it? While in many diseases
the wise body reliably warmed in response to infection, it did
not in the two great venereal diseases, syphilis and gonorrhea.
From the 1880s onward a number of physicians around the world
sought to treat these diseases, as well as other chronic diseases
that might conveniently succumb, by artificially inducing high
fever.[33] Though healing by heat had a respected history, though
spending time in a hot bath, sauna, or Turkish bath might leave
one refreshed and invigorated, pyretotherapy never acquired the
aura of the spa.[34] It remained instead at the edge of orthodoxy and
even legitimacy. Associated with morally abhorrent diseases and
an unattractive patient population of neurosyphilitics, modern
pyretotherapy would usually be carried out in the dark setting of
state psychiatric hospitals. The techniques were in fact among the
so-called shock therapies during this age of "great and desperate
cures."[35]

The few attempts to present such techniques positively drew
on the prestige of Germanic medical science.[36] At first, artificial
fevers were real infectious diseases. The pioneer was the Viennese
alienist Julius Wagner-Jauregg (1857–1940). On the basis of scat-
tered reports of severe fever relieving longstanding psychoses, he
had begun inducing malaria in 1917 (having tried other modes
of fever induction much earlier).[37] His receipt of the Nobel Prize
in medicine in 1927 sparked exploration of the range of conditions in which fever therapy might work and speculation about
its mechanisms. By 1938 fever therapy was being used for epilepsy,
asthma, multiple sclerosis, arthritis, allergies, chorea, and even
acne. Usually the basis for treatment was empirical, but for syphilis it was assumed, on the basis of in vitro experiments, that one
was, in effect, "cranking up the barbie" for a spirochete roast. Five
hours at 39°C (or three at 40°C) killed the infamous "pale spiro-

chete" of syphilis; the tougher gonococcus required a temperature of 41°C or higher for at least five hours. And needless to say, these increases in temperature had to reach the remote parts of the body where the microbes lurked.[38]

There were many ways to induce high fever. One could rely on a hot bath or hot packs or inject the patient with typhoid vaccine, some foreign protein, or malarial blood, but in America the preferred means were diathermy and fever chambers. In the former, plates placed on the chest and back zapped the patient with targeted radio waves.[39] In the latter, the patient lay in the hot and humid atmosphere of an iron-lung-like chamber for five to ten hours. Exposure to supersaturated air in this insulated environment (a process ironically known as "air-conditioning") deprived the body of its ability to cool itself since the perspiration produced could not evaporate. In effect, the device induced heatstroke.[40]

If less graphically than electroshock therapy, pyretotherapy remained vaguely punitive. It was an unpleasant and lengthy cure, usually involving many sessions. Even if used on outpatients, the equipment and expertise, not to mention the procedure itself, represented the absolute power of the total institution. As in Ken Kesey's classic *One Flew over the Cuckoo's Nest,* the embodier of that power would be female, a nurse exercising constant supervision and total control over the person in the chamber. The patient's temperature, measured via a "deep body rectal pyrometer," would be taken every five minutes during the rise to fever, and every ten during the rest of the treatment. (This would be replaced by continuous monitoring by 1950.)[41]

"While the treatment is in progress the nurse holds a human life in her hands," wrote Willa Phillips in her 1939 textbook on pyretotherapeutic nursing. This was not the most politic thing to tell the patient, she admitted, but it was important for the nurse to know.[42] Seeking to make the case for specialized training in pyretotherapy (and thus rescue it from the charlatanism associated with electrotherapies), advocates admitted dangers. "Pyretotherapy is technically difficult, arduous, and dangerous in unskilled hands," noted the authors of the 1939 text *Fever Therapy Tech-*

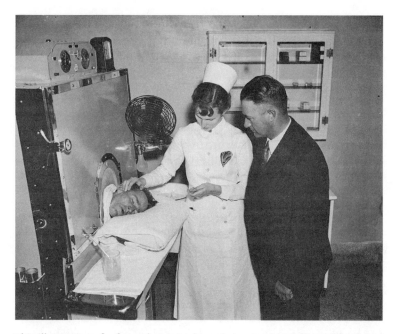

This illustration of a fever therapy cabinet being used in Gallinger Municipal Hospital in Washington, DC, highlights the nurse's role in attending to the patient's head.

Library of Congress. Photograph by Harris and Ewing.

nique. "The temperature produced is uncertain; it is very difficult, and, at times impossible, to obtain a febrile reaction . . . high enough, or long enough, to be therapeutically effective . . . [and] there is great danger of the febrile reaction getting out of control." The internal temperature might not increase steadily but instead jump suddenly. Diathermy could burn the skin. Most urged that pyretotherapy always be done in a hospital and that both the supervising physician (increasingly a member of the new specialization of physiatry) and the nurse-technician have advanced training on the apparatus in use. Ideally, the patient should give the kind of consent typical for a surgical procedure.[43]

The chief challenge of the treatment was to keep the tempera-

ture as high as possible while maintaining safety—skirting but
staying below 108°F. Patient comfort was important because patient cooperation was so necessary. Much of the concern was with
managing psychological effects. Many panicked and demanded to
be freed, usually as they crossed the 102°F–104°F range, their temperatures climbing about a degree every ten minutes on their way
to 106°F–107°F. Phillips herself had "strongly experienced that
wish to discontinue treatment" in each of her own fourteen fever
sessions. Usually the panic subsided, and Phillips advised nurses
to cajole panicky patients to commit to one more hour. But if
they would not, they were to be released. Pyretotherapy was elec-
tive, even though "they [patients] may not realize that they have
free will." A patient might react differently in subsequent sessions,
exhibiting varying tolerance to the treatment.[44]

Not everyone could handle the ordeal. While emotionally
unstable persons were not good candidates, most others, with
preparation, might be led to "grin and bear it." Phillips advised
short preliminary treatments, test drives in the box, as it were.
Or by visiting the "fever room," candidate patients might learn
that others found "the experience not extremely unpleasant." The
authors of one text suggested that the patient "imagine [lying] in
the sun on a very hot summer day with no effort to cool him." To
induce fever was, at least potentially, to induce delirium, but in
Phillips's experience this posed no serious problems for the "well
disciplined, emotionally stable patient," though "nervous symptoms, such as crying, are seen in some patients, especially those
addicted to alcohol."[45]

While Phillips does not dwell on it, pyretotherapy took a
patient to the edge of brain damage. That high heat addled the
brain was clear, but did all brains addle at the same point? In
sunstroke "too hot blood" damaged "especially" the cells "related
to consciousness." One might recover yet be " 'a little queer in the
head' for the rest of his life," noted one writer.[46] But in "Dreams
and Facts," the final chapter of his 1938 treatise *Artificial Fever
Produced by Physical Means,* the Northwestern University profes-

sor and diathermy advocate Clarence Neyman appealed to the heritage of Henderson and Cannon: pyretotherapy was simply the utilization of capacities the adapting human body had learned from nature over its eons of evolution.[47] Ultimately, the age of antibiotics would further marginalize such treatments, consigning them perhaps to the realm of mad doctoring.

FEVER AND THE MODERN MOM

The idioms of the wise body controlling its interior just as the wise homeowner controlled that body's exterior fit well into the domestication of fever, its migration from the domain of learned medicine and skilled nursing into that of the good mother, the presumptive natural nurse who handled the well-being of children and other affiliated persons when fever interrupted their normal activities.[48]

The transition was gradual. Before 1960 many fever writers were plainly confused about when the mother could decide, what she could decide, and why. There were common themes in this confusion: fever was potentially, but not usually, serious; the doctor should be called (no one was sure why); the child patient must stay in bed a long time (again, no one could say why); and finally, like the private-duty fever nurse a generation earlier, the mother was ultimately responsible for everything. "Old-fashioned" and "unsupervised home treatment" must stop, declared Dr. Herman Bundesen in the *Ladies' Home Journal* in 1943. "A child developing a fever needs immediate observation by a doctor, because there are so many dangerous ailments that begin with fever."[49] A decade earlier, Josephine Kenyon had not only insisted that the doctor be called but also set the protocol for the visit. The doctor would "examine the eyes, ears, nose, and throat, and the skin, will listen to heart and lungs and feel the abdomen, will investigate the condition of the nervous system, save the urine for analysis, look at the stool, and make a blood test if indicated." (A white-cell count, explained Dorothy Whipple in 1953, "tells the doctor something about how much of a fight the sick child is putting up against the

By the mid-twentieth century the thermometer-bearing nurse had become iconic in America, even as the brunt of temperature taking was shifting to the home.

From *Today's Health* 30 (December 1952). Photograph by A. Devaney.

infection.")[50] Well into the 1960s it was still the thermometer-wielding doctor on a house call who pronounced a child's fever safe or dangerous.

That most fevers were just concomitants of the common cold had not prevented Bundesen from insisting on medical management. And certainly, before 1950 there was still plenty to worry about. The "normal" childhood diseases sometimes killed; typhoid lingered; and polio was always a threat. A fever that continued might signify tuberculosis or some worrisome chronic condition.[51] Yet Bundesen closed with an ambivalent and sobering admission of the mother's power and responsibility: "Many persons are alive today who were saved by a mother's careful nursing."[52] For at the same time that fever was being trivialized, the expectations of motherhood were rising. Writing in *Parents* a few years later, Adeline Bullock unhelpfully amplified. The doctor's job was, in effect, to cure, "to take steps to lower a fever and bring relief." The mother's was to "face fever fearlessly by understanding it and knowing how to handle it intelligently," whatever that meant.[53] Prior to the 1960s, the mother's chief job was to keep the sick child in bed but also "comfortable, warm and happy, so as to assure the least possible physical exertion." A necessary two days (at a minimum) in bed "may seem a hardship to you and the child," admitted Kenyon, and to be "quite unnecessary. But it pays!"[54]

But whom did it pay, and who did the paying? If at the foreground of fever concern was what one's child needed, public-health considerations lay not too far behind. For bed rest was also isolation. Besides the question of a fever's relative gravity, there was the question of its contagiousness. "Because fever may be a signal for the onset of any one of the communicable diseases, [the child] should be isolated until the cause can be found," wrote Kenyon. But could amateur mothers, or any ordinary citizen, be entrusted with contagion control? To Bernard Fantus, writing in the American Medical Association's popular organ *Hygeia* in 1935, a long bed stay enforced by a doctor was ultimately a duty of medical citizenship: "As the patient . . . may give off the infectious material . . . one of the most important functions of the doctor is to protect humanity. . . . Therefore, when you have fever, you should call a doctor; but whether you do or do not, *you must go to bed and stay there* until you have completely recovered your resistance." Here the individual's needs (recovering resistance) meshed with the state's—a little too easily perhaps?[55]

These authors were braiding three legacies. The oldest was that of smallpox control: a person, however seemingly healthy, who ventured into a public space while still contagious endangered others. The other two came from typhoid nursing. Absolute rest had been seen as essential to prevent hemorrhage or perforation, and there had been much concern with adequate nourishment of the difficult-to-feed typhoidic body. As cases could last a month or longer, any unnecessary physical activities represented danger-ous and uncompensated metabolic demands.

That none of these criteria were pertinent to the vast majority of brief viral fevers suffered by American schoolchildren in the 1930s through 1950s didn't matter. Absolute bed rest was insisted upon to avoid exhaustion. The doctor would "tell the mother that every bit of energy the child has must be used to combat the disease," noted Bundesen.[56] Whipple agreed. "The child needs all the strength he can muster. . . . If . . . allowed to run around (or drag and whine around) he will waste energy by exercising and have less left for the extra pumping of his heart and lungs."[57] Fantus too, though writing for adults, insisted that "we must go to bed immediately when there is fever, even of a slight degree." The patient was in no position to know how much rest he or she needed, nor what would be the consequences of overdoing it. Awakening following a mild flu, and feeling better, "a person, just at the point of recovering his resistance, may lose it by a single chilling . . . as . . . when he leaves the bed and goes into a bath-room which may be inadequately heated. This is why absolute rest in bed, including the use of the bedpan day and night, should be insisted on even in the mildest cases of fever."[58] Bundesen simi-larly thought the mother should carry the child to the toilet.[59]

Worries about uncompensated activity joined with the ther-mostat metaphor to produce an image of the fevered body as a self-combustible furnace. Many wrote to contest the adage about starving a fever, suggesting that a fevered and starved body might burn itself up. "Fevers are wasting diseases. The body fires are raised in intensity. . . . Fires call for fuel, and in the absence . . . the stored body sugars go first, then the proteins and fats, . . . The patient literally wastes away." Longstanding concerns about di-gestibility were swept aside. Not only the Hippocratic gruels and the meat extracts of the nineteenth century but also milk and cream were recommended in *Good Housekeeping* in 1936.[60] (Here, tuberculosis treatment was very likely the model.) Remarkably, after more than two millennia writers were still understanding the treatment of a protean fever in terms of managing the non-naturals.

But real parents had to decide how much rest was enough. Identifying with those parents, Adeline Bullock in 1950 struck out in a new direction. Many had "an exaggerated fear of fever," she wrote. But like a headache or vomiting, it was merely a symptom. "We don't hop into bed every time we have a slight headache, nor do we tuck Junior snugly between covers at his slightest up-chuck." She imagined a coffee klatch of suburban mothers, with one complaining of another (presumably absent): "Ellen should throw away her thermometer. . . . She's forever taking Bob's temperature and if it's one-fifth of a degree off, she puts him to bed and calls the doctor." Moderator Bullock insists that the opposite—assuming that every fever was a normal part of childhood—was also wrong: "Every mother should have a thermometer. Taking temperatures is important." But considerable fluctuation was normal, she reminded readers. "Even the healthiest of us would probably be surprised were we to take our temperatures every two hours." Ultimately Bullock would not say where on the spectrum of concern the good mother would perch.[61] Another writer observed in 1959 that mere anxiety generated transitory fever: "Children, frightened by going to the hospital, often run fevers in the 101° to 103° range." One draft-board physician found that the temperatures of 324 presumably anxious examinees averaged nearly a degree above normal.[62]

By the end of the 1950s confidence in the child's wise body was such that what had been bad mothering two decades earlier was now good mothering. In 1930 Kenyon had hammered casual moms who let their children outdoors to play a day after having a fever. By 1959, some popularizers, citing the views of the Stanford pediatrician Alan K. Done, were recognizing a reprehensible conspiracy of parents and doctors who were denying children a natural and fulfilling fever experience. "Anti-fever therapy is often employed more for the benefit of parents, or the physician, than the child," wrote J. D. Ratcliff in *Reader's Digest.* "It is doubtful

Magazines for mothers recognized the uncertainty about how to respond to fevered children. Views were plentiful, and consensus might be difficult. This 1924 illustration suggests the involvement of many mothers and a servant.

From *Hygeia* 2 (1924): 429.

whether body temperatures in the range of 104° are harmful, even if prolonged for several days."[63] Done was still being quoted in 1981 by writers who attributed much of the fevered child's discomfort to unnatural temperature variation as antipyretic drugs wore off.[64] Another sage, Dr. John Gibson of Texas, showed that bed rest did not shorten children's fevers. "While fever may make some children feel listless or cranky, many remain as happy as clams despite fevers that would put an adult to bed."[65]

The trivialization of fever was almost complete by the mid-1970s. Fevers, authors reminded readers, were usually viral and self-limiting. Doctors had no secret remedies.[66] When should parents worry? Dr. Richard Feinbloom would bring "cool facts" to such "middle-of-the-night questions," wrote the editors of *Red-book* in 1974. Usually it would be unnecessary to whisk a baby to "an unwelcoming, frightening hospital emergency room" or to phone and risk "the wrath of [an] overburdened, perhaps unsympathetic physician."[67] There was no best advice, an anonymous 1982 *Glamour* author flatly declared: "Most experts say the best rule of thumb in treating someone with fever is to do whatever makes the person feel most comfortable."[68] Still, many authors played it safe with banal ambivalence. "Don't panic when your child has a fever, medical experts advise, but don't ignore it, either."[69] Four decades later, physicians trying to stem unnecessary emergency-room visits have recognized "fever phobia," the lasting legacy of that dependence.

DISAGREEMENTS OF DEGREES

"When the doctor comes, the first thing he does is to stick his clinical thermometer under the patient's tongue and find out how much 'fever' he has. If the fever is high—around 104 or 105 degrees, the doctor looks grave and begins to write a prescription to 'bring the fever down.' "[70] In 1929 taking and interpreting temperatures were diagnostic matters and still mainly a doctor's business. Drugstores would change that. The increasing presence in the American home of the short stick thermometer preceded consensus about its proper use. It ushered in an underappreciated medical watershed, of putting important diagnostic techniques in the hands of the patient. We now use such self-diagnosis techniques to calibrate many aspects of our bodies—to determine whether a breast or testicle feels normal, to read the color of a stool, to decide whether a skin spot is something to worry about, even to determine whether one is depressed, pregnant, or hyperglycemic.

Three problems came with the thermometer's domestication: how to use it, when to use it, and how to interpret the results. By the 1930s there was general agreement that a thermometer was a household necessity. But finding the right angle, light, and background for reading the silver sliver of mercury was no mean feat. "When you buy a thermometer, learn to read it," Kenyon counseled. "The druggist from whom you buy it, or your doctor or nurse will teach you."[71]

The matter of temperature access sites had been simplified since Wunderlich but not necessarily clarified. Vaginal temperature, except during ovulation times, was no longer an option (oddly, earlier writers seem to have overlooked its inapplicability to half the population).[72] Armpits and groins were out; to some, reliance on the undisciplined armpit was Wunderlich's great failing. This left mouth and rectum.

Recent experience with increasingly noninvasive thermometric technologies, together with the contemporary sensibilities about violation of orifices, make it hard to assess earlier sensibilities about rectal temperature taking. While rectally administered medications were once far more common than they have since become, it does not follow that rectal temperatures were culturally unproblematic in earlier decades. They were more accurate, most authors agreed, while admitting that such accuracy was superfluous. Convenience and setting were important, but so were sensibilities and power. In America, the emerging convention would be to reserve rectal temperature taking for the very young and the very old; the key mark of human competence was the ability to hold a thermometer under one's tongue for the requisite minutes. But how was that competence to be ascertained? Rarely do fever writers address these issues directly, and when they do, they often do so with evident unease.

Fantus, writing in 1935, took an extreme position: the fever thermometer was "a household necessity" to be used "preferably . . . by rectum" whenever "any member of the family does not feel well."[73] Kenyon similarly praised the superior accuracy of rectal temperatures. While she did not explicitly assert their primacy for adults, she did recognize acculturation to rectal temperature taking as an essential part of child training: "Children should become accustomed to having the thermometer inserted into the rectum as part of a routine health supervision and not mind it more than having someone look down their throat, etc."[74] But the passage implies that some children did mind. Lentz, writing about both children and elderly persons, insisted that the thermometer be "carefully inserted (one inch) and removed instantly if the patient becomes irrational or uncooperative."[75] (What, other than cooperation, rationality comprised he did not say.) Bundesen declared the minimal age of competence to be four: before that age, temperature was "always" to be taken rectally; for children above that age, temperature might be taken orally, so long as the child had been properly "trained" not to bite and to keep his or her mouth closed for at least a minute.[76] By 1980 the issue had become dignity. For the child who found, or was presumed to find, rectal temperature taking undignified but who could not hold the thermometer beneath the tongue, the armpit was again an option.[77]

There was confusion too about what an elevated temperature meant, about the border between normal and fevered. The canonical 98.6°F was actually only a rounded-off average, many triumphantly discovered. But having unfairly represented Wunderlich (who, fascinated by variation, had never claimed otherwise) as doctrinaire, they faced the problem of finding some rule a mother could apply in deciding whether to worry. Usually an elevation greater than one degree Fahrenheit was to count as fever, especially if it lasted more than a day and there were other symptoms. The site mattered, however. A normal rectal temperature was generally taken to be one degree higher than an oral temperature. Fantus had offered no wiggle room: "If the [rectal] temperature registers above 99.5 F., the patient should be put to bed immediately."[78] Since 99.6°F would have been normal according to the one-degree rule, he would evidently have had everyone always abed.

HOT BODS

As an interesting illness fever has declined. As a metaphor of desire it continues to rise. However tempting it may be to separate figurative from literal usage, one must be cautious. Modern biomedicine has no monopoly on *fever*. Even in recognizably pathological senses, it has meant many things. It has been nature's enforcer of good conduct, scourge of peoples, awful embodier of

distant place or malign wind, uniter of rich and poor or demarcator of the subhuman, door to transcendence, manifestation of the
warm-blooded body's amazing equilibrating powers, occasion of
the most difficult gift of care humans can supply to one another,
and finally, common usher of one's dissolution (which began, in
the case of "putrid" fevers, while one yet breathed).

But the term has been stretched too. There are three chief
strains of fever metaphors: (*a*) epidemic irrationality, as in "gold
fever" or a "selling fever" in the stock market; (*b*) uncontrollable
emotions; and (*c*) physical manifestations typical of fever. These
are not as distant from each other or from fever's more medical
meanings as one might think. Many earlier writers noted that intense emotion could evolve into the somatic condition of fever
and that fever undermined rationality.

Nowadays the most conspicuous fever metaphors have to
do with sex. Sex and fever do overlap physiologically—a matter
not just of heat but of breathing, perspiration, skin color, visage,
and the sensitivity of peripheral nerves. A Hippocratic metaphor
maker might have gone further to include crisis and the postcoital
or postfebrile sleep that follows. A modern public-health worker
might highlight the uncontrollable emission of fluids and the exchange of contagious material in both. And the link need not be
metaphorical at all. Ancient Greek love magic relied on the unity
of these experiences; to the Manyika of Zimbabwe, febrile and
sexual hotness were one; the parents of a fevered child must avoid
intercourse lest they increase the child's heat.[79]

None of the strains are new, but nor are they as common as
we might expect.[80] The works of Shakespeare contain twenty-
three instances of the singular *fever*.[81] Fifteen are plainly medical,
including some focusing on febrile madness. The remainder fit
these strains of metaphor. In *King John* 2.1 there is a reference to a
military assault as "a shaking fever in your walls" (*c*). In *Macbeth*
3.2, the murdered Duncan has been relieved of "life's fitful fever"
(*c*). (Here death is normality and life is disease, as in Poe's phrase
"The fever called 'Living.'") In *Measure for Measure* 3.2 there is
"a fever on goodness" (*a*). In *Henry V* 4.1 there is reference to the

"fiery fever" of love of adulation (*b*). Similarly, in *Troilus and Cressida* 1.3 there are two references to "an envious fever" among Trojan generals (*b*). More closely tied to modern uses are references in *Love's Labour's Lost* 4.3: "I would forget her; but a fever she / Reigns in my blood and will remember'd be" (*b*). The template for what is to come, however, is sonnet 147. It opens with a self-pitying lament in which passion is expressed as fever: "My love is as a fever, longing still / For that which longer nurseth the disease." The poet laments further: "My reason, the physician to my love, / Angry that his prescriptions are not kept, / Hath left me. . . . / . . . / Past cure I am, now reason is past care." It ends, however, with admission of the fevered illusion that has led to ill-advised love (and, in the views of some, to a sexually transmitted infection): "For I have sworn thee fair and thought thee bright, / Who art as black as hell, as dark as night."

However similar in substance, modern incarnations of passion as fever are generally bright and playful. That in itself is a remarkable shift. Pickup lines contain versions of "You are hot and make me hot," but for most diseases or symptoms (e.g., diarrhea or tumors) analogous metaphors are unthinkable. "You make me sick" will not work.

The index case of this shift is, I am persuaded, Peggy Lee's cool, gritty version of "Fever," icon of the smoky, jazz-club scene of the 1960s. The song was written in 1956 by Eddie J. Cooley and Otis Blackwell (a.k.a. John Davenport) for the R&B artist Little Willie John. Many will recognize its chorus: "You give me fever, / When you kiss me, Fever when you hold me tight / Fever, in the morning, fever all through the night." But beginning in 1958, Lee made it famous. She filled out the song with additional verses projecting feverish desire back through the ages, verses that many later artists have omitted. Declaring that "fever started long ago," in one verse she enlisted Romeo, who addresses Juliet: "Julie, Baby, you're my flame . . . / Thou giveth fever / When we kisseth." In another verse Pocahontas begs her angry father not to kill Captain John Smith, with whom she has had a "mad affair." "He gives me fever with his kisses," she protests. The ending, "What a lovely way to burn," jars.[82] If to some, like the early twentieth-century physiologists who studied, as thermal regulation, the fanning and banking of the fires of life, fever might conceivably be a "lovely way to burn," for any who had suffered serious fever such lines would be perverse.

By one count, the song has been recorded more than nine hundred times and by an astonishing range of artists, from Conway Twitty to James Brown, Bob Dylan to Tina Turner, the Doors to Sarah Vaughn, Elvis to Ella Fitzgerald. For Madonna and Beyonce it became a major part of their repertoire. There are versions in Chinese, Dutch, French, German, Polish, Spanish, and Swedish.[83]

Many have adapted the theme. Elvis, in "Burning Love," feels his temperature rising (to 109°F). Foreigner is famously "hot blooded": "check it and see / I got a fever of a hundred and three." In "Cat-Scratch Fever," Ted Nugent is sexually initiated by the "kitty" next door.[84] Depending on one's interpretation, "Fever" by Family Force 5 refers to ordinary sexual attraction or to possession by the Holy Spirit: "Hot to the touch / to the nth degree / Thermometer's reading / About infinity."[85] Keith's jingle "98.6" (by G. Fischoff and T. Powers) is rare in that love *cures* fever: "Hey, ninety eight point six, it's good to have you back again." And the master of the metaphor is "the Docta," the rock DJ Dr. Johnny Fever, of the 1980s television series *WKRP in Cincinnati.*

HOT MONKEY BODS

I find it easy to enjoy, but also to dismiss, the association of sex with fever—youth at play. Not so with the other main line of pop fever culture, fantasy pandemics. Here the resonances are political and disturbing. The motif—in novels, films (e.g., *Outbreak* [1995], *Doomsday* [2008], and *Contagion* [2011]), and now a video game (Plague Inc. [2013])—is familiar.[86] A new strain of fever, horrific in its effects and terrifying in the rapidity of its airborne spread, threatens to break out of its distant homeland and attack the United States. The protagonists, agonized biomedical scientists, may need to wall off those who have been exposed in order to prevent its spread. In its modern form, the scenario arose in Michael Crichton's *The Andromeda Strain* (1969).

In the journalist Richard Preston's *The Hot Zone* (1994) it crossed from fiction into fact. Preston described not only the emerging viruses of sub-Saharan Africa but also a near outbreak in the United States, in the holding pens of monkey importers in the suburbs of Washington, DC. Notably the responders were members not of the CDC, the Public Health Service, much less the World Health Organization, but of the U.S. Army's Institute of Infectious Diseases (USAMRIID). As doctors and scientists as well as warriors, the responders were appealing characters. No brutality could come from persons so noble.

Hot, in Preston's world, had nothing to do with victims' perceptions or nurses' thermometers. It referred to contagious persons or places and derived from radioactive hazard, just as the larger genre of contagion fiction drew on Cold War themes of difficult moral choices that leaders must make.[87] Some, understandably, doubted that Preston's book was nonfiction. But not only did *The Hot Zone*'s plot and characters echo earlier post-apocalyptic fiction, one of Preston's interviewees saw USAMRIID's work in terms of the fictional *Andromeda Strain.*[88]

Of course infectious diseases evolve. There have been highly deadly and easily caught airborne diseases in the past, most recently the great flu of 1919. The prospect of deadly new diseases is a disturbing one, but so too is the accompanying ideology. For these modern plague tracts are rehearsals for extreme measures. In any other sphere the unanswerable imperative fever supplies would be unacceptable, recognizable as totalitarianism. There is reason for concern. In the past, campaigns against fever have been enmeshed in imperialism or, infamously, in anti-Semitism.[89] The modern campaigns highlight place and, unavoidably, race. SARS, bird flu, and Ebola arise in remote places whose abjectness is reconfirmed in those outbreaks.

Another feature of this contagion genre has been the dehumanization of victims. Preston's gusto for describing viral ravages on the human interior rivals Homer's preoccupation with battlefield anatomy. No longer is fever a matter of unsettling sensations; it has become instead oozing slime. The most fearsome filoviruses, Ebola and Marburg, liquefy underlying tissue, withdrawing affect from the face while inflaming the brain with manic madness. All the while humans are dissolving, Preston accentuates the human characteristics of the other species, the captured and infected crab-eating monkeys. Plotters and tantrum throwers, they are like shrewd if dangerous children. What must be done to stem this plague in monkeys is precisely what we are invited to do to fictional zombies: the Army's veterinarians must kill that population (or race?) before it kills us.[90]

Zombies are the reigning fantasy of postapocalyptic contagion; to moviegoers, the genre of virus (hot-zone) films interpenetrates with zombie films.[91] Themselves victims of infection, our modern zombies signify not the needy sick but the contagious threat. We are to look past their human shape, for here vector and agent have merged. They are, in effect, the germs of an incurable disease, and like Preston's Ebola sufferers—both simian and human—they seem to want to bite deep and convert us to their biotic state. Even the CDC has recognized (and exploited) the power of the genre: zombie preparedness is pandemic preparedness.[92]

Horrific symptoms are hardly new to the history of fever. With their oozing sores and their stupor giving way to manic strength, these zombies (and hemorrhagic fever victims too) resemble the "putrid fever" victims described in eighteenth-century accounts. Yet describers of "putrid" symptoms had not sought to elicit revulsion or confrontation with apocalyptic horror. In the era before stainless steel, disinfectants, and refrigerators, putridity would have been familiar. So too would have been fever itself, in the long fevered ages that make up so much of human history. Fever is properly a unifier rather than a divider of human experience, Susruta, the Hippocratic writers, and van Swieten insist. With the exception of plague, before the nineteenth century fever had rarely been an occasion to indulge those powerful dichotomies that we see today—we and they, here and there, clean and dirty, subject and abject, progressive and stagnant, even good and evil.

And finally, last days. As Preston recognized, the end of the Cold War left pandemic fevers as the readiest idiom for contemplating the end times that the sin of the species would bring. For Preston and others the mechanisms would be ecological. As humans, pushed on by overpopulation or global capitalism invaded the habitats where viruses had been peacefully coevolving with their primary hosts, we became secondary and therefore disposable hosts. Preston represented the 90 percent fatality rate of Ebola Zaire as a triggering of Earth's own immune system: "Perhaps the biosphere does not 'like' the idea of five billion humans."[93] And yet fever outbreaks have long been understood in terms of maladaptation to or misuse of nature without eliciting chest-beating lamentations about last days. Norfolkers and early modern Italians knew that visitations would come. They might be fled from or perhaps mitigated, but they were part of the regular operation of the cosmos.

Most disturbing of all is our intoxication with these scary stories. Recently, a maladroit radio interviewer casually categorized aficionados of the (nonfictional) genre of contagion books as "disease enthusiasts." Taken aback, the interviewee, author of a new journalistic account of Ebola, proposed instead the label "health enthusiasts," which is politically correct and banal but also inaccurate.[94]

So, an indulgence: In one of these fictions as the "contaminated" die their horrible deaths and the scientists agonize, safe yet buff in their hazmat suits, I imagine the common side plot, attraction amid danger. As virologist falls for epidemiologist, imagine, as a backdrop to their passion, Peggy Lee's immortal "Fever": ". . . fever when you kiss me, / fever when you hold me tight." Bad taste? Cognitive dissonance at least? That many might see *no* irony indicates how far fever has evolved.

THE FEVER OF THE FUTURE

Lee's achievement, the metaphorical linking of two kinds of hot bodies, represents a unique moment in the history of human disease. It occurred during an age and in a place of mild and trivial fevers. Can it survive?

If fevers continue to yield to drugs and epidemiological transitions and become ever rarer, the analogy will no longer work. It will be meaningless to relate desire to fever if no one can remember what fever is like. Perhaps *fever,* like *poxy, aguish,* or *neurasthenic,* will become archaic.

Alternatively, if the prophets of Ebola are right, if our antibiotics lose to the newly robust bacteria, or if Anopheles returns to northern lands to enjoy the wetlands we have so painstakingly restored, fever's dominant links may again be to death, not life, to suffering, or care, or existential challenge, not youthful sex play.

Or finally, perhaps an exacerbation of the present, a widening gulf between lives that are effectively fever free and those in which fever pervades and always threatens. One can imagine worsening racism founded in a fever differential; the precedents are plentiful.

To expect fever to be a stable concept seems naive. If it is so fungible, what work might fever do? Can it—the idea, not the illness—even be tutelary? I am most struck by two common themes of classical medicine. First, fever as a key site for encountering and acknowledging human vulnerability, both to the coming of fever and during fever. Second, fever as an access to otherness, not just the odd transitory state of one's own fevered body and mind but also, by extension, the unique embodiment of others, whether fevered or nonfevered, precisely what the divisiveness of so much modern fever culture subverts. To bring meanings to fever seems quite right, but it is best to be careful in deciding which ones.

NOTES

Chapter 1. More than HOT

1. Y. Michael Barilan, "The Doctor by Luke Fildes: An Icon in Context," *Journal of Medical Humanities* 28 (2007): 59–80. Some say the child is a boy, others a girl.

2. Heather Birchall, *The Doctor,* accessed 3 February 2012, www.tate.org.uk/art/artworks/fildes-the-doctor-n01522.

3. Gerard van Swieten, *Commentaries upon Boerhaave's Aphorisms concerning the Knowledge and Cure of Diseases,* 15 vols. (Edinburgh: Charles Eliot, 1776), 5:3.

4. Philip Mackowiak, "Concepts of Fever," *Archives of Internal Medicine* 158 (1998): 1870–81.

5. John Cormack, *Natural History, Pathology, and Treatment of the Epidemic Fever at Present Prevailing in Edinburgh and Other Towns; Illustrated by Cases and Dissections* (London: John Churchill, 1843), 149; John Huxham, *An Essay on Fevers,* introduction by Saul Jarcho, MD (Canton, MA: Science History, 1988), 63.

6. Jane Moore, "What Sir Luke Fildes' 1887 Painting *The Doctor* Can Teach Us about the Practice of Medicine Today," *British Journal of General Practice* 58 (2008): 210–13.

7. Harriet Martineau, *Sickness and Health of the People of Bleaburn* (Boston: Crosby, Nicholls, 1853), 60. On patients' experiences, see the pioneering *Patients and Practitioners: Lay Perceptions of Medicine in Pre-industrial Society,* edited by Roy Porter (Cambridge: Cambridge University Press, 1985).

8. Margaret Currie, *Fever Hospitals and Fever Nurses: A British Social History of Fever Nursing; A National Service* (London: Routledge, 2005).

9. Sonia Shah, *The Fever: How Malaria Has Ruled Humankind for 500,000 Years* (New York: Sarah Crichton Books / Farrar, Straus, & Giroux, 2010), 36, 125.

10. Van Swieten, *Commentaries,* 5:144, 6:351–52.

11. For new diseases, see Lloyd G. Stevenson, "'New' Diseases of the Seventeenth Century," *Bulletin of the History of Medicine* 39 (1965): 1–21. For eloquent cautions, see Hans Zinsser, *Rats, Lice, and History* (Boston: Little, Brown, 1935), 57, 73, 89, 101; Mirko Grmek, *Diseases in the Ancient Greek World,* translated by Mireille Muellner and Leonard Muellner (Baltimore: Johns Hopkins University Press, 1989), 7; and Charles Creighton, *A History of Epidemics in Britain,* vol. 2, *From the Extinction of Plague to the Present Time* (Cambridge: Cambridge University Press, 1894), 4–5, 75.

12. Charles E. Rosenberg, "The Tyranny of Diagnosis: Specific Entities and

Individual Experience," *Milbank Quarterly* 80 (2002): 237–60; Walther Riese, *The Conception of Disease: Its History, Its Versions, and Its Nature* (New York: Philosophical Library, 1953), 82–90; Knud Faber, *Nosography in Modern Internal Medicine* (New York: Paul Hoeber, 1922), 66–67; Owsei Temkin, *The Double Face of Janus, and Other Essays in the History of Medicine* (Baltimore: Johns Hopkins University Press, 1977).

13. Norman D. Jewson, "The Disappearance of the Sick-Man from Medical Cosmology, 1770–1870," *Sociology* 10 (1976): 225–44; Charles E. Rosenberg, "The Therapeutic Revolution: Medicine, Meaning, and Social Change in Nineteenth-Century America," *Perspectives in Biology and Medicine* 20 (1977): 485–506; Roy Porter, "Laymen, Doctors and Medical Knowledge in the Eighteenth Century: The Evidence of the *Gentleman's Magazine*," in Porter, *Patients and Practitioners*, 283–314; Michel Foucault, *The Birth of the Clinic: An Archaeology of Medical Perception* (New York: Pantheon, 1973). The particular starting and ending points Foucault chose—from the time of Sauvages to that of Broussais—oversimplify the trajectories of European medicine, though they do not significantly impair the value of the insights. On the dictatorship of the temperature chart, see Faber, *Nosography,* 73–74. For a classic case of that transformation in American clinical medicine, see John Harley Warner, *The Therapeutic Perspective: Medical Practice, Knowledge, and Identity in America, 1820–1885* (Cambridge, MA: Harvard University Press, 1986).

14. "Gore: The Planet Has a Fever," accessed 27 January 2012, www.cbsnews .com/video/watch/?id=2592503n.

15. Geotherapeutics, accessed 14 July 2012, www.angelfire.com/mac/egmatthews /geotherapy/geotherapy.html.

Part I. The Fevers of Classical Medicines

1. Hippocrates, *Regimen III*, in *Hippocrates: Volume IV,* translated by W. H. S. Jones, Loeb Classical Library (Cambridge, MA: Harvard University Press, 1959), §§83–84. Working out looms large in Hippocratic medicine. See Vivian Nutton, *Ancient Medicine* (London: Routledge, 2004), 64.

2. Hippocrates, *Regimen II,* in *Hippocrates: Volume IV,* §46.

3. Galen, *On the Causes of Diseases* 2.2, in *Galen on Diseases and Symptoms,* translated, with introduction and notes, by Ian Johnston (Cambridge: Cambridge University Press, 2006), 161.

4. See John Huxham, *An Essay on Fevers,* introduction by Saul Jarcho, MD (Canton MA: Science History, 1988), 1; and Herman Boerhaave, *Dr. Boerhaave's Academical Lectures on the Theory of Physic,* 6 vols. (London: W. Innys, 1746), 5:394.

5. Gerard van Swieten, *Commentaries upon Boerhaave's Aphorisms concerning the Knowledge and Cure of Diseases,* 15 vols. (Edinburgh: Charles Eliot, 1776), 6:213. See also Jacques Jouanna, *Hippocrates* (Baltimore: Johns Hopkins University Press, 1999), 151.

6. Andrew Wear, *Knowledge and Practice in English Medicine, 1550–1680* (Cam-

bridge: Cambridge University Press, 2000), 147. See also Owsei Temkin, *Galenism: Rise and Decline of a Medical Philosophy* (Ithaca, NY: Cornell University Press, 1973), 130–37, 181. Classical Chinese medicine likewise accommodated substantial change, notes Marta Hanson. "Robust Northerners and Delicate Southerners: The Nineteenth-Century Invention of a Southern Medical Tradition," in *Innovation in Chinese Medicine,* edited by Elisabeth Hsu (Cambridge: Cambridge University Press, 2001), 262–91 at 262.

7. Don G. Bates, "Scholarly Ways of Knowing: An Introduction," in *Knowledge and the Scholarly Medical Traditions,* edited by Bates (Cambridge: Cambridge University Press, 1995), 1–22.

8. Walther Riese, *The Conception of Disease: Its History, Its Versions, and Its Nature* (New York: Philosophical Library, 1953), 3–4.

9. Jouanna, *Hippocrates,* 317; Margaret Trawick, "Writing the Body and Ruling the Land: Western Reflections on Chinese and Indian Medicine," in Bates, *Knowledge and the Scholarly Medical Traditions,* 279–96.

Chapter 2. Words

1. *The Susruta Samhita—An English Translation Based on Original Texts,* edited by Kaviraj Kunja Lal Bhishagratna, 3 vols. (Delhi: Kaviraj Kunja Lal Bhishagratna, 1907–16), vol. 3, *Uttara-Tantra,* ch. 39.

2. *Oxford English Dictionary,* 2nd ed., s.v. "woozy."

3. Shigehisa Kuriyama, *The Expressiveness of the Body and the Divergence of Greek and Chinese Medicine* (Cambridge, MA: MIT Press, 2006), 67–69. See also Margaret Trawick, "Writing the Body and Ruling the Land: Western Reflections on Chinese and Indian Medicine," in *Knowledge and the Scholarly Medical Traditions,* edited by Don G. Bates (Cambridge: Cambridge University Press, 1995), 279–96.

4. Paul Unschuld, translator of the *Nan-Ching,* the great book of difficult issues, outlines a body filled with "depots" and "palaces," terms that advertise analogy. The eleven or twelve vascular systems are the tracks between depots, between organs of the body, or between organ and surface. But, he notes, even when we read of organs, we should think not of dissectible entities but of functions: *liver* means what a liver does. Bian Que, *Nanjing: The Classic of Difficult Issues; with Commentaries by Chinese and Japanese Authors from the Third through the Twentieth Century,* edited by Paul Unschuld (Berkeley: University of California Press, 1986), 5–14. See also Kuriyama, *Expressiveness of the Body,* 265.

5. See Vivian Nutton, "Murders and Miracles: Lay Attitudes towards Medicine in Classical Antiquity," in *Patients and Practitioners: Lay Perceptions of Medicine in Pre-industrial Society,* edited by Roy Porter (Cambridge: Cambridge University Press, 1985), 23–53 at 31–32; and Paul F. Burke Jr., "Malaria in the Greco-Roman World: A Historical and Epidemiological Survey," in *Aufstieg und Niedergang der Romischen Welt (ANRW),* edited by Wolfgang Haase, 2nd series, vol. 37/3 (Berlin: Walter de Gruyter, 1996), 2252–81 at 2257–58.

6. *Oxford English Dictionary,* 2nd ed., s.vv. "fever" and "fevered." Analogous problems arise with other terms. In humoral theory, much would be made of differently colored biles. But were these separate things? Some ancient writers discussed the problem. See, for example, Caelius Aurelianus, *On Acute Diseases and On Chronic Diseases,* edited and translated by I. E. Drabkin (Chicago: University of Chicago Press, 1950), 9.

7. See G. E. R. Lloyd, *In the Grip of Disease: Studies in the Greek Imagination* (Oxford: Oxford University Press, 2003), 8–9; Volker Langholf, *Medical Theories in Hippocrates* (Berlin: Walter de Gruyter, 1990), 39; and Mirko Grmek, *Diseases in the Ancient Greek World,* translated by Mireille Muellner and Leonard Muellner (Baltimore: Johns Hopkins University Press, 1989), 3–7.

8. Grmek, *Diseases in the Ancient Greek World,* 1, 293; Andrew Wear, *Knowledge and Practice in English Medicine, 1550–1680* (Cambridge: Cambridge University Press, 2000), 108, 144.

9. Volker Hess, *Der wohltempeierte Mensch: Wissenschaft und Alltag des Fiebermessens (1850–1900)* (Frankfurt: Campus Verlag, 2000), 19, 39–42.

10. *Susruta Samhita,* vol. 3, ch. 39, §§4–5.

11. Catherine Despeux, "The System of Five Circulatory Phases and the Six Season Influences *(wuyan liuqui),* a Source of Innovation in Medicine under the Song (960–1279)," translated by Janet Lloyd, in *Innovation in Chinese Medicine,* edited by Elisabeth Hsu (Cambridge: Cambridge University Press, 2001) 121–65 at 143; Marta Hanson, "Robust Northerners and Delicate Southerners: The Nineteenth-Century Invention of a Southern Medical Tradition," in ibid., 262–91 at 263–64; Elisabeth Hsu, "Pulse Diagnostics in the Western Han: How *mai* and *qi* Determine *bing,*" in ibid., 51–90 at 62; Bian Que, *Nanjing,* issue 56, pp. 499–502; *Huang Di Nei Jing Su Wên: The Yellow Emperor's Classic of Internal Medicine,* translated by Ilza Veith (Berkeley: University of California Press, 1966), 51; Dean C. Epler Jr., "The Concept of Disease in an Ancient Chinese Medical Text: The *Discourse on Cold-Damage Disorders (Shang-han Lun),*" *Journal of the History of Medicine* 43 (1988): 8–35 at 15–16.

12. W. H. S. Jones and E. T. Withington, *Malaria and Greek History* (1909; reprint, New York: AMS, 1977), 41.

13. Hippocrates, *Internal Affections,* in *Hippocrates: Volume VI,* translated by Paul Potter, Loeb Classical Library (Cambridge, MA: Harvard University Press, 1988), §39.

14. Robert Christison, "Fever," in *A System of Practical Medicine: Comprised in a Series of Original Dissertations,* edited by Alexander Tweedie and W. W. Gerhard, 3 vols. (Philadelphia: Lea & Blanchard, 1842), 1:123; Gerard van Swieten, *Commentaries upon Boerhaave's Aphorisms concerning the Knowledge and Cure of Diseases,* 15 vols. (Edinburgh: Charles Eliot, 1776), 5:2–3.

15. Some have understood the appeal to polarities more in cosmological terms, as a generic way of ordering the world, than as generalized experience. See Ginnie Smith, "Prescribing the Rules of Health: Self-Help and Advice in the Later Eighteenth Century," in Porter, *Patients and Practitioners,* 249–82 at 256–58.

16. Grmek, *Diseases in the Ancient Greek World,* 289–91. Grmek writes that "Several concrete cases of *kaûsos* described in the Hippocratic corpus suggest hypothetical diagnoses that vary from one patient to the next: salmonellosis, malaria, rickettsial infection, acute food poisoning, puerperal septicemia, and, less securely, leptospirosis, relapsing fever, and acute appendicitis" (290). I would add heatstroke. Cf. Langholf, *Medical Theories in Hippocrates,* 155. Some *kaûsos* may be typhoid. See G. E. R. Lloyd, introduction to *Hippocratic Writings,* edited with an introduction by Lloyd, translated by J. Chadwick and W. N. Mann (Harmondsworth, UK: Penguin, 1978), 22; and Daniel Sennert, *Of Agues and Fevers: Their Differences, Signes, and Cures,* translated by N.D.B.M. (London: Lodowyck Loyd, 1658), 31.

17. Van Swieten, *Commentaries,* 7:73–162.

18. See, for example, Francis Adams's practices in translating, *The Seven Books of Paulus Aegineta Translated from the Greek with a Commentary Embracing the Complete View of the Knowledge Formed by the Greeks, Romans, and Arabians on All Subjects Connected with Medicine and Surgery* (London: Sydenham Society, 1844). Compare bk. 2, §5, p. 193 ("hot and ardent"), with §§29–30, pp. 260–64, on the diagnosis and cure of ardent fevers.

19. The "hunka hunka burning love" that Elvis sings of (and which raises his temperature to 109°F) is the direct descendant of *kaûsos.* Elvis Presley, "Burning Love," accessed 8 March 2012, www.youtube.com/watch?v=DcJac6OykfM.

20. Hippocrates, *Epidemics IV,* in *Hippocrates: Volume VII,* edited by Wesley D. Smith, Loeb Classical Library (Cambridge, MA: Harvard University Press, 1994), §10.

21. Ibid., 25. See also Hess, *Der wohltempeierte Mensch,* 28. For Galen, to whom preternatural heat in the heart was a central element of fever theory, *heartburn* would have different implications.

22. *Hippocrates: Volume VII.*

23. See G. E. R. Lloyd, *Adversaries and Authorities: Investigations into Ancient Greek and Chinese Science* (New York: Cambridge University Press, 1996), 112.

24. Hippocrates, *Epidemics VI,* pt. 1, §14, in *Oeuvres complètes d'Hippocrate,* translated and edited by Émile Littré, 10 vols. (1839–61; reprint, Amsterdam: Adolf Hakkert, 1962), 5:274–75.

25. Hess, *Der wohltempeierte Mensch,* 28; C. A. Wunderlich, *On the Temperature in Diseases: A Manual of Medical Thermometry,* translated by W. Bathurst Woodman (London: New Sydenham Society, 1871), 196–97; *Oxford English Dictionary,* 2nd ed., s.v. "pungent." Notably, in opposing *pungent* and *increasing,* the Hippocratic text seems at odds with the notion of a strengthening perception of heat. In *Timaeus* 65d, Plato explains the difference between *acrid* and *pungent:* acridity involves substances that actually dissolve the flesh, in this case the tongue, which senses the acridity; *pungent* refers to matters made "fiery" by the warmth of the mouth, which then rise into the head, "cleaving whatever they encounter." Francis Cornford, *Plato's Cosmology: The Timaeus of Plato Translated with a Running Commentary* (New York: Liberal Arts, 1957). See also Sennert, *Agues and Fevers,* 5, 16.

26. Hippocrates, *Regimen in Acute Diseases,* in *Hippocratic Writings,* 186–204 (§5) at 187. *Acute* refers to dangerous diseases, though at the time the antithesis was usually *pestilential,* not *chronic.* The class of *chronic* disease is more clearly associated with Aretaeus, four hundred years later. Jacques Jouanna, *Hippocrates* (Baltimore: Johns Hopkins University Press, 1999), 153. On the centrality of fever in these diseases, see Caelius Aurelianus, *On Acute Diseases,* 3, 193, 227. On *ague,* see L. J. Bruce-Chwatt, "Ague as Malaria," *Journal of Tropical Medicine and Hygiene* 79 (1976): 168–76 at 168.

27. Grmek, *Diseases in the Ancient Greek World,* 131; Jouanna, *Hippocrates,* 144.

28. Caelius Aurelianus, *On Acute Diseases,* 125–37. Phrenitis and lethargy have been ascribed to cerebral forms of malaria. See Robert Sallares, *Malaria and Rome: A History of Malaria in Ancient Italy* (Oxford: Oxford University Press, 2002), 220–21.

29. Langholf, *Medical Theories in Hippocrates,* 40–46; Hippocrates, *Diseases III,* in *Hippocrates: Volume V,* translated by Paul Potter, Loeb Classical Library (Cambridge, MA: Harvard University Press, 1988), §9; cf. M. J. Geller, "West Meets East: Early Greek and Babylonian Diagnosis," in *Magic and Rationality in Ancient Near Eastern and Graeco-Roman Medicine,* edited by H. F. J. Horstmanshoff and M. Stol (Leiden: Brill, 2004), 11–61.

30. *The Writings of Hippocrates and Galen Epitomised from the Original Latin Translations,* translated by John Redman Coxe (Philadelphia: Lindsay & Blakiston, 1846), 407.

31. A passage in Hippocrates, *Epidemics IV,* refers to "a true typhomania" (§13), which puzzled Galen, who thought of it as a mix of phrenitis and lethargy. In another case, the slave or wife of Enmyris (Myris) (§51) has "something of a typhus" (or "torpor typhöide") or "stupor" that does not involve fever. Cf. the commentary in Littré, *Oeuvres complètes d'Hippocrate,* 6:104, 145. Vivian Nutton suggests that louse-borne typhus would have been unlikely in a warm place such as Greece. *Ancient Medicine* (London: Routledge, 2004), 25.

32. Hippocrates, *Internal Affections,* §§39–44.

33. Jouanna, *Hippocrates,* 114. On dating controversies, see Langholf, *Medical Theories in Hippocrates,* 77.

34. *Epidemics VI,* pt. 3, §18, in *Hippocrates: Volume VII.* See also Wesley D. Smith, introduction to *Hippocrates: Volume VII,* 2; Jouanna, *Hippocrates,* 167; Langholf, *Medical Theories in Hippocrates,* 211–15; and Walther Riese, *The Conception of Disease: Its History, Its Versions, and Its Nature* (New York: Philosophical Library, 1953), 82.

35. W. H. S. Jones and Withington, *Malaria and Greek History,* v, 101–3, 117. Jones takes the latter quotation from John Macculloch, *Malaria: An Essay on the Production and Propagation of this Poison, and the Nature and Localities of the Places by which it is Produced: with an Enumeration of the Diseases caused by it, and of the Means of Preventing or Diminishing them, both at Home and in the Naval and Military Service* (London: Longman, Hurst, Rees, Orme, Brown, and Green, 1827), 437.

The claim is controversial, though views depend on how strong a claim Jones is presumed to be making. See also Grmek, *Diseases in the Ancient Greek World,* 33–39, 92, 97; Nutton, *Ancient Medicine,* 32; Sallares, *Malaria and Rome,* 23–39; and Burke, "Malaria in the Greco-Roman World," 2255.

36. Generally, I use Wesley Smith's translation (1994), comparing it with the mid-nineteenth-century translations of Littré and Coxe. Coxe translated from the Latin of Froesius, whereas Littré compared Greek manuscripts. Yet Littré was more interested than Coxe in imposing disease names. See also Nutton, *Ancient Medicine,* 76; Jouanna, *Hippocrates,* 389; and Wesley D. Smith, *The Hippocratic Tradition* (Ithaca, NY: Cornell University Press, 1979), 121, 149, 237.

37. Jouanna, *Hippocrates,* 115–17. Grmek, *Diseases in the Ancient Greek World,* 128, notes the frequency of trauma in ancient medicine.

38. The text begins with spring diseases and then describes two winter constitutions: at the winter solstice, a long-lasting illness of deep jaundice, burnt tongue, and chilliness (not explicitly fever); in the fluctuating weather after the solstice, an illness of jaundice, swollen glands and tonsils, foul skin (§384), and some fever. Next, after the setting of the Pleiades, comes chilliness with vomiting, appetite loss, a hard spleen, and hemorrhage (by nose and probably also by bowel). An autumnal epidemic brings "hemorrhages; short fevers that returned immediately for a little while; aversions to food; extreme languors and lassitudes; nauseas and heartburns," as well as worms, "shiverings, and bilious complaints" (§386). Last, but referring probably to the preceding spring and summer, are "erratic" and relapsing bilious fevers with appetite loss, crises on days 5–7, and "dysenteries also" (§387). On dating, see Grmek, *Diseases in the Ancient Greek World,* 315–17.

39. Hippocrates, *Epidemics IV,* §18. Hereafter, references to sections of the *Epidemics* are given in the text. I draw on both Coxe and Smith in the following passages.

40. Here there may be implications in the different translations. For example, Coxe's "slow" fever was usually the type of nervous fever that was coming to be distinguished as typhoid in the early nineteenth century.

41. *Epidemics I,* §23, in *Hippocratic Writings,* 100; *Epidemics IV,* §43. There are several similar passages in the Hippocratic corpus. See, for example, *Regimen in Acute Diseases (Appendix),* in *Hippocrates: Volume VI,* §§21–22; and *Humors,* in *Hippocrates: Volume IV,* translated by W. H. S. Jones, Loeb Classical Library (Cambridge MA: Harvard University Press, 1959), 69. See also Langholf, *Medical Theories in Hippocrates,* 53.

42. Riese, *Conception of Disease,* 87. For a shrewd and fascinating assessment of psychological aspects of fever, see Lynn LiDonnici, "Burning for It: Erotic Spells for Fever and Compulsion in the Ancient Mediterranean World," *Greek, Roman and Byzantine Studies* 39 (1998): 63–98.

43. *Epidemics IV,* §15, trans. Coxe, 385; *Epidemics VII,* §25, quoted in Wesley D. Smith, "Implicit Fever Theory in *Epidemics* 5 and 7," in *Theories of Fever from Antiq-*

uity to the Enlightenment, edited by William Bynum and Vivian Nutton (London: Wellcome Institute for the History of Medicine, 1981), 1–18 at 12.

44. Hippocrates, *Diseases II,* in *Hippocrates: Volume V,* §72. See also G. E. R. Lloyd, *Revolutions in Wisdom* (Cambridge: Cambridge University Press, 1991), ch. 1; and Hippocrates, *Internal Affections,* §39. Seeing the dead may be a trope from Mesopotamian medicine. See Langholf, *Medical Theories in Hippocrates,* 52; and Geller, "West Meets East," 43, 50.

45. Hippocrates, *Critical Days,* in *Hippocrates: Volume IX,* translated by Paul Potter, Loeb Classical Library (Cambridge: Cambridge University Press, 2010), §3. See also Jouanna, *Hippocrates,* 295, 306–7; and Caelius Aurelianus, *On Acute Diseases,* 31.

46. Hippocrates, *Regimen in Acute Diseases (Appendix),* §§25, 18, 28.

47. *Epidemics IV,* §57. See also Hippocrates, *Internal Affections,* §43.

48. Langholf, *Medical Theories in Hippocrates,* 37, 123; Geller, "West Meets East," 17, 43–44.

49. But see J. Val. de Hildenbrand, *A Treatise on the Nature, Cause, and Treatment of Contagious Typhus from the German,* translated by S. C. Gross (New York: Elam Bliss, 1829), 1–2.

50. Hippocrates, *Diseases I,* in *Hippocrates: Volume V,* §3.

51. *Susruta Samhita,* vol. 3, ch. 39; P. Kutumbiah, *Ancient Indian Medicine* (Bombay: Longmans Orient, 1962), xxv; Francis Zimmermann, "The Scholar, the Wise Man, and Universals: Three Aspects of Ayurvedic Medicine," in Bates, *Knowledge and the Scholarly Medical Traditions,* 297–319 at 303; Epler, "Concept of Disease," 15, 21.

52. Nutton, *Ancient Medicine,* 78.

53. Hippocrates, *Ancient Medicine,* in *Hippocrates: Volume I,* translated by W. H. S. Jones, Loeb Classical Library (Cambridge MA: Harvard University Press, 1922), §§16–17.

54. Hippocrates, *Regimen II,* in *Hippocrates: Volume IV,* §56.

55. Jouanna, *Hippocrates,* 310; Kuriyama, *Expressiveness of the Body,* 263.

56. Hippocrates, *Regimen III,* in *Hippocrates: Volume IV,* §83.

57. *Regimen I,* in ibid., §34; Jouanna, *Hippocrates,* 315; Epler, "Concept of Disease," 25.

58. See by Hippocrates: *Regimen in Acute Diseases,* in *Hippocrates: Volume II,* translated by W. H. S. Jones, Loeb Classical Library (Cambridge, MA: Harvard University Press, 1923), e.g., §§41, 62–63; *Regimen in Acute Diseases (Appendix),* §§13–16; *Diseases I,* §29.

59. Hippocrates, *Diseases I,* §§23–24.

60. *Susruta Samhita,* vol. 3, ch. 39, §§6, 9; *Agniveśa's Caraka Samhitā,* text with English translation and critical exposition based on *Cakrapāṇi Datta's Āyurveda Dīpikā,* edited and translated by Ram Karan Sharma and Vaidya Bhagwan Dash, 2 vols. (Varanasi: Chowkhamba Sanskrit Series Office, 1977), 2:157; Kutumbiah, *Ancient Indian Medicine,* 88.

61. Jouanna, *Hippocrates,* 319.

62. Hippocrates, *Diseases I*, §25.

63. *Regimen in Acute Diseases (Appendix)*, §32; Langholf, *Medical Theories in Hippocrates*, 80–84, 127–32. Langholf suggests that the roots of some of these understandings date from Egyptian antiquity.

64. Langholf, *Medical Theories in Hippocrates*, 79–84.

65. Epler, in "Concept of Disease," 23–30, notes the complexity of multiple levels of explanation but also the general passivity of organs. For a slightly different explanation of the relation of contraction to cooling and shivering, see Plato, *Timaeus* 62a–b.

66. *Regimen in Acute Diseases (Appendix)*, §1.

67. *Huang Di Nei Jing Su Wên* (Veith), 239–40; cf. *Huangdi nei jing su wen = Yellow Emperor's Canon of Medicine Plain Conversation*, translated by Zhaoguo Li, 3 vols. (Xian: World, 2005), 2:399–400. See also Trawick, "Writing the Body," 288.

68. *Susruta Samhita*, vol. 3, ch. 39, §§19–21.

69. Hippocrates, *Crises*, in *Hippocrates: Volume IX*, §36; see also §7.

70. Grmek, *Diseases in the Ancient Greek World*, 1–2.

71. Langholf, *Medical Theories in Hippocrates*, 93–97, 118–22, 211; Hippocrates, *Critical Days*, §11; Aulus Celsus, *De Medicina*, translated by W. G. Spencer (Cambridge, MA: Harvard University Press, 1948), bk. 1, ch. 4, §§11–16. That a seventeen-day fever might be punctuated by different schedules of stage changes suggests the importance of the number 17 rather than the regularity of febrile dynamics. The author of *Epidemics I* recognized a seventeen-day as well as a fourteen- and a twenty-day one. But its course might be punctuated in several ways. Of two brothers whose fevers start on the same day, one reaches crisis at day six, the other at day seven. After a few afebrile days, both relapse and reach a second, presumably final crisis at day seventeen. The author points out that the usual pattern of this fever is 6:6:5, but it can also be 7:7:3, 7:3:7, 6:6:3:1:1 (Evagon), or 6:7:4 (daughter of Aglaidas). See Hippocrates, *Epidemics I*, §§20–21; and Langholf, *Medical Theories in Hippocrates*, 116–17.

72. *Epidemics I*, §25; *Crises*, §7; *Epidemics III*, in *Hippocratic Writings*, §20. The even days are 4, 6, 8, 10, 14, 20, 24, 30, 40, 60, and 80; the odd days are 3, 5, 7, 9, 11, 17, 21, 27, and 31. The intervals are 2, 2, 2, 4, 6, 4, 6, 10, 20, and 20 for the even days and 2, 2, 2, 2, 6, 4, 6, and 4 for the odd days. See also Langholf, *Medical Theories in Hippocrates*, 95–97, 106, 112.

73. Langholf, *Medical Theories in Hippocrates*, 131, 209. See also *Crises*, §7; and *Epidemics I*, §3.

74. Hippocrates, *Coan Prenotions*, in *Hippocrates: Volume IX*, §134.

75. *Regimen in Acute Diseases (Appendix)*, §39. See also *Crises*, §10. What may seem, and often later became, arbitrariness or circularity—defining crisis in terms of critical day, and vice versa—may not have begun that way. *Crisis*, Langholf notes, had referred initially to a "turning point," implying no particular physiological process. Langholf, *Medical Theories in Hippocrates*, 85.

76. See Langholf, *Medical Theories in Hippocrates*, 130–33.

77. Ibid., 243–44. See also Lloyd, *In the Grip of Disease,* 9, 56.

78. Sallares, *Malaria and Rome,* 52–54; Burke, "Malaria in the Greco-Roman World," 2266.

79. Elizabeth Gaskell, *Ruth* (London: Pickering & Chatto, 2006), 62.

80. Many believe that Galen lived more than a decade longer than this, to around 217. See Nutton, *Ancient Medicine,* 226–27.

81. Ibid., 59–60, 202, 270.

82. Owsei Temkin, *Galenism: The Rise and Decline of a Medical Philosophy* (Ithaca, NY: Cornell University Press, 1973), 122–23.

83. Jouanna, *Hippocrates,* 57; Nutton, *Ancient Medicine,* 81–85, 292; W. Smith, *Hippocratic Tradition,* 210. Aristotle and many since have held that *Nature of Man* was not a work of the true Hippocrates. See Nutton, *Ancient Medicine,* 60.

84. The closest to a composite treatment of fever is the text known as *De Differentiis Febrium, Libri duo.* But there, as Burke complains, Galen offers "77 different varieties of fever . . . many are synonyms; others are merely adjectival designations." Burke, "Malaria in the Greco-Roman World," 2258.

85. Nutton, "Murders and Miracles," 32–37.

86. The four-fevers doctrine came from *Nature of Man,* §15. See Jouanna, *Hippocrates,* 61; cf. Plato, *Timaeus* 86.

87. Grmek, *Diseases in the Ancient Greek World,* 126; Hippocrates, *Diseases I,* §13.

88. Jouanna, *Hippocrates,* 315–16, 373, 384; Nutton, *Ancient Medicine,* 79; Hippocrates, *Diseases I,* §2.

89. Kutumbiah, *Ancient Indian Medicine,* xli; Kuriyama, *Expressiveness of the Body,* 49–54.

90. Plato, *Timaeus* 83b; Nutton, *Ancient Medicine,* 80, 83.

91. *Galen on Diseases and Symptoms,* translated, with introduction and notes, by Ian Johnston (Cambridge: Cambridge University Press, 2006), 115.

92. Ibid., 114.

93. For this approach in later Galenism, see Lester King, *The Road to Medical Enlightenment, 1650–1695* (London: Macdonald, 1970), 32–37.

94. Iain M. Lonie, "Fever Pathology in the Sixteenth Century: Tradition and Innovation," in Bynum and Nutton, *Theories of Fever,* 19–44. See also Sennert, *Agues and Fevers,* 12, 27–28.

95. For the former view, see Wear, *Knowledge and Practice,* 136–39. For contact putrefaction, see Sennert, *Agues and Fevers,* 13–14.

96. Galen, *On the Causes of Disease* 2.2, in *Galen on Diseases and Symptoms,* 161; cf. Sennert, *Agues and Fevers,* 13, 46.

97. Lonie, "Fever Pathology"; cf. Sennert, *Agues and Fevers,* 3. Within an Aristotelian perspective, to call heat a substance does not imply that it is a species of thing; all substances are a combination of matter and qualities. Even after thermometers became available, it was not obvious that they measured the heat of fever. See Hess, *Der wohltempeierte Mensch,* 49–54.

98. Galen, *On the Cause of Diseases* 3.4; Nutton, *Ancient Medicine,* 122.

99. Kuriyama, *Expressiveness of the Body,* 69–78; Nutton, *Ancient Medicine,* 204; Paulus Aegineta, *Seven Books,* bk. 2, §12, pp. 202–22.

100. Peter Brain, *Galen on Bloodletting: A Study of the Origins, Development, and Validity of His Opinions, with a Translation of the Three Works* (Cambridge: Cambridge University Press, 1986); Jouanna, *Hippocrates,* 160; Kuriyama, *Expressiveness of the Body,* 197–208; Sennert, *Agues and Fevers,* 18–19.

101. Celsus, *De Medicina,* preamble, §§52–54, 58–61; cf. Galen, *On Antecedent Causes,* edited by R. J. Hankinson (Cambridge: Cambridge University Press, 1998), 105, 129–33. Johnston, *Galen on Diseases and Symptoms,* 32, quoting from Galen, *On the Differentiation of Symptoms.*

102. Galen, *On the Causes of Diseases* 2.3, p. 162.

103. Galen, *On Antecedent Causes,* 143–45; see also 83, 85. Johnston, in *Galen on Diseases and Symptoms,* 113.

104. Galen, *On Antecedent Causes,* 133–35; see also 75–77, 121.

105. The nineteenth-century reconceptualization of *disease* as implying hidden pathological processes and not merely subjective and superficial symptoms would render the term superfluous.

106. Johnston, in *Galen on Diseases and Symptoms,* 108. For later use of this causal scheme, see King, *Road to Medical Enlightenment,* 33, 127–31; and Christopher Hamlin, "Predisposing Causes and Public Health in the Early Nineteenth Century Public Health Movement," *Social History of Medicine* 5 (1992): 43–70.

107. Disease as the product of multiple determinants displaced causal explanations based on a single factor, including the invoking of gods (Greece), ghosts (Mesopotamia), ancestors (China), and demons (China and India). The striking difference between Susruta's presentation of fever and the handling of the topic in the Atharvaveda is the absence of Tákman, the fever demon, who is to be placated and urged to jump to someone else, often a nubile maiden. Kenneth G. Zysk, *Medicine in the Veda: Religious Healing in the Veda:* (Delhi: Motilal Banarsidass, 1998), 41. For demon theories in the West, see Owsei Temkin, "The Scientific Approach to Disease: Specific Entity and Individual Sickness," in *The Double Face of Janus, and Other Essays in the History of Medicine* (Baltimore: Johns Hopkins University Press, 1977), 441–55 at 443; and Riese, *Conception of Disease,* 78–79.

108. Galen, *On Antecedent Causes,* 123, 137. On the importance of the legal context, see Lloyd, *Adversaries and Authorities,* 94.

Chapter 3. Books

1. James Copland, *A Dictionary of Practical Medicine,* 3 vols. (New York: Harper & Brothers, 1855), 1:1055.

2. Don Bates, "Scholarly Ways of Knowing: An Introduction," in *Knowledge and the Scholarly Medical Traditions,* edited by Bates (Cambridge: Cambridge University Press, 1995), 1–22 at 7.

3. See Ludwig Edelstein, *Ancient Medicine* (Baltimore: Johns Hopkins University Press, 1947).

4. Vivian Nutton, *Ancient Medicine* (London: Routledge, 2004), 261. See also Nutton, "Murders and Miracles: Lay Attitudes towards Medicine in Classical Antiquity," in *Patients and Practitioners: Lay Perceptions of Medicine in Pre-industrial Society*, edited by Roy Porter (Cambridge: Cambridge University Press, 1985), 23–53 at 26–37.

5. P. Kutumbiah, *Ancient Indian Medicine* (Bombay: Longmans Orient, 1962), xxv–xxvi. See also Lawrence I. Conrad, "Scholarship and Social Context: A Medical Case from the Eleventh-Century Near East," in Bates, *Knowledge and the Scholarly Medical Traditions*, 84–100, esp. 96–97.

6. Nutton, *Ancient Medicine*, 149, 168–70, 202; Wesley D. Smith, *The Hippocratic Tradition* (Ithaca, NY: Cornell University Press, 1979), 178–94.

7. Caelius Aurelianus, *On Acute Diseases and On Chronic Diseases,* edited and translated by I. E. Drabkin (Chicago: University of Chicago Press, 1950), 55, 63, 65, 159–61.

8. Owsei Temkin, *Galenism: The Rise and Decline of a Medical Philosophy* (Ithaca, NY: Cornell University Press, 1973), 67, 77, 120–21, 151.

9. Nancy Siraisi, *Taddeo Alderotti and His Pupils: Two Generations of Italian Medical Learning* (Princeton, NJ: Princeton University Press, 1981), 98–100; Cornelius O'Boyle, *The Art of Medicine: Medical Teaching at the University of Paris, 1250–1400* (Leiden: Brill, 1998), 83–84.

10. *Medieval Islamic Medicine: Ibn Riḍwān's Treatise "On the Prevention of Bodily Ills in Egypt,"* translated, with an introduction, by Michael W. Dols, Arabic text edited by Adil S. Gamal (Berkeley: University of California Press, 1984), 27–28; Temkin, *Galenism,* 69, 97–101.

11. Temkin, *Galenism,* 125. For editions of Galen, see Richard A. Durling, "Chronological Census of Renaissance Editions and Translations of Galen," *Journal of the Warburg and Courtauld Institutes* 24 (1961): 230–305.

12. Temkin, *Galenism,* 127–30.

13. Nutton, "Murders and Miracles," 45.

14. Jacques Jouanna, *Hippocrates* (Baltimore: Johns Hopkins University Press, 1999), 150–52; Nutton, *Ancient Medicine,* 21–25; Nutton, "Medical Thoughts on Urban Pollution," in *Death and Disease in the Ancient City,* edited by Valerie M. Hope and Eireann Marshall (London: Routledge, 2000), 65–73 at 71; Nutton, "Did the Greeks Have a Word for It?," in *Contagion: Perspectives from Pre-Modern Societies,* edited by Lawrence I. Conrad and Dominik Wujastyk (Aldershot, UK: Ashgate, 2000), 137–62.

15. Paulus Aegineta, *The Seven Books of Paulus Aeginata Translated from the Greek with a Commentary Embracing the Complete View of the Knowledge Formed by the Greeks, Romans, and Arabians on All Subjects Connected with Medicine and Surgery* (London: Sydenham Society, 1844), bk. 2, §36, pp. 272–77; *Agniveśa's Caraka Saṃ-*

hitā, edited and translated by Ram Karan Sharma and Vaidya Bhagwan Dash, 2 vols. (Varanasi: Chowkhamba Sanskrit Series Office, 1977), 2:142.

16. *Agniveśa's Caraka Saṃhitā,* 2:142–48. But see Peregrine Horden, "Disease, Dragons and Saints: The Management of Epidemics in the Dark Ages," in *Epidemics and Ideas: Essays on the Historical Perception of Pestilence,* edited by Terence Ranger and Paul Slack (Cambridge: Cambridge University Press, 1992), 45–76.

17. Dionysios Stathakopoulos, *Famine and Pestilence in the Late Roman and Early Byzantine Empire* (Burlington, VT: Ashgate, 2004).

18. *Medieval Islamic Medicine,* 85, 87, 97, 112–14.

19. Quoted in Paulus Aegineta, *Seven Books,* bk. 2, §36, p. 277.

20. Ann Carmichael, *Plague and the Poor in Renaissance Florence* (Cambridge: Cambridge University Press, 1986); Colin Jones, "Plague and Its Metaphors in Early Modern France," *Representations,* no. 53 (1 January 1996): 97–127; Samuel Kline Cohn, *Cultures of Plague: Medical Thinking at the End of the Renaissance* (Oxford: Oxford University Press, 2010), 77–78, 271.

21. See Lucinda McCray Beier, "In Sickness and in Health: A Seventeenth Century Family's Experience," in Porter, *Patients and Practitioners,* 101–28 at 125–26.

22. Cohn, *Cultures of Plague;* Carlo M. Cipolla, *Cristofano and the Plague: A Study in the History of Public Health in the Age of Galileo* (Berkeley: University of California Press, 1973); Cipolla, *Fighting the Plague in Seventeenth-Century Italy* (Madison: University of Wisconsin Press, 1981).

23. Andrew Wear, *Knowledge and Practice in English Medicine, 1550–1680* (Cambridge: Cambridge University Press, 2000), 280, 298–99, 340; Cohn, *Cultures of Plague,* 80, 129.

24. Wear, *Knowledge and Practice,* 280–301; Cohn, *Cultures of Plague,* 122–29; Lawrence I. Conrad and Dominik Wujastyk, introduction to Conrad and Wujastyk, *Contagion,* ix–xviii.

25. Arturo Castiglioni, quoted in Cohn, *Cultures of Plague,* 1. See also ibid., 5; Mary Fissell, "The Marketplace of Print," in *Medicine and Its Markets in England and Its Colonies, c. 1450–c. 1850,* edited by Mark S. R. Jenner and Patrick Wallis (New York: Palgrave Macmillan, 2007), 108–32; Wear, *Knowledge and Practice,* 40–41, 276; and Peter Murray Jones, "Medical Literacies and Medical Culture in Early Modern England," in *Medical Writing in Early Modern English,* edited by I. Taavitsainen and P. Pahta (Cambridge: Cambridge University Press, 2011), 30–43.

26. John Jones, *A Dial for All Agues Conteininge the Names in Greeke, Latten, and Englyshe, with the Diuersities of Them, Symple and Compounde, Proper and Accident, Definitions, Deuisions, Causes, and Signes, Comenly Hetherto Knowen* (London: William Seres, 1566).

27. Thompson Cooper, "Jones, John (*fl.* 1562–1579)," revised by Patrick Wallis, in *Oxford Dictionary of National Biography,* accessed 29 July 2012, www.oxforddnb.com/view/article/15023.

28. John Jones, "Epistle to the Reader," in Jones, *Dial for All Agues* (Jones's work

is unpaginated; chapters are cited parenthetically in the text). For this scheme of organizing fevers, see Avicenna, *Avicenae Qvarti Libri Canonis Fen Prima: De Febribus,* edited by D. Heironymo Sanctasophia (Padua: Matthaei Cadoria, 1659). In fact there were profound disagreements. See Iain M. Lonie, "Fever Pathology in the Sixteenth Century: Tradition and Innovation," in *Theories of Fever from Antiquity to the Enlightenment,* edited by William Bynum and Vivian Nutton (London: Wellcome Institute for the History of Medicine, 1981), 19–44. Charles Creighton finds Jones going far to adapt the tradition to English needs. Creighton, *A History of Epidemics in Britain,* vol. 2, *From the Extinction of Plague to the Present Time* (Cambridge: Cambridge University Press, 1894), 301.

29. P. Niebyl, "Old Age, Fever, and the Lamp Metaphor," *Journal of the History of Medicine* 26 (1971): 351–68.

30. Edward Edwards, *The Cure of All Sorts of Fevers, Both generall, and particular, with their Definition, Kindes, Differences, Causes, Signes, Prognostication, and manner of Cure, with a prespectation, their intentions curative, with their symptoms, and divers other things herein very necessarie to be judiciously observed in every Fever* (London: Thomas Harper, 1638). On the fever literature generally, see Don G. Bates, "Thomas Willis and the Fevers Literature of the Seventeenth Century," in Bynum and Nutton, *Theories of Fever,* 45–70 at 47–48.

31. Nicholas Culpeper, *Febralia: Or a Treatise on Fevers in General* (London: N. Brooke, 1656), 4, 15–16.

32. Brice Bauderon, *The expert phisician learnedly treating of all agues and feavers, whether simple or compound, shewing their different nature, causes, signes, and cure . . . ,* translated by B.W. (London: John Hancock, 1657).

33. Daniel Sennert, *Of Agues and Fevers: Their Differences, Signes, and Cures,* translated by N.D.B.M. (London: Lodowyck Loyd, 1658).

34. John Jones, "Epistle to the Reader."

35. Andrew Wear, "Epistemology and Learned Medicine in Early Modern England," in Bates, *Knowledge and the Scholarly Medical Traditions,* 159–61.

36. Lucinda McCray Beier, *Sufferers and Healers: The Experience of Disease in Seventeenth-Century England* (London: Routledge & Kegan Paul, 1987), 163.

37. There are few modern biographical studies of Santorio. See "Santorio Santorio," Galileo Project, accessed 21 February 2014, http://galileo.rice.edu/sci/santorio.html; Arturo Castiglioni, "The Life and Work of Sanctorius," *Medical Life* 38 (1931): 727–86.

38. Santorio Santorio, *Medicina Statica, or Rules of Health, in Eight Sections of Aphorisms,* translated by J.D. (London: John Starkey, 1676), 1.21–25; references are to section and aphorisms.

39. Ibid., 1.6, 51, 59; 4.2.

40. Ibid., 5.5. See also 1.46–47; 3.8, 67; and 7.30.

41. Bates, "Willis," esp. 55–56; Kenneth Dewhurst, *Willis's Oxford Casebook (1650–52)* (Oxford: Sandford, 1981); Robert Martensen, "Willis, Thomas (1621–

1675)," in *Oxford Dictionary of National Biography,* accessed 9 December 2010, www
.oxforddnb.com/view/article/29587.

42. Thomas Willis, *A Medical-philosophical Discourse of Fermentation, or, Of the Intestine Motion of Particles in Every Body by Dr. Thomas Willis . . .* (London: T. Dring, 1681); Willis, "Of Feavers," in *Dr. Willis's Practice of Physick, . . . done into English by S. P.* (London: T. Dring, 1681).

43. Willis, *Dr. Willis's Practice of Physick,* 57. Bates, "Willis," 54, identifies Cartesian influences. Robert Boyle was also a member of Willis's Oxford circle.

44. Bates, "Willis," 55.

45. Laurentius Bellini, *A Mechanical Account of Fevers* (London: A. Bell, 1720), translator's preface, v–xii; "Explanatory Introduction," ibid., xv–xxxii. See also W. Bruce Fye, "Lorenzo Bellini," *Clinical Cardiology* 20, no. 2 (1 February 1997): 181–82; Anita Guerrini, "The Varieties of Mechanical Medicine: Borelli, Malpighi, Bellini, and Pitcairne," in *Marcello Malpighi: Anatomist and Physician,* edited by Domenico Bertolini Meli (Firenze: Leo. S. Olschki, 1997), 111–28; Guerrini, "Archibald Pitcairne and Newtonian Medicine," *Medical History* 31 (1987): 70–83; and Guerrini, "Pitcairne, Archibald (1652–1713)," in *Oxford Dictionary of National Biography,* accessed 8 December 2010, www.oxforddnb.com/view/article/22320.

46. Friedrich Hoffmann, *Fundamenta Medicinae,* translated by L. S. King (London: Macdonald Scientific, 1971), 2.1, 3.1; references are to section and aphorism.

47. Thomas Gariepy, "Mechanism without Metaphysics: Henricus Regius and the Establishment of Cartesian Medicine" (PhD diss., Yale University, 1990).

48. Hoffmann, *Fundamenta Medicinae,* 3.27. See also Lester King, *The Road to Medical Enlightenment, 1650–1695* (London: Macdonald, 1970), 181–203. King reminds us that these are hypothetical Cartesian corpuscles, not what we refer to as blood cells today (189).

49. Kenneth Dewhurst, *Dr. Thomas Sydenham (1624–1689): His Life and Writings* (Berkeley: University of California Press, 1966); Wear, *Knowledge and Practice,* 450.

50. Thomas Sydenham, *Medical Observations on the History and Cure of Acute Diseases,* in *The Works of Thomas Sydenham, M.D.,* translated from the Latin of Dr. Greenhill with *A Life of the Author,* by R. G. Latham, MD, 2 vols. (London: Sydenham Society, 1848), 1:4. My reading follows Don G. Bates, "Thomas Sydenham: The Development of His Thought, 1666–1676" (PhD diss., Johns Hopkins University, 1975).

51. Sydenham, *Medical Observations,* 1:153, 36.

52. Ibid., 1:102; see also p. 40.

53. Ibid., 1:72.

54. Ibid., 1:235. See also Wear, *Knowledge and Practice,* 453–54.

55. Sydenham, *Medical Observations,* 1:102, 4.

56. Ibid., 1:44. See also King, *Road to Medical Enlightenment,* 125–26.

57. Sydenham, *Medical Observations,* 1:46.

58. Thomas Sydenham, *Methodus Curandi Febres Propriis Observationibus: The Latin Text of the 1666 and 1668 Editions with the English Translation from R. G. Latham (1848)* (Folkstone: Winterdown Books, 1987), 99.

59. Sydenham, *Medical Observations,* 1:119–20.

60. Andrew Cunningham, "Sydenham versus Newton: The Edinburgh Fever Dispute in the 1690s between Andrew Brown and Archibald Pitcarine," in Bynum and Nutton, *Theories of Fever,* 71–98.

61. Bates, "Thomas Sydenham," 88.

62. Jones, *Dial for All Agues,* ch. 2.

63. Walter Pagel, *Joan Baptista van Helmont: Reformer of Science and Medicine* (Cambridge: Cambridge University Press, 1982); Allen G. Debus, *Chemistry and Medical Debate: Van Helmont to Boerhaave* (Canton, MA: Science History, 2001); King, *Road to Medical Enlightenment,* 37–62.

64. Debus, *Chemistry and Medical Debate,* 109. Thompson draws mainly from van Helmont's "An Unheard-of Doctrine of Fevers; Wherein Is Added a Treatise Against the Four Humours of the Schooles," in *Van Helmont's Works: Containing his most Excellent Philosophy, Chirurgery, Physick, Anatomy, Wherein the Philosophy of the Schools is Examined, their Errors Refuted, and the Whole Body of Physick Reformed and Rectified,* translated by J.C. (London: Lodowick Lloyd, 1664), 932–1011; van Helmont, "A Passive Deceiving and Ignorance of the Schooles of the Humourists," in ibid., 1013–64.

65. James Thompson, *Helmont disguised, or, The vulgar errours of impericall and unskillfull practisers of physick confuted more especially as they concern the cures of the feavers, stone, plague and other diseases* (London: E. Alsop, 1657), 1.

66. Many of these criticisms vexed orthodox physicians. See Lonie, "Fever Pathology."

67. Thompson, *Helmont disguised,* 39, 43, 45, 97; van Helmont, "Unheard-of Doctrine of Fevers."

68. Thompson, *Helmont disguised,* unpaginated preface "to the ingenious reader." See also Andrew Wear, "Puritan Perceptions of Illness in Seventeenth Century England," in Porter, *Patients and Practitioners,* 55–99; and Nutton, "Murders and Miracles," 45.

69. Wear, *Knowledge and Practice,* 429.

70. Dewhurst, *Dr. Thomas Sydenham,* 41–42; Rudolph Siegel and F. N. L. Poynter, "Robert Talbor, Charles II, and Cinchona," *Medical History* 6 (1962): 82–85; M. L. Duran-Reynals, *The Fever Bark Tree: The Pageant of Quinine* (Garden City, NY: Doubleday, 1946), 95–90; Creighton, *History of Epidemics in Britain,* 2:318–26.

71. Robert Talbor, *Pyretologia, a Rational Account of the cause & cure of agues with their signes Diagnostick & Prognostick* (London: R. Robinson, 1672), 9–12, 15–20, 38.

72. Ibid., 42–44, emphasis added.

73. Saul Jarcho, *Quinine's Predecessor: Francesco Torti and the Early History of Cinchona* (Baltimore: Johns Hopkins University Press, 1993).

74. Nicholas Blégny, *The English Remedy, or, Talbor's Wonderful Secret for Cureing of Agues and Feavers . . .* (London: J. Wallis, 1682).

75. It is worth reiterating that Talbor was a practitioner as well as a huckster. Among his patients was Ralph Josselin, whose diaries have been a key source disclosing seventeenth-century sensibilities on many matters. See Beier, "In Sickness and in Health," 121. Duran-Reynals, in *Fever Bark Tree,* 85–87, believes that relations with Sydenham were more complicated.

76. G. A. Lindeboom, *Herman Boerhaave: The Man and His Work* (Rotterdam: Erasmus, 2007), 230.

77. Ibid., 24, 36–37, 98, 231–34; Herman Boerhaave, *Boerhaave's Correspondence, Part Two,* edited by G. A. Lindeboom (Leiden: Brill, 1964), 157, 247.

78. Lindeboom, *Herman Boerhaave,* 176–80; Gerard van Swieten, *Commentaries upon Boerhaave's Aphorisms concerning the Knowledge and Cure of Diseases,* 15 vols. (Edinburgh: Charles Eliot, 1776), 5:202–6.

79. Herman Boerhaave, *Dr Boerhaave's Academical Lectures on the Theory of Physic,* 6 vols. (London: W. Innys, 1746), 5:273, 329, 371; cf. Lindeboom, *Herman Boerhaave,* 45.

80. Boerhaave, *Dr. Boerhaave's Academical Lectures,* 5:266–68, 330; Lindeboom, *Herman Boerhaave,* 177–79.

81. Boerhaave, *Dr. Boerhaave's Academical Lectures,* 5:279–84.

82. Ibid., 5:493–94. See also van Swieten, *Commentaries,* 5:206–7.

83. Wear, *Knowledge and Practice,* 298–99, 303–4; Cohn, *Cultures of Plague;* Giulia Calvi, *Histories of a Plague Year: The Social and the Imaginary in Baroque Florence,* translated by Dario Biocca and Bryant T. Ragan Jr. (Berkeley: University of California Press, 1989). Followers of Paracelsus or van Helmont also adopt the appeal to arbitrarily acting poisons, if rhetorically, as a way to suggest the arbitrariness of conventional causal explanation.

84. See *Boerhaave's Correspondence, Part Two,* 145, 253, 289, 373.

85. Boerhaave, *Dr. Boerhaave's Academical Lectures,* 5:394, 437, emphasis added. Boerhaave understood his own health in this way. He said that a serious illness, retrospectively diagnosed as rheumatic fever or a slipped disk, was due to the state of his pores on an early morning visit to his garden. See Lindeboom, *Herman Boerhaave,* 84–85.

86. Lindeboom, *Herman Boerhaave,* 190–93; Volker Hess, *Der wohltempeierte Mensch: Wissenschaft und Alltag des Fiebermessens (1850–1900)* (Frankfurt: Campus Verlag, 2000), 39–44. Boerhaave did in fact incorporate thermometry into his practice, but it did not acquire the centrality it has come to have in the diagnosis of fever. His student Anton de Haen would make even more extensive use of the thermometer.

Part II. Fever as Social

1. Such a perspective follows John Simon, *English Sanitary Institutions, Reviewed in Their Course of Development, and in Some of Their Political and Social Relations*

(London: Cassell, 1890); John Pickstone, "Dearth, Dirt, and Fever Epidemics: Rewriting the History of British 'Public Health,' 1780–1850," in *Epidemics and Ideas: Essays on the Historical Perception of Pestilence,* edited by Terence Ranger and Paul Slack (Cambridge: Cambridge University Press, 1992), 125–48.

2. Adam Smith, *The Theory of Moral Sentiments* (London: A. Millar, 1759), 4–10.

3. The fullest development of this perspective is in the psychoanalyst and medical historian Walther Riese's *The Conception of Disease: Its History, Its Versions, and Its Nature* (New York: Philosophical Library, 1953).

4. This geographic orientation reflected renewed interest in Hippocrates. See David Cantor, *Reinventing Hippocrates* (Aldershot, UK: Ashgate, 2002); James C. Riley, *The Eighteenth Century Campaign to Avoid Disease* (London: Macmillan, 1987); and Wesley D. Smith, *The Hippocratic Tradition* (Ithaca, NY: Cornell University Press, 1979).

5. Margaret Pelling, *Cholera, Fever, and English Medicine, 1825–1865* (Oxford: Oxford University Press, 1978).

6. George Rosen, *From Medical Police to Social Medicine: Essays on the History of Health Care* (New York: Science History, 1974); Christopher Hamlin, "The Fate of 'The Fate of Medical Police,'" *Centaurus* 50 (2008): 63–69; Laurence Brockliss and Colin Jones, *The Medical World of Early Modern France* (Oxford: Clarendon, 1997); Dora B. Weiner, *The Citizen-Patient in Revolutionary and Imperial Paris* (Baltimore: Johns Hopkins University Press, 1993).

7. Samuel Kline Cohn, *Cultures of Plague: Medical Thinking at the End of the Renaissance* (Oxford: Oxford University Press, 2010), 7, 204–48.

8. Andrew Wear, *Knowledge and Practice in English Medicine, 1550–1680* (Cambridge: Cambridge University Press, 2000, 334–35; Patrick Wallis, "Plagues, Morality and the Place of Medicine in Early Modern England," *English Historical Review* 121 (2006): 1–24.

9. Irvine Loudon, *The Tragedy of Childbed Fever* (Oxford: Oxford University Press, 2000); Guenther B. Risse, "'Typhus' Fever in Eighteenth-Century Hospitals: New Approaches to Medical Treatment," *Bulletin of the History of Medicine* 59 (1985): 176–95; Dale C. Smith, "Medical Science, Medical Practice, and the Emergence of the Concept of Typhus in Mid-Eighteenth Century Britain," in *Theories of Fever from Antiquity to the Enlightenment,* edited by William Bynum and Vivian Nutton (London: Wellcome Institute for the History of Medicine, 1981), 121–34. For the strongest claims of novelty, see Charles Creighton, *A History of Epidemics in Britain,* vol. 2, *From the Extinction of Plague to the Present Time* (Cambridge: Cambridge University Press, 1894), 1–5.

Chapter 4. Communities

1. Marie de Rabutin-Chantal, marquise de Sévigné, *Letters from the Marchioness de Sévigné, to her daughter the Countess de Grignan,* 10 vols., 2nd ed. (London: J. Coote, 1763–68), 5:130–31; 4:150.

2. Ibid., 1:xxix–xxx.

3. Ibid., 2:33; 7:184; 2:21–22.

4. Ibid., 1:115; 2:87–88, 228–29; 6:94; 2:106–7; 6:63–64; 2:161–62.

5. Ibid., 2:12; see also 3:103–4.

6. Ibid., 6:120–22; 5:130–31, 144–45.

7. Ibid., 5:100–101.

8. Ibid., 3:29–30.

9. Ibid., 2:222–23; see also 6:169–70.

10. Ibid., 5:246–47; 6:63–64.

11. Francis R. Packard, *Guy Patin and the Medical Profession in Paris in the XVIIth Century* (New York: Paul Hoeber, 1925).

12. Sévigné, *Letters,* 1:20; 3:263–64; 6:242–43; 8:132–33, 137, 169.

13. Ibid., 5:136–37, 38–39.

14. Ibid., 6:63–64. See also 6:104–5, 120–22, 131, 242; 7:27–33; 8:68, 143–44, 180. On Talbor in the court, see M. L. Duran-Reynals, *The Fever Bark Tree: The Pageant of Quinine* (Garden City, NY: Doubleday, 1946), 76–80; and Saul Jarcho, *Quinine's Predecessor: Francesco Torti and the Early History of Cinchona* (Baltimore: Johns Hopkins University Press, 1993), 50, 64–66.

15. Sévigné, *Letters,* 10:91. See also 4:263–64; 5:144, 153–54.

16. Ibid., 8:190; 2:2–3; 5:130–31. See also 6:106.

17. Ibid., 4:284–85, 270.

18. Ibid., 1:163; 4:205–6.

19. Ibid., 2:171–72; 5:263.

20. Ibid., 5:153–54; 3:236–37; 5:44–45.

21. Ibid., 5:130–31.

22. T. A. B. Corley, "James, Robert (*bap.* 1703, *d.* 1776)," in *Oxford Dictionary of National Biography,* accessed 20 December 2010, www.oxforddnb.com/view /article/14618.

23. Charles Welsh, *A Bookseller of the Last Century: Being Some Account of the Life of John Newbery and of the Books He Published with a Notice of the Later Newberys* (1885; reprint, New York: Augustus M. Kelley, 1972), 24–27; Robert James, *Dissertation on Fevers and Inflammatory Distempers; Wherein a Method Is Proposed of Curing, or at Least of Removing the Danger Usually Attending Those Fatal Disorders* (London: John Newbery, 1748), 7; James, *Dissertation on Fevers and Inflammatory Distempers, 8th ed., to which are now first added . . . A Vindication of the Fever Powder, and a Short Treatise on the Disorders of Children* (London: Francis Newbery Jr., 1778), 100.

24. Packard, *Guy Patin.* On empirics' tactical deployment of the classical medical idiom, see Andrew Wear, *Knowledge and Practice in English Medicine, 1550–1680* (Cambridge: Cambridge University Press, 2000), 439.

25. James, *Dissertation* (1778), 126.

26. Robert James, *A Medicinal Dictionary: Including Physic, Surgery, Anatomy, Chymistry, and Botany, in all their branches relative to Medicine. Together with a History of Drugs; . . . ,* 3 vols. (London: T. Osborne, 1744–45), vol. 3, s.v. "pyrexia"; James, *Dissertation* (1778), 81–85.

27. James, *Dissertation* (1778), 85–87, 70.

28. Ibid., 88, 2–3. James did not rely solely on the powder; he also bled, blistered, applied cataplasms, ordered clysters, used oral purges, and supplemented with bark or musk (24, 127–28). Cf. William Cullen, "First Lines of the Practice of Physic," in *The Works of William Cullen . . .* , edited by John Thomson, 2 vols. (Edinburgh: Blackwood, 1827), 1:669.

29. James, *Dissertation* (1748), 46; James, *Dissertation* (1778), 40–41, 44–46, 49–52, 130, 138, 144.

30. Welsh, *Bookseller of the Last Century,* 18–22, 27, 70, 135; Ian Maxted, "Newbery, John (*bap.* 1713, *d.* 1767)," in *Oxford Dictionary of National Biography,* www.oxforddnb.com/view/article/19978; Maxted, "Newbery, Francis (1743–1818)," in ibid., www.oxforddnb.com/view/article/19977, both accessed 5 February 2011.

31. James, *Dissertation* (1778), 69, 93–94, 99.

32. Roy Porter, *Health for Sale: Quackery in England, 1660–1850* (Manchester: Manchester University Press, 1989), 52.

33. *The History of Little Goody Two-Shoes; otherwise called Mrs. Margery Two-Shoes. With the means by which she acquired her Learning and Wisdom, and, in consequence thereof, her Estate* (York: Wilson & Spence, n.d.), 5, 10, 58, 81.

34. James, *Dissertation* (1748), 44–46.

35. James, *Dissertation* (1778), iii.

36. Robert James, *Dissertation on Fevers and Inflammatory Distempers; Wherein an Expeditious Method is Proposed of Curing Those Dangerous Disorders,* 7th ed. (London: John Newbery, 1770), 74–76.

37. Welsh, *Bookseller of the Last Century,* 106.

38. James, *Dissertation* (1778), 44–46. Notably, concern would shift from general effectiveness to appropriate administration. See A Gentleman of the Faculty, *Thoughts and Observations on the Nature and Uses of Dr. James's Powder, in the Prevention and Cure of Diseases, addressed to Every One who wishes to be acquainted with the Salutary Effects of that Valuable Medicine* (Colchester: privately printed, 1790); and Malcolm Flemyng, *A Dissertation on Dr. James's Fever Powder* (London: Davis & Reymers, 1760). In the most notorious attack on James's powder, it was charged that it was responsible for Oliver Goldsmith's death. See William Hawes, *An Account of the Late Dr. Goldsmith's Illness, so far as it relates to the Exhibition of Dr. James's Powders,* 4th ed. (London: privately printed, 1780). For later uses of the powder, see James Copland, *A Dictionary of Practical Medicine, Comprising General Pathology,* 3 vols. (London: Longman, Brown, 1858), 1:1075; and Cullen, "First Lines," 1:624–25.

39. James, *Dissertation* (1748), 9–12; see also 19.

40. James, *Dissertation* (1778), 20–28.

41. Ibid., 28–36.

42. Ibid., 38–40.

43. Ibid., 70.

44. Ibid., 37, 58, 42–44.

45. Ibid., 41, 69–70, 132.

46. James, *Dissertation* (1748), 20–21, 24–25, 26, 30–31.

47. Ibid., 38–39.

48. Joseph Rogers, *Essay on Epidemic Diseases, more particularly on the Endemial Epidemics of the City of Cork, such as Fevers and Small-Pox but most professedly on the Endemial Epidemic Fever of the Year 1731* (Dublin: S. Powell, 1734), xxxix, xliv–xlv, 5–6. The first epidemic of this peculiar fever was in 1718. Mainly Rogers was writing mainly about epidemics of 1728–31.

49. Ibid., xxii–xxiii. The introduction to Rogers's book is mispaginated; these pages appear where one expects to find p. xxxviii.

50. Ibid., 3, 62.

51. Ibid., 3–4, 27, 31. Like Sydenham, Rogers saw Cork's prevailing fever as influencing or influenced by other prevailing diseases, such as smallpox.

52. Ibid., xvii, xxx–xxxiii. The English term *data* is relatively new. See *Oxford English Dictionary*, 2nd ed., s.v. "data," def. 2a.

53. Rogers, *Essay*, 5–9.

54. On intercurrency, see ibid., 5, 8, 63–64.

55. Charles Creighton, *A History of Epidemics in Britain,* vol. 2, *From the Extinction of Plague to the Present Time* (Cambridge: Cambridge University Press, 1894), 10, 234–35, 302.

56. Rogers, *Essay*, 5, 62.

57. Ibid., 20–26. Rogers looked forward to new opportunities for meteorological empiricism: thermometers, barometers, and hygrometers would clarify the relations between atmosphere and health. His source for particulate contagia was not Galen's or Fracastoro's seeds but Marcellinus's mythical history of contagion arising from the opening of a treasure chest (19–22). Cf. Vivian Nutton, "The Seeds of Disease: An Explanation of Contagion and Infection from the Greeks to the Renaissance," *Medical History* 27 (1983): 1–34.

58. Rogers, *Essay*, 36–37, 47. Water was a secondary concern. Rogers worried about geological events, which might taint a water source; conditions of heat and dryness, which might concentrate impurities; and putridity. Ibid., 44, 49–50.

59. Ibid., 38–39.

60. Ibid., 39–49.

61. Dale C. Smith, "Medical Science, Medical Practice, and the Emergence of the Concept of Typhus in Mid-Eighteenth Century Britain," in *Theories of Fever from Antiquity to the Enlightenment,* edited by William Bynum and Vivian Nutton (London: Wellcome Institute for the History of Medicine, 1981), 121–34.

62. Dorothea Singer, "Sir John Pringle and His Circle.—Part I. Life," *Annals of Science* 6 (1949): 127–80; S. Selwyn, "Sir John Pringle: Hospital Reformer, Moral Philosopher and Pioneer of Antiseptics," *Medical History* 10 (1966): 266–74; J. S. G. Blair, "Pringle, Sir John, first baronet (1707–1782)," in *Oxford Dictionary of National Biography,* accessed 4 May 2012, www.oxforddnb.com/view/article/22805.

63. John Pringle, *Observations on the Diseases of the Army* (London, 1752), 22–23, 117, 120.

64. Ibid., vii–ix.

65. Ibid., 8–11, 17, 19, 22, 27–30, 51, 58–59, 79–80, 210–14, 234–36.

66. Ibid., vii.

67. Ibid., 51–52, 129, 154, 295.

68. For a list of Assizes epidemics, see Creighton, *History of Epidemics in Britain,* 2:92–98.

69. Pringle, *Observations,* 27, 57–59, 225–26, 292. The Hungarian fever had bilious symptoms too. Pringle thought it might be a mix of his hospital fever and bilious fevers.

70. Ibid., 27.

71. Ibid., 31–32, 49, 58–59.

72. Creighton, *History of Epidemics in Britain,* 2:100.

73. Pringle, *Observations,* 339–40, 74–75, 125, 132.

74. Ibid., xiv.

75. Norman Moore, "Huxham, John (*c.* 1692–1768)," revised by Richard Hankins, in *Oxford Dictionary of National Biography,* accessed 16 March 2012, www.oxforddnb.com/view/article/14319; Michael Bartholomew, "Lind, James (1716–1794)," in ibid., accessed 5 August 2012, www.oxforddnb.com/view/article/16669. Though his book was published later, Pringle claimed to have written independently of Huxham, which was probably the case. Pringle writes: "Dr. Huxham's *Essay on Fevers* came out immediately after; in which I find so near a coincidence between his account of the malignant fever, and mine, that I imagine it must add no small weight to both our testimonies, to find two authors, in different places, without any communication, vary so little, in either the cause, description, or cure." *Observations,* xii. For present purposes I have not attempted an assessment of that convergence.

76. James, *Dissertation* (1778), 129. See also Risse, " 'Typhus' Fever," 178; Pringle, *Observations,* xiv; and Creighton, *History of Epidemics in Britain,* 2:120–30.

77. Pringle, *Observations,* 31–32; cf. 229.

78. Generally, writers downplayed the difference between small petechial spots and larger blotches characteristic of scurvy by suggesting a continuum of skin lesions. See James Lind, *Treatise of the Scurvy,* 3rd ed. (London: S. Crowder, 1772), 100–101, 114.

79. Ibid., 99–101, 130.

80. John Huxham, *An Essay on Fevers,* introduction by Saul Jarcho, MD (Canton, MA: Science History, 1988), 22–23, 27. On rapid decomposition, see James Lind, *An Essay on Diseases incidental to Europeans in hot Climates,* 5th ed. (London: J. Murray, 1792), 81; and Huxham, *Essay on Fevers,* 27. Huxham would seek to understand the unique qualities of this putrid blood in both physical and chemical terms (24–25).

81. Pringle, *Observations,* xiv, 26, 83, 102–11; cf. 215–22. Another context of pu-

trid symptoms, that of septic infection leading to gangrene, received less attention. Pringle, in *Observations,* 292, distinguishes putrid cases in which infection of a septic wound is involved. See also Lind, *Treatise,* 88.

82. Pringle, *Observations,* 211.

83. James Lind, *An Essay on the Most Effectual Means of Preserving the Health of Seamen in the Royal Navy* (London: A. Millar, 1757), 74, 88, 118, 308–9.

84. Lind, *Treatise,* 5, 35–42, 87–107, 273.

85. Ibid., 64–68, 228–29; Pringle, *Observations,* 96–97.

86. Pringle, *Observations,* 10–11, 22–24, 98–99, 116, 215–17; Lind, *Treatise,* 65.

87. Lind, *Treatise,* 139–40, 172; Mark Harrison, *Medicine in an Age of Commerce and Empire: Britain and Its Tropical Colonies, 1660–1830* (Oxford: Oxford University Press, 2010), 142–44.

88. Pringle, *Observations,* 136, 339–40; Lind, *Treatise,* 71–72. Pringle worries also about the "use of old and mouldy grain, or what has been damaged by a wet season," a recognition of what would later be recognized as ergotism, plausibly the cause of conditions understood as febrile. Mary Matossian, *Poisons of the Past: Molds, Epidemics, and History* (New Haven, CT: Yale University Press, 1989).

89. Lind, *Essay on the Most Effectual Means of Preserving the Health of Seamen,* 32–37; Lind, *Treatise,* 50–53, 78–86, 119, 131, 146–50; Pringle, *Observations,* 161; Huxham, *Essay on Fevers,* 66.

90. Huxham, *Essay on Fevers,* 29–30.

Chapter 5. Selves

1. James Adair, *Essays on Fashionable Diseases . . .* (London: Bateman, n.d.), 6. Adair blamed his own teacher, the Edinburgh professor Robert Whytt, for the change. See Christopher Lawrence, "Medicine as Culture: Edinburgh and the Scottish Enlightenment" (PhD diss., University of London, 1984), 139–43; and Lawrence, "The Nervous System and Society in the Scottish Enlightenment," in *Natural Order: Historical Studies of Scientific Culture,* edited by Barry Barnes and Steven Shapin (Beverly Hills, CA: Sage, 1979), 19–40.

2. Walther Riese, *The Conception of Disease: Its History, Its Versions, and Its Nature* (New York: Philosophical Library, 1953), 35. Riese sees this biographical approach as the apotheosis of Hippocratic medicine.

3. Charles Creighton, *A History of Epidemics in Britain,* vol. 2, *From the Extinction of Plague to the Present Time* (Cambridge: Cambridge University Press, 1894), 67–70; Anita Guerrini, *Obesity and Depression in the Enlightenment: The Life and Times of George Cheyne* (Norman: University of Oklahoma Press, 2000), 97, 108, 123, 147.

4. R. B. Todd, "On the Pathology and Treatment of Delirium and Coma: The Lumleian Lectures for 1850," *London Medical Gazette* 45 (1850): 703–8, 745–51, 789–94, 877–83, 921–25, 1031–35, at 745.

5. Gerard van Swieten, *Commentaries upon Boerhaave's Aphorisms concerning the Knowledge and Cure of Diseases,* 15 vols. (Edinburgh: Charles Eliot, 1776), 5:30.

6. Ibid., 5:30–31, 176.

7. To some theorists, the nerves were vessels subject to the same kinds of obstructions as the circulatory theorists had identified. See Creighton, *History of Epidemics in Britain,* 2:67.

8. W. F. Bynum, "Cullen and the Study of Fevers in Britain, 1760–1820," in *Theories of Fever from Antiquity to the Enlightenment,* edited by William Bynum and Vivian Nutton (London: Wellcome Institute for the History of Medicine, 1981), 135–47; Bynum, "Cullen, William (1710–1790)," in *Oxford Dictionary of National Biography,* accessed 13 May 2012, www.oxforddnb.com/view/article/6874.

9. William Cullen, "First Lines of the Practice of Physic," in *The Works of William Cullen . . . ,* edited by John Thomson, 2 vols. (Edinburgh: Blackwood, 1827), 1:480–91. Cullen, however, would take the idea of spasming capillaries from Hoffmann. James Copland, *A Dictionary of Practical Medicine, Comprising General Pathology,* 3 vols. (London: Longman, Brown, 1858), 1:1056. Even van Swieten had been converted, Cullen maintained (491).

10. Cullen, "First Lines," 492, 636; J. Val. de Hildenbrand, *A Treatise on the Nature, Cause, and Treatment of Contagious Typhus, from the German,* translated by S. C. Gross (New York: Elam Bliss, 1829), 7–8. Recognition of this dilemma helped keep vascular explanations alive. See Robert Christison, "Fever," in *A System of Practical Medicine: Comprised in a Series of Original Dissertations,* edited by Alexander Tweedie and W. W. Gerhard, 3 vols. (Philadelphia: Lea & Blanchard, 1842), 1:133.

11. Copland, *Dictionary of Practical Medicine,* 1:1046; van Swieten, *Commentaries,* 5:113, 343–45.

12. Mary Wollstonecraft Shelley, *Frankenstein; Or, The Modern Prometheus,* ch. 4, Project Gutenberg, www.gutenberg.org/files/84/84-h/84-h.htm#chap04.

13. Hildenbrand, *Treatise,* 5–6.

14. Johanna Geyer-Kordesch, "Fevers and Other Fundamentals: Dutch and German Medical Explanations c. 1680 to 1730," in Bynum and Nutton, *Theories of Fever,* 99–120; Martin Staum, *Cabanis: Enlightenment and Medical Philosophy in the French Revolution* (Princeton, NJ: Princeton University Press, 1980), 84–86; James Currie, *Medical Reports on the Effects of Water . . . ,* 4th ed. (London: Cadell & Davies, 1805), 239–40. Ascribed to the Montpellier physician Dr. Theophile de Bordeu, this union of materialism and vitalism is well developed in Denis Diderot's *d'Alembert's Dream.* See Diderot, *Rameau's Nephew and Other Works,* translated by Jacques Barzun and Ralph Bowen (Garden City, NY: Doubleday, 1956).

15. George Cheyne, *The English Malady: or, A treatise of nervous diseases of all kinds, as spleen, vapours, lowness of spirits, hypochondriacal, and hysterical distempers* (London: G. Strahan, 1733), 201–2; *Oxford English Dictionary,* 2nd ed., s.v. "valetudinarian" (the first usage was in 1703); Creighton, *History of Epidemics in Britain,* 2:67–70.

16. Guerrini, *Obesity and Depression,* esp. 119; Paul William Child, "Discourse and Practice in Eighteenth-Century Medical Literature: The Case of George

Cheyne" (PhD diss., University of Notre Dame, 1992); Bernard Mandeville, *A Treatise of the Hypochondriack and Hysterick Diseases* (1730), in *Collected Works of Bernard Mandeville,* 7 vols. (Hildesheim: Georg Olms Verlag, 1981), 2:x.

17. Christopher Lawrence, "Brown, John (*bap.* 1735, *d.* 1788)," in *Oxford Dictionary of National Biography,* accessed 13 May 2012, www.oxforddnb.com/view /article/3623. See also the several chapters in W. F. Bynum and Roy Porter, *Brunonianism in Britain and Europe* (London: Wellcome Institute for the History of Medicine, 1988), esp. Christopher Lawrence, "Cullen, Brown, and the Poverty of Essentialism," 1–21; Nelly Tsouyopoulos, "The Influence of John Brown's Ideas in Germany," 63–73; and Ramunas Kondratas, "The Brunonian Influence on the Medical Thought and Practice of Joseph Frank," 75–88.

18. Currie, *Medical Reports,* 233–34.

19. Van Swieten, *Commentaries,* 6:208–9.

20. Bartholomew Parr, *London Medical Dictionary; Including under Distinct Heads, Every Branch of Medicine, Viz. Anatomy, Physiology, and Pathology, the Practice of Physick and Surgery, Therapeutics and Materia Medica; with Whatever Relates to Medicine in Natural Philosophy, Chemistry, and Natural History* (Philadelphia: Mitchell, Ames, and White, 1819), s.v. "Febres."

21. Copland, *Dictionary of Practical Medicine,* 1:1040.

22. Cullen, "First Lines," 482 In Hippocratic medicine, notes Riese, disease was conceived in terms of music. Riese, *Conception of Disease,* 89, 32–35.

23. See Cullen, "First Lines," 502, 492.

24. Henry Cohen, "The Evolution of the Concept of Disease," *Proceedings of the Royal Society of Medicine* 48 (1955): 155–60 at 55. See also Owsei Temkin, *The Double Face of Janus, and Other Essays in the History of Medicine* (Baltimore: Johns Hopkins University Press, 1977). On the German approach, see Geyer-Kordesch, "Fevers and Other Fundamentals."

25. Christison, "Fever," 175, 181–82; Riese, *Conception of Disease,* 33–38.

26. Van Swieten, *Commentaries,* 5:202, 191; cf. 5:7, 10, 2, 166–69.

27. Copland, *Dictionary of Practical Medicine,* 1:1039, 1046; Riese, *Conception of Disease,* 37.

28. Cullen, "First Lines," 482, 492.

29. William Wallace Currie, ed., *Memoir of the Life, Writings, and Correspondence of James Currie, M.D., FRS,* 2 vols. (London: Longmans, 1831), 1:4–8, 11, 32–33, 37–43.

30. Currie, *Medical Reports,* 3–4.

31. Ibid., 88, 104. See also Cullen, "First Lines," 637–38. A decade later Vincent Priessnitz, founder of hydropathy, would espouse similar views in calling for a much more general water cure. Hydropathy would acquire its own theories and cadres of experts; Currie's deeper radicalism championed the victim's perspective. Another precursor was John Hancocke, author of *Febrifugiam Magnum* (1722). See Ginnie Smith, "Prescribing the Rules of Health: Self-Help and Advice in the Later

Eighteenth Century," in *Patients and Practitioners: Lay Perceptions of Medicine in Pre-industrial Society,* edited by Roy Porter (Cambridge: Cambridge University Press, 1985), 249–82 at 261.

32. Currie, *Medical Reports,* 243–49.

33. Ibid., 248–49.

34. Ibid., 257–58; cf. Cullen, "First Lines," 493–96, 561. Currie misrepresents Cullen, who writes that "the operations of nature are very precarious" (599). See also Christison, "Fever," 126.

35. Volker Hess, *Der wohltemperierte Mensch: Wissenschaft und Alltag des Fieber-messens (1850–1900)* (Frankfurt: Campus Verlag, 2000), 46–48.

36. Copland, *Dictionary of Practical Medicine,* 1:1045–47.

37. Van Swieten, *Commentaries,* 5:122–24; Cullen, "First Lines," 595–97; Hildenbrand, *Treatise,* 23–24.

38. Riese, *Conception of Disease,* 43–44.

39. Cyril Pearl, *Spectral Visitants, or a Journal of a Fever, by a Convalescent* (Portland, ME: S. H. Colesworthy, 1845). Galen, however, had recognized when his mind was affected by fever. Van Swieten, *Commentaries,* 6:220.

40. Copland, *Dictionary of Practical Medicine,* 1:1041, 1045–46. See also Cullen, "First Lines," 653; and van Swieten, *Commentaries,* 5:122–24.

41. Van Swieten, *Commentaries,* 5:69, 233.

42. Ibid.; Christison, "Fever," 179–80, 263.

43. Van Swieten, *Commentaries,* 5:253–54; 6:232–33, 246.

44. Robert James, *Dissertation on Fevers and Inflammatory Distempers; Wherein a Method Is Proposed of Curing, or at Least of Removing the Danger Usually Attending Those Fatal Disorders* (London: John Newbery, 1748), 23; see also the cases of Martha Bucktrout and Robert Horsborough (25–27).

45. Ibid., 14–15, 44–46.

46. Copland, *Dictionary of Practical Medicine,* 1:1041.

47. Van Swieten, *Commentaries,* 5:129. Only after 1850 would an inherently "relapsing fever" be recognized as accounting for some such cases. See Christison, "Fever," 154.

48. Timothy Flint, quoted in *Memoir of the Life and Medical Opinions of John Armstrong, M.D. to Which Is Added an Inquiry into the Facts Connected with Those Forms of Fever Attributed to Malaria or Marsh Effluvium,* by Francis Boott, 2 vols. (London: Baldwin & Craddock, 1833), 1:323n.

49. Christison, "Fever," 156; see also 202.

50. Emil Krapelin, quoted in C. B. Farrar, "The Mental Complications of Typhoid Fever," *Maryland Medical Journal* 43 (1902): 330–33 at 331; Henry Berkley, *A Treatise on Mental Diseases, Based Upon the Lecture Course at the Johns Hopkins University, 1899, and Designed for the Use of Practitioners and Students of Medicine* (New York: Appleton, 1900), 38–60.

51. E. J. Edwards, "Mental Derangements Following Influenza," *London Medi-*

cal Recorder, 20 August, 1890, 283–84; William K. Anderson, *Malarial Psychoses and Neuroses with Chapters Medico-Legal, and on History, Race Degeneration, Alcohol, and Surgery in Relation to Malaria* (London: Humphrey Milford / Oxford University Press, 1927).

52. G. E. R. Lloyd, *In the Grip of Disease: Studies in the Greek Imagination* (Oxford: Oxford University Press, 2003), 9; Riese, *Conception of Disease,* 35.

53. See Marie de Rabutin-Chantal, marquise de Sévigné, *Letters from the Marchioness de Sévigné, to her daughter the Countess de Grignan,* 10 vols., 2nd ed. (London: J. Coote, 1763–68), 4:284–85.

54. Parr, *London Medical Dictionary,* s.v. "Febres."

55. Van Swieten, *Commentaries,* 6:202–8, 217–19; Cullen, "First Lines," 498.

56. James, *Dissertation* (1748), 17; see also 11–12, 21.

57. Robert James, *Dissertation on Fevers and Inflammatory Distempers. 8th ed. To which are now first added . . . A Vindication of the Fever Powder, and a Short Treatise on the Disorders of Children* (London: Francis Newbery Jr., 1778), 29, 56–59. James assumed that foaming at the mouth indicated rabies.

58. John Pringle, *Observations on the Diseases of the Army* (London, 1752), 163, 207–8, 299–300.

59. Van Swieten, *Commentaries,* 6:217–31.

60. Ibid., 6:179; Cullen, "First Lines," 498, 603.

61. Charles Murchison, *A Treatise on the Continued Fevers of Great Britain* (London: Parker, Son, & Bourn, 1862), 487–88, 151.

62. Ibid., 152–53; Robert Lyon, *A Treatise on Fever* (Philadelphia: Lea & Blanchard, 1861), 101, 113–15; Christison, "Fever," 175, 195.

63. Murchison, *Treatise,* 153; Lyon, *Treatise on Fever,* 114; Christison, "Fever," 195, 259. On the windows episode, see W. B. Carpenter, *Principles of Mental Physiology with Their Application to the Training and Discipline of the Mind, and the Study of Its Morbid Conditions* (New York: D. Appleton, 1889), 654–66.

64. Murchison, *Treatise,* 151; Lyon, *Treatise on Fever,* 113.

65. Murchison, *Treatise,* 151–53.

66. Edgar Allan Poe, "For Annie," accessed 22 February 2014, www.daypoems .net/poems/648.html.

67. Lyon, *Treatise on Fever,* 50; see also 62.

68. Jean-Jacques Rousseau, *Confessions* (Chicago, 1856), pt. 2, bk. 7.

69. Flint, quoted in Boott, *John Armstrong, M.D.,* 1:320n.

70. Pearl, *Spectral Visitants,* 65.

71. Alfred Russel Wallace, *Alfred Russel Wallace: Letters and Reminiscences,* edited by James Marchant (New York: Harper, 1916), 88–89. I thank Gene Cittadino for reminding me of this singular case.

72. For authorship, see Joseph Williamson, *A Bibliography of the State of Maine from the earliest period to 1891,* 2 vols. (Portland, ME: Thurston Print, 1896). Pearl cites Hitchcock's report of his own hallucinations and Thomas Upham, whose *Out-*

lines of Imperfect and Disordered Mental Action (New York: Harper, 1840) recognized fever as a common cause of such "spectral visitants." *Spectral Visitants,* 129–32.

73. Pearl, *Spectral Visitants,* 3, 64.

74. Ibid., 32–47.

75. Ibid., 49–53.

76. Quoted in Boott, *John Armstrong, M.D.,* 1:321n.

77. Edward Hitchcock, "Case Of Optical Illusion in Sickness, with an Attempt to Explain Its Psychology," *New Englander* 3 (1845): 192–213 at 194. Hitchcock also notes a tendency of shapes to transform into figures.

78. Guerrini, *Obesity and Depression,* 164–67, 180.

79. Laura Jean Libbey, *Parted by Fate* (New York: Robert Bonner, 1890), 227–28.

80. Fanny Burney, *Cecilia, or Memoirs of an Heiress,* 3 vols., 2nd ed. (Dublin: Price, Moncrieffe, 1783), 3:315–42.

81. Emily Brontë, *Wuthering Heights* (1847; London: Penguin, 1990), 113–29.

82. Elizabeth Gaskell, *Ruth* (London: Pickering & Chatto, 2006), 79–86, 219.

83. Ibid., 315, 327. Easing her decision is the willingness of the local surgeon to take her son as an apprentice (323).

84. Ibid., 327–32.

85. The novel, as Bailin notes, has twelve significant fevers. Miriam Bailin, *The Sickroom in Victorian Fiction* (Cambridge: Cambridge University Press, 1994); Gaskell, *Ruth,* 59, 62, 182, 330.

86. Riese, *Conception of Disease,* 74–75.

87. Gaskell, *Ruth,* 334.

88. Christison, "Fever," 167.

89. *Health of Towns: Report of the Speeches of E. Chadwick, Esq., Dr. Southwood Smith, Richard Taylor, Esq., James Anderton, Esq. and Others . . . to Promote a Subscription in Behalf of the Widow and Children of Dr. J.R. Lynch, Who Died of Fever, Caught in the Course of Exertions to Alleviate the Sufferings of the Poor . . .* (London: Chapman, Elcoate, 1847), 4–5; J. W. Cusack and William Stokes, "On the Mortality of Medical Practitioners from Fever in Ireland," *Dublin Quarterly Journal of Medical Sciences,* n.s., 4 (1847): 134–45. There were links to military sacrifice; attending fever was even more dangerous.

90. Gaskell, *Ruth,* 313–14.

91. The case of fever does not fit the well-known Ackerknecht thesis, long a staple of public-health history. E. H. Ackerknecht associated an emerging orientation toward the public's health with the rise to power of trade and manufacture in liberal states and the associated ascendency of anticontagionist models of disease. Contagionism, by contrast, persisted in autocratic states trying to preserve economic and demographic security by tight control of borders. There, public health would be mainly a matter of preventing the arrival of infected foreign bodies. Yet the persistence of the idea that contagiousness could arise from fevers owing to debility, together with the conceptual overlap between contagia and miasms (the main agent

of anticontagionism), undermines the distinctions on which the model depends. See E. H. Ackerknecht, "Anticontagionism between 1821 and 1867," *Bulletin of the History of Medicine* 22 (1948): 562–93, reprinted with commentaries in the *International Journal of Epidemiology* 38, no. 1 (2009): 7–33. See also Peter Baldwin, *Contagion and the State in Europe, 1830–1930* (New York: Cambridge University Press, 1999); and Charles E. Rosenberg, "Erwin H. Ackerknecht, Social Medicine, and the History of Medicine," *Bulletin of the History of Medicine* 81 (2007): 511–32.

92. Thomas Carlyle, *Past and Present,* edited by R. D. Altick (New York: Houghton Mifflin, 1965), 150–51.

93. Charles Kinglsey, *Yeast—A Problem* (New York: J. F. Taylor, 1903), 304.

94. William Pulteney Alison, *Observations on the Management of the Poor in Scotland, and Its Effects on the Health of Great Towns,* 2nd ed. (Edinburgh: Blackwood, 1840), 18; Christopher Hamlin, "William Pulteney Alison, the Scottish Philosophy, and the Making of a Political Medicine," *Journal of the History of Medicine and Allied Sciences* 61 (2006): 144–86; Hamlin, "The 'Necessaries of Life' in British Political Medicine, 1750–1850," *Journal of Consumer Policy* 29 (2006): 373–97; Creighton, *History of Epidemics in Britain,* 2:99–102.

95. Mary J. Dobson, *Contours of Death and Disease in Early Modern England* (Cambridge: Cambridge University Press, 1997), 383ff.

96. *Some Account of the Origin and Plan of An Association, formed for the Establishment of a House of Recovery, or Fever Hospital, in the city of Dublin, with extracts shewing the necessity and utility of such an institution* (Dublin: Bates, 1801); Hildenbrand, *Treatise,* 174–75; John Ferriar, "Account of the Establishment of Fever-Wards in Manchester," in Ferriar, *Medical Histories and Reflections,* 1st American ed. 4 vols. in 1 (Philadelphia: Dobson, 1816), 3:316–40; John V. Pickstone, "Ferriar's Fever to Kay's Cholera: Disease and Social Structure in Cottonopolis," *History of Science* 22 (1984): 401–19. Creighton, in *History of Epidemics in Britain,* 2:175–80, noted coercive attempts to hospitalize fever victims in some places.

97. Henry Kennedy, *A Few Observations on the Nature and Effect of Fever, to which the Poor of Dublin are liable: and on the Proposed Plan of a Receiving House for the Accommodation of those laboring under it* (Dublin: Gilbert & Hodges, 1801), 7–12.

Part III. Fever Becomes Modern

1. Charles E. Rosenberg, "The Tyranny of Diagnosis: Specific Entities and Individual Experience," *Milbank Quarterly* 80 (2002): 237–60.

Chapter 6. Facts

1. Elisha Bartlett, *The History, Diagnosis, and Treatment of Typhoid and Typhus Fever; with an Essay on the Diagnosis of Bilious Remittent, and of Yellow Fever* (Philadelphia: Lea & Blanchard, 1842), 137. On Bartlett, see *Elisha Bartlett's Philosophy of Medicine,* edited by William Stempsey, SJ (Dordrecht: Springer, 2005).

2. Bartlett, *History, Diagnosis, and Treatment,* 137–38.

3. Robert Christison, "Fever," in *A System of Practical Medicine: Comprised in a Series of Original Dissertations,* edited by Alexander Tweedie and W. W. Gerhard, 3 vols. (Philadelphia: Lea & Blanchard, 1842), 1:128.

4. Walther Riese, *The Conception of Disease: Its History, Its Versions, and Its Nature* (New York: Philosophical Library, 1953), 87–88.

5. Caroline Hannaway and Ann La Berge, "Paris Medicine: Perspectives Past and Present," in *Constructing Paris Medicine,* edited by Hannaway and La Berge (Amsterdam: Rodopi, 1998), 1–69; Edwin Ackerknecht, *Medicine at the Paris Hospital, 1794–1848* (Baltimore: Johns Hopkins Press, 1967), 9–10. The transformation was the subject of Michel Foucault's famous book *The Birth of the Clinic: An Archaeology of Medical Perception* (New York: Pantheon, 1973). Its impact on American medicine is the subject of John Harley Warner's *The Therapeutic Perspective: Medical Practice, Knowledge, and Identity in America, 1820–1885* (Cambridge, MA: Harvard University Press, 1986).

6. Charles E. Rosenberg, "The Tyranny of Diagnosis: Specific Entities and Individual Experience," *Milbank Quarterly* 80 (2002): 237–60.

7. The opposing perspective is well developed in Knud Faber, *Nosography in Modern Internal Medicine* (New York: Paul Hoeber, 1922); and Riese, *Conception of Disease,* 7–8, 96.

8. Ackerknecht, *Medicine at the Paris Hospital,* 6–11; Martin Staum, *Cabanis: Enlightenment and Medical Philosophy in the French Revolution* (Princeton, NJ: Princeton University Press, 1980), 103–8.

9. Ackerknecht, *Medicine at the Paris Hospital,* 32–33.

10. Faber, *Nosography,* 22–29; Lester King, "Boissier de Sauvages and 18th Century Nosology," *Bulletin of the History of Medicine* 40 (1966): 43–51; Julian Martin, "Sauvages's Nosology: Medical Enlightenment in Montpellier," in *The Medical Enlightenment of the Eighteenth Century,* edited by Andrew Cunningham and Roger French (Cambridge: Cambridge University Press, 1990), 111–37.

11. William Cullen, "First Lines of the Practice of Physic," in *The Works of William Cullen . . . ,* edited by John Thomson, 2 vols. (Edinburgh: Blackwood, 1827), 1:516–25, 534; Christison, "Fever," 124, 137.

12. James Currie, *Medical Reports on the Effects of Water . . . ,* 4th ed. (London: Cadell & Davies, 1805), 44–47 at 45.

13. P. C. A. Louis, *Anatomical, Pathological, and Therapeutic Researches upon the Disease known under the Name of Gastro-enterite, Putrid, Adynamic, Ataxic, or Typhoid Fever, etc. compared with the most common acute diseases,* translated by Henry Bowditch, 2 vols. (Boston: Isaac Butts, 1836). The work was originally published as *Recherches anatomiques, pathologiques et thérapeutiques sur la maladie connue sous les noms de gastro-entérite, fièvre putride, adynamique, ataxique, typhoïde, etc.; comparée avec les maladies aiguës les plus ordinaires,* 2 vols. (Paris: Bailliere, 1829).

14. Charles Murchison, *A Treatise on the Continued Fevers of Great Britain* (London: Parker, Son, & Bourn, 1862), 517–20.

15. Christison, "Fever," 129–30.

16. E. H. Ackerknecht, "Elisha Bartlett and the Philosophy of the Paris Clinical School," *Bulletin of the History of Medicine* 24 (1950): 43–60 at 55; Jacalyn Duffin, "Laennec and Broussais: The 'Sympathetic' Duel," in Hannaway and La Berge, *Constructing Paris Medicine,* 251–74.

17. Thomas Sydenham, *Medical Observations on the History and Cure of Acute Diseases,* in *The Works of Thomas Sydenham, M.D.,* Translated from the Latin of Dr. Greenhill with *A Life of the Author* by R. G. Latham, MD, 2 vols. (London: Sydenham Society, 1848), 1:235–50; Gerard van Swieten, *Commentaries upon Boerhaave's Aphorisms concerning the Knowledge and Cure of Diseases,* 15 vols. (Edinburgh: Charles Eliot, 1776), 5:166–69.

18. Dora B. Weiner, *The Citizen-Patient in Revolutionary and Imperial Paris* (Baltimore: Johns Hopkins University Press, 1993), 181–83.

19. King, "Boissier de Sauvages," 46–47.

20. Cullen, "First Lines," 525, 541.

21. Van Swieten, *Commentaries,* 6:350.

22. Ackerknecht, "Elisha Bartlett," 50. See also Hannaway and La Berge, "Paris Medicine: Perspectives Past and Present."

23. Faber, *Nosography,* 42–49.

24. F. J. V. Broussais, *Conversations on the Theory and Practice of Physiological Medicine . . .* (London: Burgess & Hill, 1825), 258–59, 266–69; Jean-François Braunstein, *Broussais et le matérialisime: Médecine et philosophie au xix siècle* (Paris: Méridiens Klincksieck, 1986). See also Ackerknecht, *Medicine at the Paris Hospital,* 61–80; Ackerknecht, "Broussais, or a Forgotten Medical Revolution," *Bulletin of the History of Medicine* 27 (1953): 320–43; Duffin, "Laennec and Broussais"; W. R. Albury, "Corvisart and Broussais: Human Individuality and Medical Dominance," in Hannaway and La Berge, *Constructing Paris Medicine,* 221–50; and Faber, *Nosography,* 47–49.

25. Broussais, *Examen de la doctrine médicale génralment adoptée et des systèmes modernes de nosologie . . .* (Paris: Moronval, 1816), 17–24.

26. Broussais, *Examen,* 17.

27. Broussais, *Conversations,* 13–15.

28. Ibid., 173, 266–68. Broussais believed that all inflammations moved.

29. Ibid., 7–32, 44–47.

30. Ackerknecht notes that Paris physicians were well trained in the medical classics, though Broussais was unusually vigorous in rejecting antiquity. *Medicine at the Paris Hospital,* 5–6.

31. Broussais, *Conversations,* 31, 140, 153–56.

32. Christison, "Fever," 124–25, 129.

33. James Copland, *A Dictionary of Practical Medicine, Comprising General Pathology,* 3 vols. (London: Longman, Brown, 1858), 1:1044.

34. Broussais, *Conversations,* 96 (puerperal fever), 56–57 (tuberculosis), 319–21 (surgical inflammations).

35. Louis, *Anatomical, Pathological, and Therapeutic Researches.* See also Christison, "Fever," 129–30; and Faber, *Nosography,* 44. The discovery was sometimes credited to Johann Georg Roederer and Carl Gottlieb Wagler, of Göttingen, in 1762. For Louis's explanation of the concept and the term, see the preface to the much enlarged second edition of his *Anatomical Researches: Recherches anatomiques, pathologiques et thérapeutiques sur la maladie connue sous les noms de fièvre typhoide, putride, adynamique, ataxique, bilieuse, muqueuse, gastro entérite, entérite folliculeuse, dothinentérie, etc., comparée avec les maladies aiguës les plus ordinaire,* 2nd ed., 2 vols. (Paris: Ballière, 1841), 1:xv–xxiii.

36. Ackerknecht, who admired him enormously, astutely concluded that Broussais had become, "medically speaking, an abortion." Ackerknecht, "Elisha Bartlett," 53.

37. See, for example, Marianna Karamanou et al., "P. C. A. Louis (1787–1872): Introducing Medical Statistics in Pneumonology," *American Journal of Respiratory and Critical Care Medicine* 182 (2010): 1569.

38. Louis, *Anatomical, Pathological, and Therapeutic Researches,* 1:xii; Faber, *Nosography,* 42–44.

39. Louis, *Anatomical, Pathological, and Therapeutic Researches,* 2:266.

40. Ibid., 2:388–94.

41. William Osler, *Principles and Practice of Medicine,* 7th ed. (New York: D. Appleton, 1910), 57.

42. Judith Leavitt, *Typhoid Mary: Captive to the Public's Health* (Boston: Beacon, 1996).

43. John Harley Warner definitively depicts the scene in *Against the Spirit of System: The French Impulse in Nineteenth-Century American Medicine* (Princeton, NJ: Princeton University Press, 1998), 3ff.

44. Austin Flint, *Clinical Reports on Continued Fever* (Philadelphia: Lindsay & Blakiston, 1855), 9–15. On American typhoid research, see Dale C. Smith, "Gerhard's Distinction between Typhoid and Typhus and Its Reception in America, 1833–1860," *Bulletin of the History of Medicine* 54 (1980): 368–84.

45. Bartlett, *Typhoid and Typhus Fever,* 109; see also 4, 110–13.

46. Dale C. Smith, "The Rise and Fall of Typhomalarial Fever: I. Origins," *Journal of the History of Medicine and Allied Sciences* 37 (1982): 182–219.

47. Bartlett, *Typhoid and Typhus Fever,* 41.

48. Ibid., 74–75.

49. Ibid.; Lloyd G. Stevenson, "A Pox on the Ileum: Typhoid Fever among the Exanthemata," *Bulletin of the History of Medicine* 51 (1977): 496–504. Bartlett is following Bretonneau. See Faber, *Nosography,* 44.

50. Elisha Bartlett, *An Essay on the Philosophy of Medical Science* (Philadelphia: Lea & Blanchard, 1844), 76–120.

51. Ibid., 132.

52. Ibid., 127–40.

53. Ibid., 257–66 at 259–60; Broussais, quoted in Braunstein, *Broussais et le matérialisime*, 31.

54. Daniel Drake, *A Systematic Treatise Historical, Etiological, and Practical of the Principal Diseases of the Interior Valley of North America as they appear in the Caucasian, African, Indian, and Esquimaux Varieties of its Population*, edited by Hanbury Smith and Francis Smith, 2 vols. (1854; reprint, New York: Burt Franklin, 1971), 2:478–79.

55. Ibid., 2:430–35 at 433.

56. *The Sketches of Erinensis: Selection of Irish Medical Satire, 1824–1836*, edited by Martin Fallon (London: Skilton & Shaw, 1979); Faber, *Nosography*, 51–53.

57. Parisian physicians pioneered too in charting the demographic effects of poverty. But tendencies toward specialization of academic labor in France encouraged divergence between the research trajectories of these *hygienistes* and those of hospital-based clinician-dissectors like Louis and Auguste Chomel. See William Coleman, *Death Is a Social Disease: Public Health and Political Economy in Early Industrial France* (Madison: University of Wisconsin Press, 1982); Ackerknecht, *Medicine at the Paris Hospital;* and Weiner, *Citizen-Patient.*

58. Henry MacCormac, *An Exposition of the Nature, Treatment, and Prevention of Continued Fever* (London: Longman, 1835), 49.

59. See, for example, the astonishment of Joseph Lindwurm in *Der Typhus in Irland beobachtet im Sommer 1852* (Erlangen, Germany: Enke, 1853), 70. Creighton agreed: there was "something special and peculiar in the . . . epidemiology of Ireland." *A History of Epidemics in Britain*, vol. 2, *From the Extinction of Plague to the Present Time* (Cambridge: Cambridge University Press, 1894), 224.

60. William Stokes, *Lectures in the Theory and Practice of Physic* (Philadelphia: Waldie, 1837), 328; Robert Graves, *A System of Clinical Medicine* (Dublin: Fannin, 1843), 45.

61. Richard Grattan, "Medical Report of the Fever Hospital in Cork Street, Dublin, for the Year 1815," *Transactions of the Association of Fellows and Licentiates of the King and Queen's College of Physicians in Ireland* 1 (1817): 435–94, 439–40, 443.

62. Richard Grattan, "Medical Report of the Fever Hospital in Cork Street, Dublin, for the Year Ending 4th January 1819," *Transactions of the Association of Fellows and Licentiates of the King and Queen's College of Physicians in Ireland* 3 (1820): 316–447 at 365. See also William Stoker, "A Sketch of the Medical and Statistical History of Epidemic Diseases in Ireland, from 1798 to 1835, with the Method of Prevention and Cure, as Practised in the Fever Hospital, Cork Street, Dublin," *London Medical and Surgical Journal* 8 (1835): 422–27, 455–60, 495–96, 518–21, at 423–25 and 458.

63. Grattan, "Medical Report . . . 1819," 341, 343. See also Creighton, *History of Epidemics of Britain*, 2:278.

64. William Stoker, "Medical Report of the Fever Hospital and House of Recovery, Cork Street, for the Year 1816," *Transactions of the Association of Fellows and Licentiates of the King and Queen's College of Physicians in Ireland* 2 (1818): 397–471 at

451; John O'Brien, "Medical Report of the Sick Poor Institution for the Year 1817," ibid., 472–511 at 477; Edward Percival, *Practical Observations on the Treatment, Pathology, and Prevention of Typhous Fever* (London: Longman, Hurst, Rees, Orme, & Browne, 1819), 41. See also Christison, "Fever," 158–59.

65. Grattan, "Medical Report . . . 1815," 484–85. See also Christison, "Fever," 124, 141.

66. John O'Brien, "Medical Report of the House of Recovery, and Fever Hospital, Cork Street Dublin, for the Year 1829," *Transactions of the Association of Fellows and Licentiates of the King and Queen's College of Physicians in Ireland*, n.s., 1 (1830): 256–373 at 310. Only a few saw petechiae as distinctive. See Francis Barker, "Medical Report of the Fever Hospital, Cork-Street, Dublin; Containing an Account of the Progress of the Present Epidemic," ibid. 2 (1818): 512–602 at 520, 543, 557–60; and Robert Graves, *Clinical Lectures on the Practice of Medicine, To Which Is Prefixed a Criticism by Professor Trousseau, Reprinted from the Second Edition, Edited by the Late Dr. Neligan,* 2 vols. (1848; reprint, London: New Sydenham Society, 1884), 1:263–64.

67. In Francis Barker and John Cheyne, *An Account of the Rise, Progress, and Decline of the Fever lately Epidemical in Ireland,* 2 vols. (London: Baldwin, Craddock, & Joy, 1821), 1:371.

68. John O'Brien, "Medical Report of the Fever Hospital and House of Industry, Cork Street Dublin, for 1820," *Transactions of the Association of Fellows and Licentiates of the King and Queen's College of Physicians in Ireland* 3 (1820): 448–505 at 492–94; Barker and Cheyne, *Fever lately Epidemical in Ireland,* 1:452–55, 464; Mr. Parr, "Remarks on the Late Epidemics—Fever and Dysentery," *Dublin Medical Press* 19 (1848): 226–30 at 226; Christison, "Fever," 179.

69. Cullen had made much of cold as a cause of fever, but he had considered the effects of hunger indirect. "First Lines," 520, 550–56.

70. Graves, *Clinical Lectures,* 1:137–38. See also Stoker, "Medical Report . . . 1816," 432, 449.

71. Jones Lamprey, in "Report upon the Recent Epidemic Fever in Ireland," *Dublin Quarterly Journal of Medical Science,* n.s., 7 (1849): 100–104.

72. John Cheyne, "Report of the Hardwicke Fever Hospital for the Year ending on the 31st March, 1817," *Dublin Hospital Reports* 1 (1817): 1–116 at 4; Cheyne, "Report of the Hardwicke Fever Hospital for the Year ending on the 31st March, 1818," ibid. 2 (1818): 1–146 at 46.

73. Barker, "Medical Report [1818]," 523; O'Brien, "Medical Report . . . 1817," 482–83.

74. Christison, "Fever," 131.

75. O'Brien, "Medical Report . . . 1829," 297–99; cf. Barker and Cheyne, *Fever lately Epidemical in Ireland,* 1:231–41, 482.

76. J. O. Curran, "Observations on Scurvy As It Has Lately Appeared throughout Ireland and in the Several Parts of the British Isles," *Dublin Quarterly Journal of Medical Sciences,* n.s., 4 (1847): 83–134; Charles Ritchie, "Contributions to

the Pathology and Treatment of the Scorbutus, Which Is at Present Prevalent in Various Parts of Scotland," *Edinburgh Monthly Journal of Medical Science* 8 (1847): 38–49, 76–86; Thomas Shapter, "On the Recent Occurrence of Scurvy in Exeter and the Neighbourhood," *Provincial Medical and Surgical Journal,* no. 11 (2 June 1847): 281–85; Graves, *Clinical Lectures,* 2:415–27; William P. MacArthur, "Medical History of the Famine," in *The Great Famine: Studies in Irish History, 1845–52,* edited by R. Dudley Edwards and T. Desmond Williams (Dublin: Irish Committee of Historical Sciences / Browne & Nolan, 1956), 263–315 at 287–89.

77. H. Vandyke Carter, *Spirillum Fever: Synonyms; Famine or Relapsing Fever As Seen in Western India* (London: J. & A. Churchill, 1882), 300–301.

78. Laurence Geary, *Medicine and Charity in Ireland, 1718–1851* (Dublin: University College Dublin Press, 2004).

79. D. Smith, "Gerhard's Distinction."

80. Creighton, *History of Epidemics of Britain,* 2:202.

81. W. I. McDonald, "Jenner, Sir William, first baronet (1815–1898)," in *Oxford Dictionary of National Biography,* accessed 7 June 2012, www.oxforddnb.com /view/article/14754. Jenner was appointed to the staff of the London Fever Hospital in 1852 following his celebrated study on the typhus-typhoid distinction. On the hospital, see William F. Bynum, "Hospital, Disease, and Community: The London Fever Hospital, 1801–1850," in *Healing and History: Essays for George Rosen,* edited by Charles E. Rosenberg (New York: Science History, 1979), 97–115.

82. Cullen, "First Lines," 528. See also Christison, "Fever," 151–52.

83. William Jenner, *On the Identity or Non-Identity of Typhoid and Typhus Fevers* (London: John Churchill, 1850), 2–4.

84. Ibid., 55–62.

85. Jenner wanted readers to regard these ulcers as the work of the "fever-poison" and not products of "inflammation." The shade of Broussais still affected medical debate. Ibid., 97–98.

86. "Review of Jenner, *On the Identity or Non Identity of Typhoid and Typhus Fevers,*" *Dublin Medical Press* 25 (1851): 89–90.

87. August Hirsch, whose magisterial *Handbook of Geographical and Historical Pathology,* 2 vols. (London: Sydenham Society, 1883–85), is often the conduit to subsequent medical literature, often cites Murchison respectfully. Creighton's relation is more complicated: he is simultaneously critical of Murchison and heavily dependent on him, particularly with regard to typhus. Creighton, *History of Epidemics of Britain,* 2:4.

88. Murchison, *Treatise,* v; Christopher Hamlin, "Murchison, Charles (1830–1879)," in *Oxford Dictionary of National Biography,* accessed 22 July 2011, www.ox forddnb.com/view/article/19554.

89. Some note Bartlett's approving allusion to the insect hypothesis of Henry Holland. *Essay,* 223. Bartlett lists this proto-germ-theory concept as an example of appropriate hypothesizing but does not explain what makes it so.

90. John O'Brien, "Medical Report of the House of Recovery, and Fever Hospital, Cork Street Dublin, for the Year Ending 4th of January 1827," *Transactions of the Association of Fellows and Licentiates of the King and Queen's College of Physicians in Ireland* 5 (1828): 512–71 at 526–36; Stoker, "Medical Report . . . 1816," 453; William Pickells, "Report of the South Fever Asylum, Cork," *Transactions of the Association of Fellows and Licentiates of the King and Queen's College of Physicians in Ireland* 3 (1820): 194–235 at 211–16; Robert Reid, "Clinical Observations Made during the Epidemic Fever of 1826," ibid. 5 (1828): 266–302 at 267–83. Relapsing fever would later be ascribed to various species of louse- or tick-borne *Borrelia,* a group of organisms that includes the agent of lyme disease.

91. John King Bracken of Waterford, quoted in Barker and Cheyne, *Fever lately Epidemical,* 1:202.

92. Ibid., 1:440; William Scott, "Observations on Fever," *Dublin Medical Press* 11 (1844): 69–70. Cf. John Wardell, *The Scotch Epidemic Fever of 1843–44: Its History, Pathology, and Treatment, in Which It Is Maintained That, the Disease Essentially Differed from the Ordinary Forms of Continued Fever, Witnessed in This Country* (London: John Churchill, 1848); John Cormack, *Natural History, Pathology, and Treatment of the Epidemic Fever at Present Prevailing in Edinburgh and Other Towns; Illustrated by Cases and Dissections* (London: John Churchill, 1843), 87; Murchison, *Treatise,* 47–48; Creighton, *History of Epidemics of Britain,* 2:57, 150, 156, 172–78, 204.

93. Claude Bernard, *Introduction to the Study of Experimental Medicine,* translated by H. C. Greene (New York: Dover, 1957), 116–37.

94. Murchison, *Treatise,* 10–13.

95. Broussais, *Conversations,* 40–41.

96. José Ramón Bertomeu-Sánchez and Agustí Nieto-Galan, eds., *Chemistry, Medicine, and Crime: Mateu J. B. Orfila (1787–1853) and His Times* (Sagamore Beach, MA: Science History, 2006).

97. Broussais develops these links explicitly in *Conversations,* 45–47, 291–93. See also Christison, "Fever," 172.

98. Margaret Pelling, *Cholera, Fever, and English Medicine, 1825–1865* (Oxford: Oxford University Press, 1978).

99. John M. Eyler, *Victorian Social Medicine: The Ideas and Methods of William Farr* (Baltimore: Johns Hopkins University Press, 1979); Christopher Hamlin, "Providence and Putrefaction: Victorian Sanitarians and the Natural Theology of Health and Disease," *Victorian Studies* 28 (1985): 381–411.

Chapter 7. Naming the Wild

1. William Coleman, *Yellow Fever in the North: The Methods of Early Epidemiology* (Madison: University of Wisconsin Press, 1987), 3–4; cf. Lloyd G. Stevenson, "Exemplary Disease: The Typhoid Pattern," *Journal of the History of Medicine and Allied Sciences* 37 (1982): 159–82.

2. Alexander Collie, *On Fevers: Their History, Etiology, Diagnosis, Prognosis, and*

Treatment (Philadelphia: P. Blakiston, Son, 1887). On the seduction of diagnosis, see Charles E. Rosenberg, "The Tyranny of Diagnosis: Specific Entities and Individual Experience," *Milbank Quarterly* 80 (2002): 237–60.

3. Burke Cunha, *Fever of Unknown Origin* (New York: Informa Health Care, 2007).

4. Bruno Latour, *The Pasteurization of France,* translated by Alan Sheridan and John Law (Cambridge, MA: Harvard University Press, 1988); Kim Pelis, *Charles Nicolle, Pasteur's Imperial Missionary: Typhus and Tunisia* (Rochester, NY: University of Rochester Press, 2006), xvii.

5. Coleman, *Yellow Fever in the North,* 31, 44–45.

6. Mark Harrison finds that in many respects the experience and concepts of colonial medicine, including naval and military medicine, underlay the very medical revolution we associate with Paris. *Medicine in an Age of Commerce and Empire: Britain and Its Tropical Colonies, 1660–1830* (Oxford: Oxford University Press, 2010).

7. Sir Joseph Fayrer, *On the Climate and Fevers of India: Being the Croonian Lectures Delivered at the Royal College of Physicians in March 1882* (London: Churchill, 1882), 217–18.

8. On the dialectic between civilization and wildness, see Alan Bewell, *Romanticism and Colonial Disease* (Baltimore: Johns Hopkins University Press, 1999), 19, 49.

9. Patrick Manson, *Tropical Diseases: A Manual of the Diseases of Warm Climates* (New York: William Wood, 1898), xv.

10. Ibid., 91–92. On Manson, see J. W. W. Stephens, "Manson, Sir Patrick (1844–1922)," revised by Mary P. Sutphen, in *Oxford Dictionary of National Biography,* accessed 22 March 2013, www.oxforddnb.com/view/article/34865.

11. Fayrer, *Climate and Fevers of India,* 61–62, 151, 179. See also August Hirsch, *A Handbook of Geographical and Historical Pathology,* vol. 1, *Acute Diseases* (London: Sydenham Society, 1883). Ultimately geographic designators would largely give way to pathogen-based distinctions. For comparison with modern practices, see "WHO International Classification of Diseases," accessed 23 February 2014, www.who.int /classifications/icd/en/.

12. "Here are the names of some emerging viruses: Lassa. Rift Valley. Oropouche. Rocio. Q. Guanarito. VEE. Monkeypox. Dengue. Chikungunya. The hantaviruses. Machupo. Junin. The rabieslike strains Mokola and Duvenhage. LeDantec. The Kyasnur Forest brain virus. . . . The Semliki Forest agent. Crimean-Congo. Sindbis. O'nyongnyong. Nameless Sao Paulo. Marburg. Ebola Sudan. Ebola Zaire. Ebola Reston." Richard Preston, *The Hot Zone* (New York: Random House, 1994), 287.

13. Literary motifs, including Conradesque allusions, still shape representation of tropical fevers. Describing Ebola along the Bumba River in Congo, Preston refers to the isolation of the outbreak in the "silent heart of darkness" and to Ebola Zaire's seeming "to emerge out of the stillness of an implacable force brooding on an inscrutable intention." Ibid., 69–70, 78.

14. Ronald Ross, quoted in *The Conquest of Malaria: Italy, 1900–1962,* by Frank M.

Snowden (New Haven, CT: Yale University Press, 2006), 87. See also Ronald Ross, *The Prevention of Malaria* (New York: E. P. Dutton, 1910).

15. Mariola Espinosa, *Epidemic Invasions: Yellow Fever and the Limits of Cuban Independence, 1878–1930* (Chicago: University of Chicago Press, 2009); Nancy Stepan, "The Interplay between Socio-Economic Factors and Medical Science: Yellow Fever Research, Cuba and the United States," *Social Studies of Science* 8 (1978): 397–423. See also Alison Bashford, *Medicine at the Border: Disease, Globalization, and Security, 1850 to the Present* (New York: Palgrave Macmillan, 2006).

16. For an exemplary illumination of these issues, see Kenneth Kiple and Virginia King, *Another Dimension to the Black Diaspora: Diet, Disease, and Racism* (Cambridge: Cambridge University Press, 1981).

17. Fayrer, *Climate and Fevers of India,* 13.

18. Manson, *Tropical Diseases,* xi–xii, xv, 15–19; James L. A. Webb Jr., *Humanity's Burden: A Global History of Malaria* (Cambridge: Cambridge University Press, 2009), 131–32. On race more generally, see Warwick Anderson, *Colonial Pathologies: American Tropical Medicine, Race, and Hygiene in the Philippines* (Durham, NC: Duke University Press, 2006), 47, 59–61, 75; Paul Weindling, *Epidemics and Genocide in Eastern Europe, 1890–1945* (Oxford: Oxford University Press, 1999), 33; and Coleman, *Yellow Fever in the North,* 96.

19. Sonia Shah, *The Fever: How Malaria Has Ruled Humankind for 500,000 Years* (New York: Sarah Crichton Books / Farrar, Straus, & Giroux, 2010), 125, 231–32.

20. Philip Curtin, *Death by Migration: Europe's Encounter with the Tropical World in the Nineteenth Century* (Cambridge: Cambridge University Press, 1986), 8, 33 (tables 1.1 and 1.6); J. R. McNeill, *Mosquito Empires: Ecology and War in the Greater Caribbean, 1620–1914* (Cambridge: Cambridge University Press, 2010).

21. Fayrer, *Climate and Fevers of India,* 13–17. For the persistence of the plague trope, see Preston, *Hot Zone,* 68, 254, 261.

22. McNeill, *Mosquito Empires,* 64.

23. James Lind, *An Essay on Diseases Incidental to Europeans in Hot Climates,* 5th ed. (London: J. Murray, 1792), 9.

24. Bewell, *Romanticism and Colonial Disease,* 11, 74–79. Novelists too were important recognizers of tropical febrility. See Mrs. Aubin, *The Life of Charlotta Du Pont, an English Lady; taken from her own Memoirs* (London: Bettesworth, 1723); and [W. R. Chetwood], *The Voyages and Adventures of Captain Robert Boyle in several parts of the World,* 2nd ed. (London: Millar, 1728).

25. McNeill, *Mosquito Empires,* 66–67.

26. Henry Warren, *A Treatise Concerning the Malignant Fever in Barbados* (London: Fletcher Gyles, 1740), 3–4.

27. Ibid., 2; Lind, *Essay on Diseases Incidental to Europeans,* 11.

28. Warren, *Malignant Fever in Barbados,* 20–21. In keeping with contemporary iatromechanical views, Warren views these contagia as particulate.

29. James Lind, *A Treatise on the Putrid and Remitting Fen Fever Which Raged at Bengal in the Year 1762* (London: C. Dilley, 1772), 2–3, 22–34. This Lind was a cousin of the better-known expert on scurvy.

30. Conevery Bolton Valenčius, *The Health of the Country: How American Settlers Understood Themselves and Their Land* (New York: Basic Books, 2002); Karen Ordahl Kupperman, "Fear of Hot Climates in the Anglo-American Colonial Experience," *William and Mary Quarterly*, 3rd ser., 41 (1984): 213–40; Manson, *Tropical Diseases*, 124.

31. Francis Boott, *Memoir of the Life and Medical Opinions of John Armstrong, M.D. to Which Is Added an Inquiry into the Facts Connected with Those Forms of Fever Attributed to Malaria or Marsh Effluvium*, 2 vols. (London: Baldwin & Craddock, 1833), 1:300.

32. John Macculloch, *Malaria: An Essay on the Production and Propagation of this Poison, and the Nature and Localities of the Places by which it is Produced: with an Enumeration of the Diseases caused by it, and of the Means of Preventing or Diminishing them, both at Home and in the Naval and Military Service* (London: Longman, Hurst, Rees, Orme, Brown, & Green, 1827), 280.

33. L. J. Bruce-Chwatt, "John Macculloch, M.D., F.R.S. (1773–1835) (The Precursor of the Discipline of Malariology)," *Medical History* 21 (1977): 156–65. Later malariologists would recognize malaria's subacute effects and its complicity in other diseases through its weakening of the immune system. See Manson, *Tropical Diseases*, 89; Shah, *Fever*, 222; and Snowden, *Conquest of Malaria: Italy*, 13–15.

34. Macculloch, *Malaria*, 2–3, 20–23, 443; Macculloch, *An Essay on the Remittent and Intermittent Diseases, Including, Generically Marsh Fever and Neuralgia . . . ,* 2 vols. (London: Longman, Rees, Orme, Brown, & Green, 1828), 1:48–51, 73, 130, 220–26. Macculloch maintained that much "typhus" was really malarial disease. *Malaria*, 3, 39. See also Charles Creighton, "Malaria," *Encyclopedia Britannica*, 9th ed.

35. The increasing use of *miasm* as a general causal term in the nineteenth century allowed *malaria* to become a disease name. See Macculloch, *Remittent and Intermittent Diseases*, 1:xviii.

36. Macculloch, *Malaria*, 3, 7–8.

37. Ibid., 279–80, 381, 99–101. A sympathetic reviewer, the American Charles Caldwell, nevertheless pronounced Macculloch a "hydrophobe." *Essays on Malaria and Temperament* (Lexington, KY: Finnell & Hernden, 1831), 162–69.

38. Macculloch, *Malaria*, 246–47, 258–60, 391.

39. Ibid., 413–14.

40. Ibid., 9–11.

41. Ibid., 232–33; see also 403–4.

42. Macculloch, *Remittent and Intermittent Diseases*, 1: 195–96, 10–11.

43. Macculloch, *Malaria*, 111–12, 207–11, 363, 393, 415–16. Or it might be, noted Macculloch, that certain places would be habitable only by certain races. In the

southern United States this would be a population of freed African slaves. For modern expressions of fever as a population check, see Preston, *Hot Zone,* 83, 287–88; and Shah, *Fever,* 213–14.

44. In the view of one contemporary reviewer, the chemist-economist Thomas Cooper, Macculloch's catalog of the conditions of malaria clearly indicated mosquito transfer. See [Cooper], "Malaria: Essay on the Production and Propagation of This Poison," *American Quarterly Review* 4 (1828): 206–308.

45. W. H. S. Jones and E. T. Withington, *Malaria and Greek History* (1909; reprint, New York: AMS, 1977), 6–8; Leonard Jan Bruce-Chwatt and Julian de Zuleta, *The Rise and Fall of Malaria in Europe: A Historico-Epidemiological Study* (Oxford: Clarendon, 1980), 135.

46. Boott, *John Armstrong, M.D.,* 2:572–83; F. J. V. Broussais, *Conversations on the Theory and Practice of Physiological Medicine . . .* (London: Burgess & Hill, 1825), 181–200.

47. Manson, *Tropical Diseases,* 37–41, 105.

48. G. C. Boase, "Boott, Francis (1792–1863)," revised by Stephanie J. Snow, in *Oxford Dictionary of National Biography,* accessed 29 October 2011, www.oxforddnb .com/view/article/2901.

49. Boott, *John Armstrong, M.D.,* 2:233–60.

50. Anita McConnell, "Armstrong, John (1784–1829)," in *Oxford Dictionary of National Biography,* accessed 6 December 2011, www.oxforddnb.com/view/article/662.

51. Boott, *John Armstrong, M.D.,* 1:17–18, 24–26, 31.

52. Ibid., 1:200, 114–24, 128–33. The symptoms—"confusion of mind or stupor, a dull intoxicated look, blanched conjunctiva, dimness of sight, weakness of the muscular power, attended by an unsteadiness of gait or inability to stand or move, weakness or oppression of the pulse, and by slow, laborious respiration" (1:129)—resembled shock, and Boott uses the term (1:128, 133, 135). Boott also notes coldness of skin and weak pulse.

53. Macculloch, *Remittent and Intermittent Diseases,* 1:2–3.

54. Fayrer, *Climate and Fevers of India,* 58–59. Opposition to dissection was an issue (16).

55. Dale C. Smith, "The Rise and Fall of Typhomalarial Fever: I. Origins," *Journal of the History of Medicine and Allied Sciences* 37 (1982): 182–219. On the use of the term in tropical medicine, see Manson, *Tropical Diseases,* 109, 196; and Fayrer, *Climate and Fevers of India,* 215.

56. Lind, *Essay on Diseases Incidental to Europeans,* 118–19. West African "bulam fever" would later be included as the Atlantic yellow fever.

57. Boott, *John Armstrong, M.D.,* 1:239; see also 118–24, 151–55, 167, 215, 235–39, 259, 341.

58. Ibid., 1:307, 357; cf. John Cormack, *Natural History, Pathology, and Treatment*

of the Epidemic Fever at Present Prevailing in Edinburgh and Other Towns; Illustrated by Cases and Dissections (London: John Churchill, 1843), 89–91.

59. Boott, *John Armstrong, M.D.,* 1:338, 300–301, 341, 351, 391, 513.

60. Ibid., 1:401–2; see also 443. Boott is quoting a Dr. Duvall and a Dr. Speed.

61. On Lyons, see D. G. Crawford, *Roll Call of the Indian Medical Service, 1615–1930,* 2 vols. (London: W. Thacker, 1930), 1:167.

62. Macculloch, *Remittent and Intermittent Diseases,* 1:108–10; C. A. Wunderlich, *On the Temperature in Diseases: A Manual of Medical Thermometry,* translated by W. Bathurst Woodman (London: New Sydenham Society, 1871), 5, 16, 241. The clinical distinction remains problematic. Sally J. Cutler, "Possibilities for Relapsing Fever Reemergence," *Emerging Infectious Diseases* 12 (2006): 369–74.

63. R. T. Lyons, *A Treatise on Relapsing or Famine Fever* (London: Henry S. King, 1872), 1; cf. Charles Murchison, *A Treatise on the Continued Fevers of Great Britain* (London: Parker, Son, & Bourn, 1862).

64. Lyons, *Relapsing or Famine Fever,* 108, 171–72; cf. Murchison, *Treatise,* 290.

65. Lyons, *Relapsing or Famine Fever,* v–vi, 1, 3–4, 108, 112–13, 141.

66. Manson, *Tropical Diseases,* 41, 55, 67, 87.

67. Lyons, *Relapsing or Famine Fever,* viii.

68. Ibid., 205, 3–4.

69. Ibid., v–viii, 230–33.

70. Ibid., 248–52, 260–80.

71. Ibid., 107, 117, 123, 139, 148–57, 161–65, 183–84, 243–45; cf. Manson, *Tropical Diseases,* 89, for relapsing fever's overlap with malarial cachexia.

72. "Review of Lyons, Relapsing Fever," *Indian Medical Gazette* 8 (1873): 74–76, 106–8.

73. Lyons, *Relapsing or Famine Fever,* 173–77.

74. H. Vandyke Carter, *Spirillum Fever: Synonyms; Famine or Relapsing Fever As Seen in Western India* (London: J. & A. Churchill, 1882). On the politics of hunger in India during the period, see David Hall-Matthews, "Inaccurate Conceptions: Disputed Measures of Nutritional Needs and Famine Deaths in Colonial India," *Modern Asian Studies* 42 (2008): 1189–1212. For later assessments of the issues and of Lyons, see Leo Popoff, "Relapsing Fever," in *Twentieth-Century Practice: An International Encyclopedia of Modern Medical Science, by Leading Authorities of Europe and America,* edited by Thomas L. Stedman, vol. 16, *Infectious Diseases* (New York: William Wood, 1899), 455–548; and Oscar Felsenfeld, *Borrelia: Strains, Vectors, Human and Animal Borreliosis* (St. Louis: Warren H. Green, 1971).

75. Charles Creighton, *A History of Epidemics in Britain,* vol. 2, *From the Extinction of Plague to the Present Time* (Cambridge: Cambridge University Press, 1894), 69n.

76. Henry D. Shapiro, "Drake, Daniel," *American National Biography Online,* accessed 4 June 2012, www.anb.org/articles/12/12-00229.html; Shapiro, "Daniel

Drake and the Crisis in American Medicine of the 19th Century," *Journal of the American Medical Association* 254 (1985): 2113–16; John Harley Warner, "The Idea of Southern Medical Distinctiveness: Medical Knowledge and Practice in the Old South," in *Sickness and Health in America: Readings in the History of Medicine and Public Health,* edited by Judith Walzer Leavitt and Ronald L. Numbers, 2nd ed. (Madison: University of Wisconsin Press, 1985), 53–70.

77. Daniel Drake, *A Systematic Treatise Historical, Etiological, and Practical of the Principal Diseases of the Interior Valley of North America as they appear in the Caucasian, African, Indian, and Esquimaux Varieties of its Population,* edited by Hanbury Smith and Francis Smith, 2 vols. (1850; reprint, New York: Burt Franklin, 1971), 2:17–18.

78. Ibid., 2:17, 49, 72, 94–95, 186; cf. Macculloch, *Remittent and Intermittent Diseases,* 1:19, 26, 139, 197, 201.

79. Drake, *Systematic Treatise,* 2:22–25, 31–41, 151, at 39. Drake referred to alluvial streams and to neuralgia as a result of malarial exposure, suggesting familiarity with Macculloch's works. Henry Warren too had compared the effects of tropical diseases to those of toxins from insect and snake bites and from poisonous plants. *Malignant Fever in Barbados,* 22.

80. H. P. Cholmeley, "Fayrer, Sir Joseph, first baronet (1824–1907)," revised by W. F. Bynum, in *Oxford Dictionary of National Biography,* accessed 20 December 2011, www.oxforddnb.com/view/article/33099.

81. Fayrer, *Climate and Fevers of India,* 24, 34–37, 179–85.

82. Ibid., 26–31.

83. Ibid., 24–25, 34, 179–82, 185, 211, 215–25.

84. Cf. Manson, *Tropical Diseases,* 29, 59–60, 105, 108; and Ross, *Prevention of Malaria,* 218–34.

85. Samuel Kline Cohn, *Cultures of Plague: Medical Thinking at the End of the Renaissance* (Oxford: Oxford University Press, 2010), 140.

86. Coleman, *Yellow Fever in the North,* 3, 96.

87. Manson, *Tropical Diseases,* 139–41.

88. Yellow fever epidemics remain a popular subject. For a review, see Margaret Humphreys, *Yellow Fever and the South* (New Brunswick, NJ: Rutgers University Press, 1992), 11–12. More recently, see Khaled J. Bloom, *The Mississippi Valley's Great Yellow Fever Epidemic of 1878* (Baton Rouge: Louisiana State University Press, 1993); Deanne Nuwer, *Plague among the Magnolias: The 1878 Yellow Fever Epidemic in Mississippi* (Tuscaloosa: University of Alabama Press, 2009); and J. Worth Estes and Billy G. Smith, *A Melancholy Scene of Devastation: The Public Response to the 1793 Philadelphia Yellow Fever Epidemic* (Canton, MA: Science History, 1997).

89. The best-known exemplar of village epidemiology was William Budd, who tracked typhoid fever through the Devon village of North Tawton. See Coleman, *Yellow Fever in the North,* 173–77; and C.-E. A. Winslow, *The Conquest of Epidemic Disease* (Princeton, NJ: Princeton University Press, 1943).

90. Thomas Miner and William Tully, *Essays on Fevers, and Other Medical Subjects* (Middletown, CT: E. & H. Clark, 1823), 365–402.

91. Ibid., 305, 356, 366.

92. Coleman, *Yellow Fever in the North,* 67, 77, 181.

93. "Review of Miner and Tully," *North American Review* 17 (1823): 323–40 at 336–37.

94. Miner and Tully, *Essays on Fevers,* 356; Humphreys, *Yellow Fever and the South,* 22–25.

95. "Review of Miner and Tully," 336–37; cf. Miner and Tully, *Essays on Fevers,* 373.

96. David R. Goldfield, "Disease and Urban Image: Yellow Fever in Norfolk, 1855," *Virginia Cavalcade* 23, no. 2 (1973): 34–41. See also Humphreys, *Yellow Fever and the South,* 8–9.

97. *Report of the Howard Association of Norfolk, Va., to All Contributors Who Gave Their Valuable Aid in Behalf of the Sufferers from Epidemic Yellow Fever during the Summer of 1855* (Philadelphia: Inquirer Print Office, 1857) (hereafter cited as *Howard Association Report*); *Report of the Portsmouth Relief Association to the Contributors of the Fund for the Relief of Portsmouth, Virginia, during the Prevalence of the Yellow Fever in That Town in 1855 . . .* (Richmond: H. K. Ellyson, 1856), 136 (hereafter cited as *Portsmouth Report*); *Report of the Philadelphia Relief Committee, Appointed to Collect Funds for the Sufferers by Yellow Fever, at Norfolk & Portsmouth, Va.* (Philadelphia: Inquirer Office, 1856), 114 (hereafter cited as *Philadelphia Report*).

98. For the views of Norfolk physicians, see *Howard Association Report,* 95–112; for Portsmouth, see J. N. Schoolfield, "Sketch of the Yellow Fever," in *Portsmouth Report,* 75–121.

99. George Armstrong, *The Summer of the Pestilence: A History of the Ravages of the Yellow Fever in Norfolk, Virginia, AD 1855,* 2nd ed. (Philadelphia: Lippincott, 1856), 175–88; William S. Forrest, *The Great Pestilence in Virginia; Being an Historical Account of the Origin, General Character, and Ravages of the Yellow Fever in Norfolk and Portsmouth in 1855 . . .* (Philadelphia: Lippincott, 1856), 8–9, 26.

100. Armstrong, *Summer of the Pestilence,* 17–20; Forrest, *Great Pestilence in Virginia,* 8–9, 128, 283. Cf. Daniel Goldberg, "On Ideas as Actors: How Ideas about Yellow Fever Causality Shaped Public Health Policy Responses in 19th-Century Galveston," *Canadian Journal of Medical History* 29 (2012): 351–71.

101. On Armstrong (1813–1899), see, Gary Scott Smith, "Armstrong, George Dod," *American National Biography Online,* www.anb.org/articles/08/08-00055 .html. Armstrong's twelve "letters," conveying his "feelings and impressions" during the epidemic, were in fact cobbled together in its aftermath. *Summer of the Pestilence,* 7–8, 11. For Armstrong's motif, see Sally F. Griffith, "'A Total Dissolution of the Bonds of Society': Community Death and Regeneration in Matthew Carey's Short Account of the Malignant Fever," in Estes and Smith, *Melancholy Scene of Devastation,* 45–60.

102. Forrest, *Great Pestilence in Virginia*, 176–77; *Philadelphia Report*, 6, 10–11.

103. Forrest, *Great Pestilence in Virginia*, 67–68, 318; *Philadelphia Report*, 17–18.

104. Compare *Portsmouth Report*, 49, with *Howard Association Report*, 4–5.

105. *Portsmouth Report*, 14–15.

106. Ibid., 240.

107. Forrest, *Great Pestilence in Virginia*, 87–88, 105; *Portsmouth Report*, 139; *Philadelphia Report*, 121.

108. Armstrong, *Summer of the Pestilence*, 62, 70–71, 153–54; Forrest, *Great Pestilence in Virginia*, 40–41, 57–58, 173–74; *Portsmouth Report*, 248; *Philadelphia Report*, 89, 93, 96–97, 100.

109. Armstrong, *Summer of the Pestilence*, 54–55.

110. Forrest, *Great Pestilence in Virginia*, 163, 273; *Portsmouth Report*, 15.

111. *Portsmouth Report*, 19, 47; *Philadelphia Report*, 88–89.

112. Humphreys, *Yellow Fever and the South*, 49.

113. Armstrong, *Summer of the Pestilence*, 32.

114. Forrest, *Great Pestilence in Virginia*, 197.

115. Armstrong, *Summer of the Pestilence*, 84–86, 122–23; cf. 140.

116. *Portsmouth Report*, 244, 255; *Philadelphia Report*, 4, 19–20, 25–33, 85–86, 93, 98, 117, 120.

117. Armstrong, *Summer of the Pestilence*, 45–47; Forrest, *Great Pestilence in Virginia*, 39.

118. Goldfield, "Disease and Urban Image," 34–41.

119. *Philadelphia Report*, 23, 31–32.

120. *Portsmouth Report*, 312–16; *Philadelphia Report*, 14–15, 101–5, 108. The Philadelphians insisted that their volunteers were acclimated.

121. Snowden, *Conquest of Malaria: Italy*, 72–75 at 72. Snowden is quoting Gustavo Foa.

122. Bewell, *Romanticism and Colonial Disease*, 47–49; see also 77–78.

123. Anderson, *Colonial Pathologies*, 90–93.

124. Shah, *Fever*, 24, 93. See also Snowden, *Conquest of Malaria: Italy*, 93.

125. Drake, *Systematic Treatise*, 2:150–51.

126. W. H. S. Jones and Withington, *Malaria and Greek History*, 100–117. See also Fayrer, *Climate and Fevers of India*, 100–101; Macculloch, *Remittent and Intermittent Diseases*, 1:59–70, 88–91, 247–51, 321–23; and Snowden, *Conquest of Malaria: Italy*, 13–15, 74.

127. William K. Anderson, *Malarial Psychoses and Neuroses with Chapters Medicolegal, and on History, Race Degeneration, Alcohol, and Surgery in Relation to Malaria* (London: Humphrey Milford / Oxford University Press, 1927), 133.

128. Martin Willis notes Stoker's medical allusions but not any uniquely malarial concerns. "'The Invisible Giant,' 'Dracula,' and Disease," *Studies in the Novel* 39 (2007): 301–25.

129. Armstrong, *Summer of the Pestilence,* 169–70, 175, 178, 180–81, 191–92; *Portsmouth Report,* 100.

130. Leon Warshaw, *Malaria: The Biography of a Killer* (New York: Rinehart, 1949), 60–64; Forrest, *Great Pestilence in Virginia,* 288–90.

131. Armstrong, *Summer of the Pestilence,* 160–63; Forrest, *Great Pestilence in Virginia,* 32–33. However tempting it might be to think that the allusion is to the small Aedes mosquito, descriptions suggest that it was a blowfly feasting on corpses. "Insects and Pestilence: Cholera and Yellow Fever," *Scientific American* 11 (1855): 30;

132. Forrest, *Great Pestilence in Virginia,* 133, 143, 286.

133. Macculloch, *Malaria,* 143, 150, 266–68, 299–303. Others did speculate. See [Cooper], "Malaria"; and George H. F. Nuttall, "Upon the Part Played by Mosquitoes in the Propagation of Malaria: A Historical and Critical Study," *Journal of Tropical Medicine* 2 (1900): 198–200, 231–33, 245–47.

134. Armstrong, *Summer of the Pestilence,* 53, 93, 122–25. On natural-historical method, see Humphreys, *Yellow Fever and the South,* 35–44.

135. Warshaw, *Malaria,* 45–76.

136. George B. Wood, *A Treatise on the Practice of Medicine,* 2nd ed. (Philadelphia: Grigg, Elliott, 1849); Thomas Watson, *Lectures on the Principles and Practice of Physic,* 4th ed., 2 vols. (London: Parker, 1857).

137. Walther Riese, *The Conception of Disease: Its History, Its Versions, and Its Nature* (New York: Philosophical Library, 1953), 88. Drake, *Systematic Treatise,* 2:42; see also 50.

138. Macculloch, *Malaria,* 474–75.

139. Boott, *John Armstrong, M.D.,* 1:233–34.

140. Ibid., 1:156–64 at 156–57. Boott, following Benjamin Rush, sometimes suggests that predispositions like fatigue will play a larger role than toxins. See ibid., 1:169–70. See also Drake, *Systematic Treatise,* 2:129; and Macculloch, *Malaria,* 280–89, 404, 468–70.

141. Coleman, *Yellow Fever in the North,* 129.

142. Macculloch, *Malaria,* 53–55.

143. Ibid., 381; cf. Coleman, *Yellow Fever in the North,* 55, 135.

144. Drake, *Systematic Treatise,* 2:42; see also 134.

145. Manson, *Tropical Medicine,* 122.

146. Macculloch, *Malaria,* 128–33, 208–9.

147. Boott, *John Armstrong, M.D.,* 1:164–65, 170–71. See also Macculloch, *Malaria,* 86–92.

148. Macculloch, *Malaria,* 390, 130–31; see also 12, 225, 231, 233–34, 403.

149. Ibid., 466.

150. Weindling, *Epidemics and Genocide,* 9.

151. Fayrer, *Climate and Fevers of India,* 27–32; Humphreys, *Yellow Fever and the South,* 33–34; Warshaw, *Malaria,* 43; Manson, *Tropical Diseases,* 229.

152. Deborah J. Neill, *Networks in Tropical Medicine: Internationalism, Colonialism, and the Rise of a Medical Specialty, 1890–1930* (Stanford, CA: Stanford University Press, 2012).

153. William Osler, *Principles and Practice of Medicine,* 7th ed. (New York: D. Appleton, 1910), 64.

154. Carter, *Spirillum Fever,* 19, 34, 316.

155. Snowden, *Conquest of Malaria: Italy,* 149–50, 173–74.

156. Pelis, *Charles Nicolle,* 53–54, 69.

157. Weindling, *Epidemics and Genocide,* xv.

158. George Rosen, "What Is Social Medicine? A Genetic Analysis of the Concept," *Bulletin of the History of Medicine* 21 (1947): 674–733.

159. K. C. Carter, "Koch's Postulates in Relation to the Work of Jacob Henle and Edwin Klebs," *Medical History* 29 (1985): 353–74.

160. In 1875 Ferdinand Cohn would name the spiral organism that Obermaier found *Spirochaeta obermaieri.* Konrad Birkhaug, "Otto H. F. Obermeier, February 13, 1843–August 20, 1873," in *A Symposium on Relapsing Fever in the Americas,* edited by F. R. Moulton (Washington, DC: AAAS, 1942), 7–14 at 9–10. At the end of the century it still would not be possible to reproduce the fever. Popoff, "Relapsing Fever," 463–65.

161. Popoff, "Relapsing Fever," 511, 525–28; H. L. Wynns, "The Epidemiology of Relapsing Fever," in Moulton, *Symposium on Relapsing Fever,* 100–105 at 100. See also Richard Strong et al., *Typhus Fever with Particular Reference to the Serbian Epidemic* (Cambridge, MA: Harvard University Press / American Red Cross, 1920), 3–4.

162. George Magee, "Symptomatology, Clinical Diagnosis, and Therapy of Relapsing Fever," in Moulton, *Symposium on Relapsing Fever,* 106–8 at 106–7; Cutler, "Possibilities for Relapsing Fever Reemergence," 369; Shah, *Fever,* 132–33.

163. Rudolph Virchow, *On Famine Fever and Some of the Other Cognate Forms of Typhus: A Lecture Held for the Benefit of the Sufferers of East Prussia, February 9, 1868* (London: Williams & Norgate, 1868), 13, 34–41, 44–47. See also Weindling, *Epidemics and Genocide,* 25.

164. M. V. Assous and A. Wilamowski, "Relapsing Fever Borreliosis in Eurasia—Forgotten, but Certainly Not Gone!," *Clinical Microbiology and Infection* 15 (2009): 407–14; Cutler, "Possibilities for Relapsing Fever Reemergence."

165. Pelis, *Charles Nicolle,* 69–70.

166. Strong et al., *Typhus Fever,* 18–20; cf. Hans Zinsser, *Rats, Lice, and History* (Boston: Little, Brown, 1935), 225–26. The Central Powers were no less assiduous in mounting an alliance to combat typhus. Weindling, *Epidemics and Genocide,* 73–74.

167. Strong et al., *Typhus Fever,* 97–100 at 97.

168. Ibid., 21–28 at 27; for the Polish expedition, see 97–106, 109. Any compliments were usually minimal and backhanded. See also Alfred E. Cornebise, *Typhus and Doughboys: The American Polish Typhus Relief Expedition, 1919–1921* (Newark: University of Delaware Press, 1982); and Pelis, *Charles Nicolle,* 63–70.

169. Strong et al., *Typhus Fever,* 8, 111–22, 143, 163–65, 170–71, 178, 181.

170. Ibid., 161–66.

171. Ibid., 111–12, 122–27, 168. The effectiveness of quinine, enlargement of the spleen in so-called relapsing fever (ibid., 163), and the failure to base diagnosis on any test all suggest that malarial diseases were being diagnosed as relapsing fever. Historians of malaria note that it too was epidemic in the Balkans during the First World War. Bruce-Chwatt and Zuleta, *Rise and Fall of Malaria in Europe,* 46–47.

172. Strong et al., *Typhus Fever,* 147–53. Cf. Riese, *Conception of Disease,* 93–94.

173. Shah, *Fever,* 134, 229, 238–39.

174. Fergus Walsh, "Global Malaria Death Rate Falling," 13 December 2011, www .bbc.co.uk/news/health-16161907; Neil Bowdler, "Malaria Deaths Hugely Underestimated—Lancet Study," 2 February 2012, www.bbc.co.uk/news/health-16854026; and Michelle Roberts, "Third of Malaria Drugs Are Fake," 21 May 2012, www.bbc .co.uk/news/health-18147085, all from *BBC Health,* accessed 21 June 2012.

Chapter 8. Numbers and Nurses

1. The quotations in the first three paragraphs are from J. B. Coleman, "Proper Nursing Absolutely Necessary for the Successful Treatment of Typhoid Fever," *Medical Council,* 4 November 4 1899, 433–34. My discussion focuses mainly on American private-duty fever nursing. Fever nursing in Europe remained more public, concerned with controlling epidemic disease. See Margaret Currie, *Fever Hospitals and Fever Nurses: A British Social History of Fever Nursing; A National Service* (London: Routledge, 2005).

2. Walther Riese, *The Conception of Disease: Its History, Its Versions, and Its Nature* (New York: Philosophical Library, 1953), 96.

3. William Coleman, *Yellow Fever in the North: The Methods of Early Epidemiology* (Madison: University of Wisconsin Press, 1987), 51.

4. Volker Hess, *Der wohltempeierte Mensch: Wissenschaft und Alltag des Fiebermessens (1850–1900)* (Frankfurt: Campus Verlag, 2000), 20. The key work is C. A. Wunderlich, *On the Temperature in Diseases: A Manual of Medical Thermometry,* translated by W. Bathurst Woodman (London: New Sydenham Society, 1871); see 21–41.

5. J. Worth Estes, "Quantitative Observation of Fever and Its Treatment before the Advent of Short Clinical Thermometers," *Medical History* 35 (1991): 189–216; Daniel Sennert, *Of Agues and Fevers: Their Differences, Signes, and Cures,* translated by N.D.B.M. (London: Lodowyck Loyd, 1658), 1.

6. Usually the recognition is credited to Boerhaave's protégé Anton de Haen. See Wunderlich, *On the Temperature in Diseases,* 21, 30; and James Currie, *Medical Reports on the Effects of Water . . . ,* 4th ed. (London: Cadell & Davies, 1805), 244. Wunderlich interpreted these perceptions as reflecting differences between the trunk and the extremities.

7. Patrick Manson, *Tropical Diseases: A Manual of the Diseases of Warm Cli-*

mates (New York: William Wood, 1898), 39; Edward Register, *Practical Fever Nursing* (Philadelphia: W. B. Saunders, 1907), 192; George P. Paul, *Nursing in the Acute Infectious Fevers*, 3rd ed. (Philadelphia: W. B. Saunders, 1917), 199.

8. Wunderlich, *On the Temperature in Diseases,* 21–27.

9. Currie, *Medical Reports,* iii–iv, 126–29.

10. Ibid., 198–220.

11. Ibid., 243, 248–50, at 243.

12. John E. Lesch, *Science and Medicine in France: The Emergence of Experimental Physiology, 1790–1855* (Cambridge, MA: Harvard University Press, 1984); Frederick Gregory, *Scientific Materialism in Nineteenth Century Germany* (Dordrecht: D. Reidel, 1977); Lynn K. Nyhart, *Modern Nature: The Rise of the Biological Perspective in Germany* (Chicago: University of Chicago Press, 2009).

13. Knud Faber, *Nosography in Modern Internal Medicine* (New York: Paul Hoeber, 1922), 59–70, 88, 147–48, 186–87, at 65–67; Virchow is quoted on 70, Faber from 66.

14. Wunderlich, *On the Temperature in Diseases,* v. Wunderlich's book was translated into English in 1871 and French in 1872. There was also an 1871 American abridgment. Temperature studies were increasingly common in the 1850s and 1860s.

15. Ibid., 68–71. Wunderlich did recommend taking the rectal temperatures for infants and some others. Register worried that rectal thermometers might break off or disappear into the bowel and "be difficult to extract." *Practical Fever Nursing,* 26–27. Urine-stream temperatures were also mentioned, as were vaginal temperatures. Register saw these as most accurate. Later authors would argue that one should add up to a degree to axillary temperatures.

16. Wunderlich, *On the Temperature in Diseases,* 71–78 at 75.

17. Ibid., 77–78.

18. Ibid., 48–49.

19. Ibid., 99–100. Philip Mackowiak and Gretchen Worden, "Carl Reinhold August Wunderlich and the Evolution of Clinical Thermometry," *Clinical Infectious Diseases* 18, no. 3 (1994): 462. Given the number of observations Wunderlich claimed to have made, Mackowiak and Worden questioned the means by which he could reach such generalizations in a precomputer and prestatistics age.

20. Wunderlich, *On the Temperature in Diseases,* v–vii, 5, 39; Faber, *Nosography,* 73–74.

21. Wunderlich, *On the Temperature in Diseases,* 48.

22. Ibid., 79.

23. Mackowiak and Worden, "Wunderlich and the Evolution of Clinical Thermometry," 458, emphasis added; Wunderlich, *On the Temperature in Diseases,* vi.

24. Robert Graves, *Clinical Lectures on the Practice of Medicine, To Which Is Prefixed a Criticism by Professor Trousseau, Reprinted from the Second Edition, Edited by the Late Dr. Neligan,* 2 vols. (1848; reprint, London: New Sydenham Society, 1884), 1:134.

25. Watching fits well into Ulrich's category of social medicine. See Laurel Ul-

rich, *A Midwife's Tale: The Life of Martha Ballard, Based on Her Diary, 1785–1812* (New York: Knopf, 1990).

26. Cyril Pearl, *Spectral Visitants, or Journal of a Fever, by a Convalescent* (Portland, ME: S. H. Colesworthy, 1845), 68–70.

27. Alfred Holbrook, *Reminiscences of the Happy Life of a Teacher* (Cincinnati: Elm Street Printing, 1885), 336–41.

28. Elizabeth Hill, *The Widow's Offering: An Authentic Narrative of the Parentage, Life, Trials and Travels of Mrs. Elizabeth Hill* (New London, CT: D. S. Ruddock, 1852), 154–58.

29. "The Sick Man's Remonstrance," *Northern Monthly* 1 (1864): 106–7. On acuteness of hearing during fever, see Register, *Practical Fever Nursing,* 59.

30. Register, *Practical Fever Nursing,* 191, 279.

31. Lecture notes of Edith Lewis Palmer Griscom, 1909–12, Barbara Bates Center for the History of Nursing, University of Pennsylvania, MC 12/3.

32. Register devotes 81 pages of his 1907 *Practical Fever Nursing* to typhoid (compared with 157 pages for 19 other fevers), and Paul devotes 23 pages of his 1917 *Nursing in the Acute Infectious Fevers* (compared with 100 for 16 others).

33. Paul, *Nursing in the Acute Infectious Fevers,* 29; Register, *Practical Fever Nursing,* 75.

34. Register, *Practical Fever Nursing,* 15. Strikingly, many elements of the rest cure for neurasthenia, famously depicted in Charlotte Perkins Gilman's novella *Yellow Wall Paper,* carried over to fever nursing generally.

35. Bedside Stories series, Chatauqua School of Nursing, Barbara Bates Center for the History of Nursing, University of Pennsylvania, MC 22. Whatever their origin, the pamphlets bear an editor's toolmarks.

36. Paul, *Nursing in the Acute Infectious Fevers,* 100.

37. Kay Brune, "The Early History of Non-opioid Analgesics," *Acute Pain* 1, no. 1 (December 1997): 33–40.

38. Hobart Amoy Hare, *Fever: Its Pathology and Treatment by Antipyretics* (Philadelphia: F. A. Davis, 1891), 1. See also Faber, *Nosography,* 110.

39. Hare, *Fever,* 2–4, 97, 36.

40. Ibid., 60–62, 92–95, 121, at 60, 62.

41. Ibid., 159–60; cf. T. J. Maclagan, *Fever: A Clinical Study* (London: J. & A. Churchill, 1888), 148, 162; Register, *Practical Fever Nursing,* 164; Paul, *Nursing in the Acute Infectious Fevers,* 99.

42. Register, *Practical Fever Nursing,* 48–49.

43. Wunderlich, *On the Temperature in Diseases,* 48–52, 35, 106–7.

44. Ibid., 35, 81–82, 125, 131–32, 196, at 81.

45. Ibid., 53, 124, 170, 174–78.

46. Ibid., 120, 170.

47. Ibid., 177–78.

48. Ibid., 178.

49. Ibid., 81.

50. John Harley Warner, *The Therapeutic Perspective: Medical Practice, Knowledge, and Identity in America, 1820–1885* (Cambridge, MA: Harvard University Press, 1986), 101. And patients too: Volker Hess, "Standardizing Body Temperature: Quantification in Hospitals and Daily Life, 1850–1900," in *Body Counts: Medical Quantification in Historical and Sociological Perspectives,* edited by Gérard Jorland, Annick Opinel, and George Weisz (Montréal: McGill-Queen's University Press, 2005), 109–26.

51. Manson, *Tropical Diseases,* xi–xii, emphasis added.

52. H. C. Wood Jr., *Thermic Fever, or Sunstroke* (Philadelphia: J. B. Lippincott, 1872). Wood had suggested the term *thermic fever* in 1863, when he conceived the fever process to involve a toxin as well as heat. He later became convinced that heat alone was sufficient (103). On Wood, see John P. Swann, "Horatio C. Wood Jr.," *American National Biography Online,* accessed 24 February 2013, www.anb.org /articles/12/12-00998.html.

53. Wood, *Thermic Fever,* 7, 12–15, 20–21, 33.

54. Ibid., 37.

55. Ibid., 52–79. In a similar vein, see Charles Reber, *Paresis of the Sympathetic Centers from over Excitation by High Solar Heat, Long Continued and Suddenly Withdrawn, Etc.; So-called Malaria, Its Etiology, Pathogenesis, Pathology and Treatment* (St. Louis: Geo. Rumbold, 1879). On nineteenth-century notions of negative feedback, see M. N. Wise, "Work and Waste," *History of Science* 27 (1989): 221–61, 263–301, 391–449.

56. Maclagan, *Fever: A Clinical Study,* 28–29; Hare, *Fever,* 63; James M. Gully, *A Monograph on Fever and Its Treatment by Hydro-Therapeutic Means* (London: Simpkin, Marshall, 1885), 10–11.

57. William Osler used furnace analogies, among a mix of metaphors, in his lectures to nursing students at the University of Pennsylvania in 1887. Mary Clymer Ward Papers, Barbara Bates Center for the History of Nursing, University of Pennsylvania, MC 16.

58. Maclagan, *Fever: A Clinical Study,* 24.

59. Ibid.

60. On Maclagan, see W. Broadbent, "Dr. Maclagan and His Great Work," *Nineteenth Century and After* 55 (1904): 734–40.

61. Maclagan, *Fever: A Clinical Study,* 1.

62. Ibid., 36–46 at 41, 46.

63. Ibid., 48, 49.

64. Ibid., 68–69.

65. Hare, *Fever,* 61–62, 67, 103.

66. Isabel Leighton, ed., *The Aspirin Age, 1919–41* (Harmondsworth, UK: Penguin, 1964).

67. Notably, Register, *Practical Fever Nursing* (1907), covers yellow fever, relapsing fever, plague, and dengue, whereas Paul, *Nursing in the Acute Infectious Fevers* (1917), covers none of these and omits typhus and tuberculosis as well.

68. Maclagan, *Fever: A Clinical Study,* 116.

Part IV. Fever, Modern and Post-Modern

1. Robert Graves, *Clinical Lectures on the Practice of Medicine, To Which Is Prefixed a Criticism by Professor Trousseau, Reprinted from the Second Edition, Edited by the Late Dr. Neligan*, 2 vols. (1848; reprint, London: New Sydenham Society, 1884).

2. Helen Clapesattle, *The Doctors Mayo* (New York: Pocket Books, 1956), 6–13.

3. "Malaria Cases in U.S. Reach 40-Year High," accessed 21 November 2013, www.cdc.gov/media/releases/2013/p1031-malaria-cases.html; Randall Packard, *The Making of a Tropical Disease: A Short History of Malaria* (Baltimore: Johns Hopkins University Press, 2007); James L. A. Webb Jr., *Humanity's Burden: A Global History of Malaria* (Cambridge: Cambridge University Press, 2009); Sonia Shah, *The Fever: How Malaria Has Ruled Humankind for 500,000 Years* (New York: Sarah Crichton Books / Farrar, Straus, & Giroux, 2010).

Chapter 9. Machines, Mothers, Sex, and Zombies

1. World Health Organization, Initiative for Vaccine Research (IVR): Diarrhoeal Diseases (updated February 2009), accessed 28 January 2012, www.who.int /vaccine_research/diseases/diarrhoeal/en/index7.html.

2. K. David Patterson, "Typhus and Its Control in Russia, 1870–1940," *Medical History* 37 (1993): 361–81.

3. Paul Weindling, *Epidemics and Genocide in Eastern Europe, 1890–1945* (Oxford: Oxford University Press, 1999), xv, 3, 10.

4. "Typhus," accessed 31 January 2012, www.whale.to/b/typhus_h.html.

5. Beulah France, "So Your Child Has a Fever," *Country Gentleman* 124 (1954): 86; see also J. D. Ratcliff, "What It Means When You Have a Fever," *Reader's Digest*, December 1959, 60–63.

6. "Modern Views of Fever," *Literary Digest* 92 (1927): 18.

7. See Norman Plummer, "Fever," *Hygeia* 9 (1931): 1032–34; Iris Fry, "On the Biological Significance of the Properties of Matter: L. J. Henderson's Theory of the Fitness of the Environment," *Journal of the History of Biology* 29 (1996): 155–96; John Parascondola, "Organismic and Holistic Concepts in the Thought of L. J. Henderson," ibid. 4 (1971): 63–113.

8. Philip J. Pauly, "The Development of High School Biology," *Isis* 82 (1991): 662–88.

9. "Modern Views of Fever," 18.

10. "Why We Get Feverish," *Literary Digest* 103 (1929): 52–53. See also "Anhydremia and Fever," *Journal of the American Medical Association* 79 (1922): 218–19; Plummer, "Fever," 1033; and Henry G. Barbour, "The Heat Regulating Mechanism of the Body," *Physiological Reviews* 1 (1921): 295–326 at 309–10. Other possible master controllers included the adrenal glands and even, sometimes, the mind, through the power of suggestion.

11. John Lentz, "The Tale of the Thermometer," *Today's Health,* May 1957, 40–41, 49–50, at 41.

12. Theodore Irwin, "Fever: How to Play It Cool," ibid., December 1968, 52–55 at 53. See also Ratcliff, "What It Means When You Have a Fever," 61; and "Fever Fact and Fancy," *Current Health,* December 1978, 18.

13. Irwin, "Fever: How To Play It Cool," 52.

14. "Fever Fact and Fancy," 18; Dr. Morris A. Wessel, "Fever: What Does It Mean?," *Parents,* October 1981, 38.

15. France, "So Your Child Has a Fever," 86.

16. C. W. Ramsey, "When You're Too Hot . . . ," *McCall's,* February 1980, 106.

17. Dr. Herman N. Bundesen, "The Problem of Fever," *Ladies' Home Journal,* February 1943, 115–16.

18. Adeline Bullock, "Fever: Good or Bad?," *Parents,* March 1950, 50, 83–85, at 50.

19. Dr. Dorothy Whipple, "What You Should Know about Fever," ibid., August 1953, 32–33, 78, at 32.

20. Bullock, "Fever: Good or Bad?," 50.

21. "What You Should Know about Sore Throat and Fever," *Better Homes and Gardens,* February 1979, 13; Irwin, "Fever: How To Play It Cool," 52.

22. Irwin, "Fever: How To Play It Cool," 55.

23. "Why We Get Feverish," 52–53.

24. Whipple, "What You Should Know about Fever," 78.

25. See, for example, www.sansscience.wordpress.com/2011/09/11/video-of-pacman-t-cells-running-around-inside-your-body/; and www.nature.com/nri/journal/vii/nii/full/nri3102.html, both accessed 29 April 2013.

26. Bernard Fantus, "What to Do and What Not to Do for Fever," *Hygeia* 13 (March 1935): 259–60; Ratcliff, "What It Means When You Have a Fever," 61–62 at 62.

27. Wessel, "Fever: What Does It Mean?"; Karen Leila Woodworth, "Fever Dos and Don'ts for Anxious Parents," *Parents,* June 1980, 68–72 at 69; Irwin, "Fever: How to Play It Cool," 52–55; Ratcliff, "What It Means When You Have a Fever," 61–62.

28. "Why We Get Feverish," 53.

29. Lentz, "Tale of the Thermometer," 41, 49.

30. Fantus, "What to Do and What Not to Do for Fever," 259–60.

31. Dr. L. E. Holt Jr., "Fever: What Causes It and What to Do about It," *Good Housekeeping,* May 1957, 30–31, 155, at 155.

32. Susan Sontag, *"Illness as Metaphor" and "AIDS and Its Metaphors"* (New York: Picador, 2001).

33. By the mid-1930s the phrase *fever therapy* might refer to either a means of curing fever or a means of creating it. For clarity, the latter was sometimes called *pyrotherapy* or *pyretotherapy.*

34. Willa Phillips, *Textbook of Pyretotherapy* (Ann Arbor: Edwards Bros., 1939), 2–3.

35. Elliot S. Wallenstein, *Great and Desperate Cures: The Rise and Decline of Psychosurgery and Other Radical Treatments for Mental Illness* (New York: Basic Books, 1986).

36. "Why We Get Feverish," 52–53.

37. Magda Whitlow, "Wagner-Jauregg and Fever Therapy," *Medical History* 34 (1990): 294–310; Gladys Terry, *Fever and Psychoses: A Study of the Literature and Current Opinion on the Effects of Fever on Certain Psychoses and Epilepsy* (New York: Paul Hoeber, 1939).

38. Michael Pijoan and Claude Payzant, *A Handbook of Artificial Fever Therapy* (Boston: Little, Brown, 1938), 2–3, 18–24.

39. One advocate, Jacques d'Arsonval, considered Lee Deforest's invention of the triode as one of the great advances of medical instrumentation. *Fever Therapy: Abstracts and Discussions of Papers presented at the First International Conference on Fever Therapy, College of Physicians and Surgeons, Columbia University, New York City, March, 29, 30, 31, 1937* (New York: Paul Hoeber, 1937), 5–6.

40. Phillips, *Textbook of Pyretotherapy*, 8–13; H. Worley Kendell, *Fever Therapy* (Springfield, IL: Charles C. Thomas, 1951), 4. A number of manufacturers were involved. Besides General Electric's Inductotherm, the Humidtherm Therapeutic fever cabinet was manufactured by the Therapeutic Appliance Corporation of Los Angeles, and the L-F Hypertherm, by the Liebel-Flarshiem Company of Cincinnati.

41. Phillips, *Textbook of Pyretotherapy*, 4–5, 15–16, 55.

42. Ibid., 1.

43. Jack Ewalt, Ernest H. Parsons, Stafford Warren, and Stafford Osborne, *Fever Therapy Technique* (New York: Paul Hoeber, 1939), 2–5 at 5; Pijoan and Payzant, *Handbook of Artificial Fever Therapy*, 11, 13, 15. On physiatry, see Glenn Gritzer and Arnold Arluke, *The Making of Rehabilitation: A Political Economy of Medical Specialization, 1890–1980* (Berkeley: University of California Press, 1985).

44. Phillips, *Textbook of Pyretotherapy*, 7, 14; cf. Joel T. Braslow, "The Influence of a Biological Therapy on Physicians' Narratives and Interrogations: The Case of General Paralysis of the Insane and Malaria Fever Therapy, 1910–1950," *Bulletin of the History of Medicine* 70 (1996): 577–608. Pijoan and Payzant, writing about the Boston area, assumed that patients would be sedated, often with morphine (*Handbook of Artificial Fever Therapy*, 14); this does not appear to have been common practice elsewhere.

45. Phillips, *Textbook of Pyretotherapy*, 5–6.

46. "Modern Views of Fever," 18.

47. Clarence Neyman, *Artificial Fever Produced by Physical Means: Its Development and Application* (New York: Paul Hoeber, 1938), 245.

48. Rima B. Apple, *Perfect Motherhood: Science and Childrearing in America* (New Brunswick, NJ: Rutgers University Press, 2006).

49. Bundesen, "Problem of Fever," 115–16.

50. Dr. Josephine Hemenway Kenyon, "Facts about Fever," *Good Housekeeping,* April 1930, 106; Whipple, "What You Should Know about Fever," 32.

51. Tuberculosis was a concern in *Better Homes and Gardens* as late as 1966. See "When Someone in Your Family Has a Fever," April 1966, 126.

52. Bundesen, "Problem of Fever," 116.

53. Bullock, "Fever: Good or Bad?," 85.

54. Bundesen, "Problem of Fever," 116; Kenyon, "Facts about Fever," 106.

55. Kenyon, "Facts about Fever"; Fantus, "What to Do and What Not to Do for Fever," 260, emphasis added.

56. Bundesen, "Problem of Fever," 115.

57. Whipple, "What You Should Know about Fever," 33.

58. Fantus, "What to Do and What Not to Do for Fever," 260.

59. Bundesen, "Problem of Fever," 115.

60. Dr. Walter H. Eddy, "Stuff a Cold and Starve a Fever and You Won't Get Well Fast," *Good Housekeeping,* January 1936, 95.

61. Bullock, "Fever: Good or Bad?," 50.

62. Ratcliff, "What It Means When You Have a Fever," 61.

63. Ibid., 62; cf. "Friendly Fever," *Time,* 27 April 1959, 77.

64. See, for example, Wessel, "Fever: What Does It Mean?"

65. Woodworth, "Fever Dos and Don'ts for Anxious Parents," 70–72 at 71–72.

66. Ramsey, "When You're Too Hot . . . ," 56.

67. Editors' introduction to Dr. Richard Feinbloom, "Fevers," *Redbook,* March 1974, 87, 160–62, at 87.

68. "When Fever Strikes: How to Fight Back," *Glamour,* January 1982, 162, 165, at 165.

69. Woodworth, "Fever Dos and Don'ts for Anxious Parents," 72.

70. "Why We Get Feverish," 52–53.

71. Kenyon, "Facts about Fever." See also Bullock, "Fever: Good or Bad?," 83.

72. Lentz, "Tale of the Thermometer," 50; Irwin, "Fever: How To Play It Cool," 55.

73. Fantus, "What to Do and What Not to Do for Fever," 259.

74. Kenyon, "Facts about Fever."

75. Lentz, "Tale of the Thermometer," 50.

76. Herman Bundesen, "Fever," *Ladies' Home Journal,* September 1943, 154.

77. Woodworth, "Fever Dos and Don'ts for Anxious Parents," 70.

78. Kenyon, "Facts about Fever"; Bundesen " Fever," 154; Bullock, "Fever: Good or Bad?," 50; Fantus, "What to Do and What Not to Do for Fever," 259.

79. Lynn LiDonnici, "Burning for It: Erotic Spells for Fever and Compulsion in the Ancient Mediterranean World," *Greek, Roman and Byzantine Studies* 39 (1998): 63–98; Anita Jacobson-Widding, "Notions of Heat and Fever among the Manyika of Zimbabwe," in *Culture, Experience and Pluralism: Essays on African Ideas of Ill-*

ness and Healing, edited by Anita Jacobson-Widding and David Westerlund (Stockholm: Almqvist & Wiksell, 1999), 27–44.

80. See *Oxford English Dictionary,* 2nd ed., s.vv. "fever," "fevered," and "feverish."

81. Concordance of Shakespeare's complete works, accessed 27 January 2012, www.opensourceshakespeare.org/concordance.

82. See www.stlyrics.com/lyrics/madhotballroom/fever.htm, accessed 27 January 2012.

83. See www.rintelen.ch/konzept_und_text/index.php?sparte=fever/db.php&lang =en. For the author's preferred version, see www.youtube.com/watch?v=z5wOd 2aGxdU. Both accessed 13 July 2012.

84. See www.lyricsfreak.com/e/elvis+presley/burning+love_20047794.html; www .lyricsfreak.com/f/foreigner/hot+blooded_20054836.html; and www.lyricsfreak.com /t/ted+nugent/cat+scratch+fever_20551632.html, all accessed 29 April 2013.

85. See www.lyricsmode.com/lyrics/f/family_force_5/fever_lyrics.html, accessed 27 January 2012. Thus in the "Song Meanings" section of this website: "This song can be taken two ways. For the non-believer it could be like whenever a guys around a girl they really like. For the believer it can be that the holy-spirit fills you and it can spread, like a fever. The latter is what ff5 has said it to be" (November 2011). "Fever is about a guy who thinks a girl is hot and its catchy" (January 2012).

86. Plague Inc., accessed 24 February 2014, www.ndemiccreations.com/en/22 -plague-inc. Here the object is to help the plague spread. On this form of popular culture see Priscilla Wald, *Contagions: Cultures, Carriers, and the Outbreak Narrative* (Durham, NC: Duke University Press, 2008). My thanks to David Barnes for this reference.

87. Richard Preston, *The Hot Zone* (New York: Random House, 1994), 13, 32–33, 44, 50, 69, 156.

88. Ibid., 83.

89. Weindling, *Epidemics and Genocide.* The most thoughtful explorations of these issues are Greg Bear's *Darwin's Radio* (New York: Del Ray, 1999) and *Darwin's Children* (New York: Del Ray, 2003).

90. Preston, *Hot Zone,* 11, 47, 55, 68, 72.

91. See, for example, "The Good, The Bad & The Disgusting: Virus Movies," accessed 24 February 2014, http://skymovies.sky.com/contagion-2011/best-virus -movies.

92. See "Preparedness 101: Zombie Pandemic," www.cdc.gov/phpr/zombies /#/page/1; and Office of Public Health Preparedness and Response, "Zombie Preparedness," www.cdc.gov/phpr/zombies.htm, both accessed 29 April 2013.

93. Preston, *Hot Zone,* 287.

94. See "When Infections 'Spillover,'" accessed 19 December 2012, www.npr .org/2012/10/19/163245528/when-infections-spillover.

INDEX

acclimatization, 138, 181, 209, 212, 215, 223, 237, 254, 274, 335n88, 356n120
acetaminophen, 81, 269
acidity, 114, 177
Ackerknecht, E. H., 175, 340n91, 343n30, 344n36
acridity, 28, 36, 46, 100, 177, 317n25
Adair, J. M., 125
adaptation, fever as, 283–84, 288–91. *See also* wise body
adynamique fever, 172–73
Africa, West, 92, 211, 213, 240–41
African Americans, 251–52
Agniveśa, 60
ague, 29, 63–64, 80, 98, 235, 318n26
alchemical elements and principles, 24, 71, 82. *See also* iatrochemists
alcohol, 91, 95, 115, 147, 159, 212, 296. *See also* wine, as therapy
Alexander of Tralles, 56
Alexandria, 51, 55–58
Alison, William Pulteney, 161–62
alkalinity, 114, 177
American Medical Association, 299
analgesics, 268–69, 278
anatomico–clinical correlation, 14, 169, 173–74, 178–181, 198–203, 255
anatomy, 20, 49, 51, 56, 68, 70–71, 74, 80, 82–83, 216. *See also* anatomico-clinical correlation; dissection; pathological anatomy
Andromeda Strain, 309
animals, experiments on, 224, 244–45, 273–75. *See also* postulates, Henle-Koch
anthrax, 207
antibiotics, 3, 10, 43, 282–85, 292, 297, 312

antimonial drugs, 96–97, 100, 104. *See also* James's powder; sudorifics; therapeutics
antipyretics, ix, 3, 81, 302. *See also* febrifuges
anxiety, 95, 100, 129, 157, 182, 301. *See also* mind, state of
apocalyptic and postapocalyptic literature, 14–15, 309–10
archeus, 79, 129
ardent fever, 27, 29, 41, 115, 275. See also *kaûsos*
Aristotelianism, 44. *See also* nature, concept of
Aristotle, 58, 63, 290
Armstrong, George Dod, 230–33, 236–37
Armstrong, John, 196, 217–18, 224
Asia, 61, 215, 220–22, 243, 285. *See also* South Asia
aspirin, ix, 139, 269, 279, 292
Assize fevers, 105, 108, 116, 334n68. *See also* jail fever
ataxique fever, 172–73
atmospheric constitutions, 19, 111, 220, 240. *See also* Sydenham, Thomas
atmospheric variables, 223, 239–40, 333n57; chemical, 111–12, 239–40; particulate, 214, 220; physical, 111, 294
Avicenna, 58, 63
Ayurvedic medicine. *See* Indian classical concepts of fever

Baconianism, 74, 79, 81
bacteria, 312
bacteriology, 84, 163, 168, 183, 199–200, 206, 211, 242, 245, 263–64, 275, 291
Balkans, 236, 359n171
Barbados, 211–12

Snowden, Frank, 243

soil, 115, 121, 214–15, 223

soldiers, 13, 84, 90, 92, 114–18, 121–25, 137, 146, 152, 176, 178, 210–11, 241, 274, 290

Sontag, Susan, 292

South Asia, 207–10, 216, 218–24, 243–44, 274–75, 350n21, 352n54, 353n74

Southwood Smith, Thomas, 196

specificity. *See* disease specificity, concept of

spirits, concept of, 45, 48–49, 56, 58, 63, 67, 71–72, 79, 94, 97, 141–42, 170. *See also* Galenism

spleen, 32, 46, 80, 131, 208, 213, 235, 237, 359n171

spots, 197, 247, 334n78; blotches, 119–20, 195; rose-colored, 180–83, 186, 195, 197, 208. *See also* petechiae; rashes

stages. *See* course of fever

Stahl, Georg Ernst, 74, 129, 135

starvation, 64, 124, 194, 244, 272. *See also* famine fever; hunger

state, interest of, 4–5, 10–12, 55, 62, 162, 182, 184, 221, 225, 234, 242, 246–47, 260–61, 282, 308–10, 340n91

statistics: analytical, 198, 225, 360n19; descriptive, 13, 175, 182, 200, 202; vital, 55, 58, 225, 245, 247, 249. *See also* numerical method

Stoker, Bram, 236, 356n128

Stoker, William, 190–91

Stokes, William, 190

straitjackets, 148

stupor, 7, 29, 34, 110, 154, 158, 172–73, 181, 186, 192, 250, 310, 318n31. *See also* typhoid

subjectivity, 4, 7, 9, 11, 18, 24, 26, 46, 126–27, 132–33, 135–36, 139, 182, 202, 248, 253–55, 323n105. *See also* coldness, sense of; hotness, sense of; romanticism

sudorifics, 3, 10, 100–101, 177. *See also* James's powder; therapeutics

suicidality, 148, 236

sulfa drugs, 292

sun, 9, 51, 123, 146, 274–75, 296

surfeit as cause, 37, 49–50, 100, 141. *See also* non-naturals

surgeons, 58, 189, 197, 211, 274

Susruta, 23, 25, 35, 37, 40, 44, 46, 310, 323n107

swamps. *See* marshes

sweating, 14, 19, 24, 31–32, 37–38, 41, 61, 64, 69, 86, 100, 119, 124, 128, 131, 180, 271. *See also* perspiration; sudorifics

Sydenham, Thomas, 69, 74–77, 92–93, 261, 327n38; on disease specificity, 75–76, 101, 135, 145, 169, 171, 174, 201, 240; empiricism of, 74, 76, 112, 114, 217, 219, 333n51; on etiology, 75; on theory, 76

sympathy, doctrine of, 90, 158–59, 163

synocha and synochus, 63–65, 172, 175, 192

syphilis, 92, 144, 293–94

Talbor, Robert, 69, 79–82, 97, 329n75

telephones, 251–52

temperature: by age, 272, 289, 301; as definition, 5–6, 14, 18, 25, 137–39, 166, 202, 249, 251, 276–80; as key indicator, 6, 9, 11, 250–55, 258, 261, 265, 268–74, 276, 301–4, 308, 360n14; by race, 258, 272; by sex, 258, 272; standardization of, 256–59, 288

temperature charts, 11, 163, 166, 259–61, 271, 314n13

temperature regulation, 3, 251, 271–78, 288, 291–92. *See also* pathophysiological processes

temperature taking, 69, 84–86, 136, 202, 251–53, 256–58, 298–301; axillary, 257, 304; groin, 304; oral, 257, 304–5; rectal, 242, 294, 304–5; vaginal, 304, 360n27

tertian fever, 2, 23, 34, 42, 46, 64–65, 74, 94, 115, 134, 171, 218, 222. *See also* intermittent fever

therapeutic nihilism, 251, 263

therapeutics, 3, 10, 38–40, 43–45, 51–52, 58, 63, 68, 70, 74–75, 96–97, 108, 115–16, 123, 132–35, 139, 166, 171–74, 177, 179–81, 187–88, 192, 198, 228, 240, 248, 251, 259, 284, 301–2, 364n33. *See also* antipyretics; aspirin; bloodletting; febrifuges; opium; purging; quinine; sudorifics; wine, as therapy